Lecture Notes in Physics

The Lecture Notes in Physics

The series Lecture Notes in Physics (LNP), founded in 1969, reports new developments in physics research and teaching – quickly and informally, but with a high quality and the explicit aim to summarize and communicate current knowledge in an accessible way. Books published in this series are conceived as bridging material between advanced graduate textbooks and the forefront of research and to serve three purposes:

- to be a compact and modern up-to-date source of reference on a well-defined topic
- to serve as an accessible introduction to the field to postgraduate students and nonspecialist researchers from related areas
- to be a source of advanced teaching material for specialized seminars, courses and schools

Both monographs and multi-author volumes will be considered for publication. Edited volumes should, however, consist of a very limited number of contributions only. Proceedings will not be considered for LNP.

Volumes published in LNP are disseminated both in print and in electronic formats, the electronic archive being available at springerlink.com. The series content is indexed, abstracted and referenced by many abstracting and information services, bibliographic networks, subscription agencies, library networks, and consortia.

Proposals should be sent to a member of the Editorial Board, or directly to the managing editor at Springer:

Christian Caron
Springer Heidelberg
Physics Editorial Department I
Tiergartenstrasse 17
69121 Heidelberg / Germany
christian.caron@springer.com

M. Damnjanović
I. Milošević

Line Groups in Physics

Theory and Applications to Nanotubes and Polymers

Springer

Milan Damnjanović
University of Belgrade
Fac. Physics
Studentski trg 12
11001 Beograd
Serbia
yqoq@rcub.bg.ac.rs

Ivanka Milošević
University of Belgrade
Fac. Physics
Studentski trg 12
11001 Beograd
Serbia
ivag@rcub.bg.ac.rs

Damnjanović, M., Milošević, I., *Line Groups in Physics*, Lect. Notes Phys. 801 (Springer, Berlin Heidelberg 2010), DOI 10.1007/978-3-642-11172-3

Lecture Notes in Physics ISSN 0075-8450 e-ISSN 1616-6361
ISBN 978-3-642-11171-6 e-ISBN 978-3-642-11172-3
DOI 10.1007/978-3-642-11172-3
Springer Heidelberg Dordrecht London New York

Library of Congress Control Number: 2010921144

Cover design: Integra Software Services Pvt. Ltd., Pondicherry

Printed on acid-free paper

Springer is part of Springer Science+Business Media (www.springer.com)

To our parents

Preface

Over last decades low-dimensional materials are in focus of physics and chemistry as well as of material and other natural sciences. Like Vitaly Ginzburg has foreseen 30 years ago, low dimensionality offers physical phenomena and properties unseen in three-dimensional world. To see how thin films and monomolecular layers realize such a prediction it suffices only to observe intensity of research devoted to recently synthesized graphene. Still, quasi-one-dimensional compounds are over long period established as the origin of the most important and most interesting discoveries of material science and solid state physics. To mention only deoxyribonucleic acid, the most important molecule in nature, and diversity of nanotubes and nanowires, the cornerstones of the present and future nanotechnology.

Line groups, describing symmetry of quasi-one-dimensional materials, offer the deepest insight to their characteristic properties. Underlying many of the laws, they are very useful, but far from simple. This book is intended to explain them, their properties, and their most common applications. In particular, it is important to understand that the line groups are much wider class of symmetries than the well-known rod groups. While the latter describe only translationally periodical objects, line groups include symmetries of incommensurate periodical structures. Indeed, just the incommensurability is one of new features related to quasi-one-dimensionality. Being not subgroups of the space groups (like rod groups), line groups are beyond the scope of the standard crystallography. New techniques are needed to understand and use line groups. It follows that the book is devoted both to mathematicians and physicists, chemists, and other material scientists. We hope that it will be interesting for crystallographers, too, at least for the comparison of the techniques.

The purpose of the book is twofold. First, it gives overview of the state of art of the line groups, at the level enabling reader to understand and possibly further develop or apply the theory. Besides this, it should serve as a more or less exhaustive manual or reference book. Therefore, together with discussion of various notions and properties (with brief but mathematically rigorous explanations), there are many tables and figures, summarizing results in a comprehensive way. Still, there are some especially lengthy data, not suitable to present them in a text of this type, despite their importance in applications (e.g., Clebsch–Gordan coefficients). Many of them, as well as concrete calculations of the most of the theoretical results given here in a general form, are automatically generated by the computer code POLSym and online available through our Website www.nanolab.rs. (where also this book can be commented).

While the first half of the text is focused on the preparation of the main result, and therefore with prevailing mathematical discussions and seldom examples on their usage, the last part is completely devoted to applications in two levels of elaboration. First, the most important applications are considered generally, in order to provide understanding and derive necessary data. Finally, all the results are applied to nanotubes. Having in mind how deeply symmetry determines the properties of nanotubes, the last chapter can serve as a sort of the comprehensive textbook of these important (and also fashionable) systems.

We want to acknowledge help of many people during several years of preparation of this book. Younger members of our research team at NanoLab, Faculty of Physics, Belgrade, contributed by their research to the material of the book, like B. Dakić with new results on the invariant and covariant functions, T. Vuković with some of the number theoretical results, as well as with the most of the theory related to diffraction, E. Dobardžić performed much of the phonon work, while B. Nikolić was involved in the derivation of parities of nanotubes rolled up from arbitrary layer. Collaboration with Professor Thomsen's group, Institute for Solid State Physics, Technical University Berlin, was crucial in motivation to implement symmetry in various problems related to nanotubes; discussions with Christian Thomsen, Janina Maultzsch, Stephanie Reich, and Marcel Mohr resulted in several papers which are in some form reflected in this book. In the final stage of preparation of manuscript Professor Evarestov, Department of Chemistry, St. Petersburg University, made several suggestions important to make text closer to people already familiar with standard crystallographic methods and terminology. Professor Stergios Logothetidis, Aristotle University Thessaloniki, and students of the master program Nanoscience and Nanotechnology, organized by him, by putting lot of questions on nanotubes, convinced us to make the last chapter more extensive. In addition, the first lines of the book (in spring of 2005) were written and the last revision (in 2008) of the text was done in Stergios' and Alcestis' Halkidiki house, inspiring quiet spot among the pine trees overlooking Egéo Pélagos, with Óros Ólimbos on the horizon. And last but not least, we want to express our gratitude to many other friends and collaborators which helped us in technical and computer matter. Especially, we are grateful to our invincible young hacker I. Y. Perushka.

Belgrade,
June, 2009

Milan Damnjanović
Ivanka Milošević

Contents

Acronyms

$m, n = \mathrm{GCD}(m, n)$	Greatest common divisor of m and n
$\overline{m, n} = \mathrm{LCM}(m, n)$	Least common multiple of m and n
$[x]$	Integer part of x (maximal integer less then x)
$\{x\}$	Fractional part of x ($x - [x]$)
$\overline{x}, \underline{x}$	Numerator and denominator of the rational, i.e., $x = \overline{x}/\underline{x}$
$\tilde{x}$	x/n (division by the line group parameter n); exceptionally, $\tilde{k}$ and $\tilde{m}$ denote helical quantum numbers.
$x^{-1}_{(x')}$	Inverse of x modulo x', i.e., $xx^{-1}_{(x')} = 1 + zx'$
I	Identity matrix, identity orthogonal transformation
$\lvert X\rvert$	Absolute value (when X is a number), number of elements in X (when X is a set or a group), dimension of X (when X is a vector space or a representation)
$A \overset{n}{=} B$	Equality modulo n, i.e., $A = B + zn$
$A \overset{\circ}{=} B$	Equality modulo known interval $(x, y]$, i.e. $A = B + z(y - x)$
$\boldsymbol{L}, \boldsymbol{L}^{(F)}$	Line group, Fth family
$\boldsymbol{Z}$	Generalized translations of a line group
$\boldsymbol{P}$	Point factor of a line group
$\boldsymbol{Y}$	Transversal
S_x	Orbit of the point x
$\boldsymbol{P}_{\mathrm{I}}$	Isogonal point group
$\boldsymbol{P}_{\mathrm{M}}$	Symmetry group of monomer
$\boldsymbol{M}$	Intersection of $\boldsymbol{L}$ and $\boldsymbol{P}_{\mathrm{M}}$
$D, D^{(\lambda)}$	Representation, irreducible representation
$D^{(\mathrm{id})}$	Identical representation of the group: $D^{(\mathrm{id})}(\ell) = 1$
SAB	Symmetry-adapted basis
$(A\lvert \boldsymbol{f})$	Koster–Seitz symbol of transformation $(A\lvert \boldsymbol{f})\boldsymbol{r} \overset{\mathrm{def}}{=} A\boldsymbol{r} + \boldsymbol{f}$
$e = (I\lvert 0)$	Identical transformation

Chapter 1
Introduction

Polymers, due to palette of remarkable and applicable properties, attract interest of physicists, chemists, and biologists over decades. Most of these extraordinary characteristics originate from their reduced dimensionality and regular structure. Discovery of carbon nanotubes by Iijima in 1991, and the revolution they caused in material science, additionally stressed out that quasi-one-dimensionality was crucially responsible for the peculiarities of these systems. The well-established notions of nanoscience, nanotechnology, and/or nanobiotechnology illustrate impact of these systems on both the fundamental science and technology. Actually, the acronym N&N, stressing out only the nanoscale, is the best description of the whole bunch of interrelated classical sciences and high-tech breakthroughs of the fast-growing field initialized by the discovery of nanotubes. This is probably the most remarkable example how development of the fundamental and applied science is interrelated through endless series of feedbacks.

Quantum mechanics proved to be the key to the nano-world. Over seven decades before the dawn of N&N quantum mechanical formalism and techniques have been developing intensively and successfully through many challenges. Various approaches to many-body problems, correlation and interference effects, advanced numerical algorithms, and many other quantum mechanical achievements have been already at nano-researchers disposal.

However, symmetry, one of the deepest concepts in science and philosophy, which after the work of Wigner in the second quarter of the previous century became one of the roots as well as a powerful technical tool of quantum mechanics, has not been fully exploited. This is a bit strange in a view of its cornerstone status in particle physics and its long and extensive use in solid state physics over more than half a century, to mention only the Bloch theorem, through which symmetry underlies all the results for crystals. Nevertheless, in nanoscience symmetry is not used in a systematic way. This is even more surprising since the translational invariance from the very beginning proved to be extremely fruitful in the theoretical studies of nanotubes and stereo-regular polymers, enabling many important predictions. In particular, due to the symmetry electronic bands of carbon nanotubes are easily found analytically (in a simplified model though). Prediction of a wide variety of the conducting properties was probably the first major result, which paved the way

Damnjanović, M., Milošević, I.: *Introduction*. Lect. Notes Phys. **801**, 1–5 (2010)
DOI 10.1007/978-3-642-11172-3_1

for N&N. And almost a decade after the discovery of nanotubes, their full symmetry was reported and the symmetry-based results started to appear, although still rarely enough.

One possible reason for such a delay is that there is no systematic monograph on the symmetry of quasi-one-dimensional structures, unlike the vast number of books devoted to the space groups of crystals. Even more, there is a frequent belief that the symmetry of the systems periodical in one dimension is described by the rod groups, for which there exist exhaustive reviews [1]. However, rod groups are subgroups of the space groups, thus subdued to the crystallographic restrictions applicable for the two- and three-dimensional crystals. Thus all of 75 rod groups are only a small subset of the continuously many line groups. They are not applicable to the carbon nanotubes, and this prevented standard full symmetry considerations from the very beginning.

The other reason may be the experience from the physics of two- and three-dimensional crystals. The translation group with associated conserved quasi-momentum is simple to understand and use, in contrast to the 230 space groups and their irreducible representations. On the other hand, the translations are a major part of the symmetry: the translational subgroup index of the space group is small and theoretically never greater than 32. Moreover, for the most of the typical problems in which the usage of the remaining symmetry proved to be fruitful, ad hoc recipes are more or less well elaborated. Thus, the idea of direct application of the translational symmetry only is quite natural and seemingly equally good enough for the quasi-one-dimensional physics. However, it is just the opposite.

First, line groups and related conserved quantum numbers (i.e., irreducible representations) are by far simpler than the space groups. Second, lack of the crystallographic restrictions makes translations just a small part of the full symmetry, in contrast to the higher dimensional crystals. In fact, pure translations are completely absent in the incommensurate structures. Therefore the considerations, including only translational symmetry (and Bloch theorem in its original form), are inefficient, and in the incommensurate cases conceptually inapplicable. While in the unit cell of a three-dimensional crystal there are typically several to hundred atoms, unit cell of a typical chiral nanotube includes hundreds or thousands of atoms. This is the source of the major technical obstacle for theoretical predictions: existing ab initio numerical codes do not implement line group symmetry, but only space groups, meaning that the computer capacity restricts calculations to the rare nanotubes and polymers with small number of atoms per unit cell, excluding automatically the incommensurate systems from the studies. To summarize, with a smaller effort invested to understand line groups (in comparison to the space groups), the gain is much larger, opening the single way out in some cases. Actually, structural differences between various line groups are in many aspects larger than those between the space groups, and this is a sort of hallmark of the variety of properties of quasi-one-dimensional crystals.

First notions of the symmetry of mono-periodic systems appeared in the second quarter of the last century in the works of Speiser (stripe groups) [2], Hermann (commensurate line groups) [3], Alexander (incommensurate groups are mentioned for the first time) [4], Shubnikov [5], and Belov [6] (rod groups).

In the late 1950s the relevance of the line groups for polymers was discussed by B. K. Vainshtein [7]. Shortly after, the symmetry of time reversal was included, yielding a structure of the magnetic rod groups (Neronova and Belov [8]). In the first volume of the series *Modern Crystallography*, among other types of symmetry groups Vainshtein reviewed line groups under the name spiral groups [9]. Finally, in the volume E of *International Tables of Crystallography*, many data about rod groups are given [1].

Another derivation of the commensurate line groups, based on the theory of extensions of the translational group, was performed in the late 1970s in the works of Vujičić, Božović, and Herbut [10]. Also, they systematically constructed irreducible representations of the line groups just 1 year later [11, 12].

Later on the factorization of the line groups into week direct product of cyclic groups proved to be essential property of both commensurate and incommensurate line groups. It was accomplished in the context of construction of the magnetic line groups (describing the symmetry of the quasi-1D systems with spin ordering) by Damnjanović and Vujičić [13]. Later on it has been extensively used in many applications of the line groups. In the early 1990s the line group formalism has been applied to general classification of monoperiodic systems and analysis of their normal vibrations and vibronic instability. Jahn–Teller theorem, originally formulated and proved for molecules (i.e., for systems with point group symmetry), has been reformulated and proved for polymers and other systems with line group symmetry by Milošević and Damnjanović [14].

In 1999, in the paper "Full symmetry, optical activity, and potentials of single-wall and multi-wall nanotubes" by Damnjanović et al. [15] elucidated that symmetry of single wall carbon nanotubes was described by non-symmorphic line groups. This had many important consequences, to mention here a dramatic effect on the understanding of vibrational spectra of achiral tubes, almost negligible interaction between the layers in multi-wall tubes, selection rules for the processes in tubes, etc. From that time on, line group symmetry and methods based on it have been extensively applied in the studies of various quasi-one-dimensional compounds: carbon and transition metal chalcogenide nanotubes, ZnO nanorods, nanosprings and NTs, and many other (in)organic nanostructures [16–18].

Quite recently, generalized Bloch theorem [19] for line groups and general result on the symmetry of nanotubes rolled up from arbitrary two-dimensional lattices along an arbitrary chiral vector [20] completed the base for full application of the line groups in the physical problems.

Chapters 2, 3, 4, 5, and 6 are more mathematical and give a systematic overview of the current state of the theory of line groups. The specific structure of the line groups (Chap. 2) has been used to derive quite general classification of the systems with line group symmetry (Chap. 3), irreducible representations (Chap. 4), tensors characterizing such systems (Chap. 5), and magnetic line groups (Chap. 6). In the last three chapters we treat a number of physical properties of the systems with line group symmetry. Normal modes and vibronic instability are derived in Chap. 7, while Chap. 8 reviews basic results on the applications of the line groups in various physical problems. Finally, Chap. 9 is completely devoted to nanotubes and their properties; besides overview of such an interesting and modern subject, this chapter can serve to reader as a detailed illustration of the previously developed methods. Appendices are quite brief remainders on the relevant well-known mathematical topics, aimed to enable interested readers and researchers to fully verify the conclusions of the main text and to introduce a particular notation. To facilitate reading and later usage of the book, we extracted the most of the symbols in a separate list.

Therefore, the reader interested only in mathematical topics is referred to Chaps. 2, 3, 4, 5, and 6, while the last three are mainly devoted to physicists. However, for this physical part some knowledge of the material of the previous chapters is assumed. Generally, these prerequisites depend on the problem considered and are roughly sketched in the scheme of the interdependencies of the chapters below. Still, most of Chap. 2 is indispensable for understanding of the later ones.

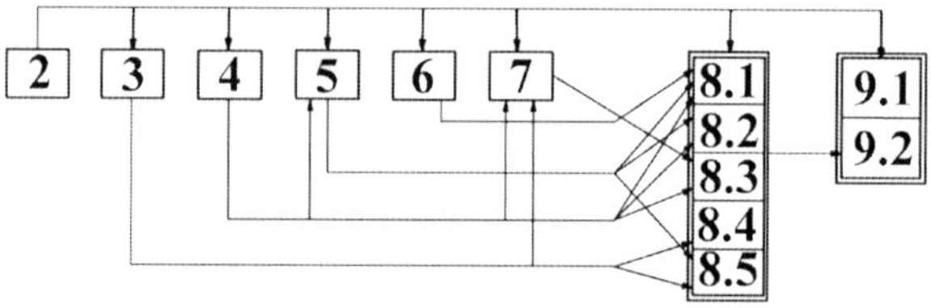

Most of the contents exist in the literature, dispersed in many articles with different points of view and notation; it is exposed here in a systematic and consistent way. Still, there are several original results, to mention only incommensurate line groups and generalizations of the line groups (in particular bihelical line groups and their relation to the symmetry breaking), general treatment of the chirality of the physical systems, list of the symmetries and symmetry cells for the line group orbits, completion of the epikernels (Landau's problem in the phase transitions), etc.

We found that the most natural way to expose the material is to follow specific, factorized, structure of the line groups. For example, we systematically use helical momenta and their relationship to the helical coordinate system. In fact, it turns out that linear momenta cannot be introduced for incommensurate systems, and even for commensurate ones they are not fully conserved, making their application more cumbersome; however, linear momenta may be preferable when system is influenced by an external field. For the same reason we generally use the term *polymer* for all the considered systems generalizing the notion usually related to polymers. Similarly, although we introduced both factorized and international notation, we prefer the former, not only because international refers only to the commensurate groups.

References

1. V. Kopsky, D. Litvin, *Subperiodic Groups, International Tables for Crystallography*, vol. E (Kluwer, Dordrecht, 2003)
2. A. Speiser, *Die Theorie der Gruppen von endlicher Ordnung, mit Anwendungen auf algebraische Zahlen und Gleichungen sowie auf die Kristallographie* (Springer, 1923)
3. C. Herman, Z. Kristallogr. **69**, 250 (1928)
4. E. Alexander, Z. Kristallogr. **70**, 367 (1929)
5. A.V. Shubnikovr, *Symmetry (The Laws of Symmetry and Their Application in Science, Technology, and Applied Art)* (Akademii Nauk SSSR, Moskva, 1940). In Russian
6. N.V. Belov, Kristallografiya **1**, 474 (1956)
7. B.K. Vainshtein, Kristallografiya **4**, 842 (1959)
8. N.N. Neronova, N.V. Belov, Kristallografiya **6**, 3 (1961)
9. B.K. Vainshtein, *Modern Crystallography: Fundamentals of Crystals, Symmetry and Methods of Structural Crystallography*, vol. 1 (Springer, Berlin, 1994)
10. M. Vujičić, I. Božović, F. Herbut, J. Phys. A **10**, 1271 (1977)
11. I. Božović, M. Vujičić, F. Herbut, J. Phys. A **11**, 2133 (1978)
12. I. Božović, M. Vujičić, J. Phys. A **14**, 777 (1981)
13. M. Damnjanović, M. Vujičić, Phys. Rev. B **25**, 6987 (1982)
14. I. Milošević, M. Damnjanović, Phys. Rev. B **47**, 7805 (1993)
15. M. Damnjanović, I. Milošević, T. Vuković, R. Sredanović, Phys. Rev. B **60**, 2728 (1999)
16. M. Damnjanović, I. Milošević, E. Dobardžić, T. Vuković, B. Nikolić, in *Applied Physics of Nanotubes: Fundamentals of Theory, Optics and Transport Devices*, ed. by S.V. Rotkin, S. Subramoney (Springer-Verlag, Berlin-Heidelberg-New York, 2005), pp. 41–88
17. S. Reich, C. Thomsen, J. Maultzsch, *Carbon Nanotubes — Basic Concepts and Physical Properties* (Wiley-VCH, Weinheim, 2004)
18. E.B. Barros, A. Jorio, G.G. Samsonidze, R.B. Capaz, A.G.S. Filho, J.M. Filho, G. Dresselhaus, Physics Reports **431**, 261–302 (2006)
19. I. Milošević, B. Dakić, M. Damnjanović, J. Phys. A: Math. Gen. **39**, 11833 (2006)
20. M. Damnjanović, B. Nikolić, I. Milošević, Phys. Rev. B **75**, 033404 (2007)

Chapter 2
Line Groups Structure

Abstract Line groups are introduced as symmetry groups of the system periodic in a single direction, with periodicity being not restricted to the translational one. Their structure is a weak direct product of the intrinsic symmetry of monomer and the group of generalized translations, arranging these monomers along the direction of periodicity. Continuously many of these groups are classified into 13 infinite families. Only 75 of the line groups are subgroups of the space groups and they are known as rod groups.

2.1 Factorization of the Line Groups

It is physical point of view which gives the natural and straightforward insight into the line group structure. When we analyze stereo-regular polymers, nanotubes, nano-rods, nano-springs, and quasi-one-dimensional subsystems in three-dimensional solids, we first notice their large longitudinal-lateral aspect ratio which singles out the direction along which basic constituents, *monomers*, are repeated regularly. Therefore their symmetries, being defined as the geometrical transformations leaving the compound unchanged, arise in two different ways: as periodicity of the regular arrangement of monomers and as an intrinsic symmetry of a single monomer. However, only if the intrinsic symmetry leaves simultaneously all the monomers unchanged, i.e., if it is compatible with the periodical arrangement, it is symmetry of the whole structure. Consequently, the symmetries of the total compound are combined from the symmetries of the arrangement and the intrinsic symmetries of monomers. This fact gives a hint to classification of the line groups: after independent classification of all possible arrangements and their symmetries, as well as of the symmetries of monomers, one should combine these two in all compatible ways and will thus get [1] all the line groups. Besides, in such a way important information on the specific structure of the line groups is directly obtained and it will be used extensively in order to derive many significant physical consequences.

Damnjanović, M., Milošević, I.: *Line Groups Structure*. Lect. Notes Phys. **801**, 7–27 (2010)
DOI 10.1007/978-3-642-11172-3_2

2.1.1 Generalized Translations: Symmetry of Arrangements

Here we analyze the symmetries coming from the periodical arrangement of the monomers along one direction which by the convention is denoted as the z-axis. Note that such a system should necessarily be infinite along z-axis. Therefore, the considered symmetries only approximately describe real compounds.

We shall slightly generalize the notion of periodicity. Usually translational periodicity is considered only and it is sufficient for the compounds which can be obtained by successive translations from some minimal part, called *elementary cell*. Generalizing this, we recall that polymer is an infinite chain of the identical monomers, regularly arranged along the z-axis. Although this regular arrangement is in some cases realized by translations, other situations may occur as well. Thus, instead of pure translations $(I|\boldsymbol{f})$, we allow more general geometrical transformation $Z = (X|\boldsymbol{f})$ (see Appendix A for notation), mapping each monomer of the chain onto the adjacent one.

Such a periodicity implies that element Z generates infinite cyclic group, $\mathbf{Z}$. Translational part $\boldsymbol{f}$ enables to move from one monomer to the adjacent one, and this way it determines the direction of the system. Therefore, it is a vector along the axis of the system, $\boldsymbol{f} = f\boldsymbol{e}_z$; knowing this, we shorten notation to $(X|f)$, explicating only the length f. Finally, conveniently choosing the orientation of z-axis we take f positive.

The orthogonal part X of the generator Z must leave z-axis invariant. Otherwise, after applying Z our system would not be directed along z-axis, and Z could not be a symmetry. This leaves two possibilities: $X\boldsymbol{e}_z = \boldsymbol{e}_z$ and $X\boldsymbol{e}_z = -\boldsymbol{e}_z$. However, as square of Z is $(X|f)^2 = (X^2|fX\boldsymbol{e}_z + f\boldsymbol{e}_z)$, in the latter case the translational part vanishes and infinite system cannot be generated. Therefore, X leaves invariant also the orientation of the z-axis. Hence X can be either a rotation around z-axis or a reflection in the vertical mirror plane (composition of these two is a mirror plane), i.e., the possibilities are $X = C_Q$ (rotation for $\phi = 2\pi/Q$ around z-axis) and $X = \sigma_{\mathrm{v}}$ (vertical mirror plane). Altogether, the group $\mathbf{Z}$ is infinite and cyclic, being *screw-axis* (alternatively called *helical*) group generated by $Z = (C_Q|f)$ or *glide plane* group, with $Z = (\sigma_{\mathrm{v}}|f)$. Here, Q may be any real number and for the sake of uniqueness we conveniently take it to be greater than or equal to one.

Now we examine when the arrangement generated by the above defined $Z = (X|f)$ has *translational periodicity*, i.e., we search for the conditions under which in the cyclic group $\mathbf{Z}$ there is a subgroup $\boldsymbol{T}$ of the pure translations. Such arrangements and groups will be called *commensurate*, to distinguish from the *incommensurate* ones, having no translations leaving the system invariant. Translational subgroup is cyclic, and it is generated by $(I|A)$, where the *translational period* A is the minimal among the pure translations in $\mathbf{Z}$. As an element of $\mathbf{Z}$, this minimal translation is obtained as a successive application of Z, i.e., there is a natural number q such that $(X|f)^q = (X^q|qf) = (I|A)$. This gives the condition $X^q = I$. For glide planes this is always automatically fulfilled for $q = 2$ ($\sigma_{\mathrm{v}}^2 = I$). As for the screw-axes, the condition reads $Q = q/r$, where r is a natural and not greater than q (as $Q \geq 1$).

So, translational periodicity appears only when Q is rational, i.e., when $\phi = 2\pi r/q$ is rational multiple of 2π. Conveniently, we assume that r and q are co-primes, $\mathrm{GCD}(q, r) = 1$. Particularly, if $q = 1$ (i.e., $\phi = 0$) screw-axis is pure *translational group*; formally we take $\phi = 2\pi$, with $Q = q = r = 1$. The obtained period[1] A is multiple of the fractional translation of the generator, $A = qf$, meaning that the elementary cell contains q monomers. Thus the helical generator in this case is $(C_q^r|A/q)$.

To summarize, there are two types of periodical arrangements, corresponding to the generalized translational groups:

1. *screw-axis* group, $\boldsymbol{T}_Q(f)$ generated by $(C_Q|f)$. In the special cases when $Q = q/r$, with positive co-prime integers $r \leq q$, screw-axis $\boldsymbol{T}_q^r(A/q)$ generated by $(C_q^r|A/q)$ is commensurate, with index q subgroup of pure translations with period $A = qf$; particularly, for $q = 1$ and $q = 2$, screw-axis degenerates to the translational and zigzag group, respectively.
2. *glide plane* group, $\boldsymbol{T}'(A/2)$, generated by $(\sigma_{\mathrm{v}}|f)$, which has a halving subgroup of pure translations with period $A = 2f$.

These groups are schematically depicted in Fig. 2.1.

Knowing the group of generalized translations, we arbitrary chose an initial monomer and denote it by M_0. Then applying Z successively t times on monomer M_0, i.e., acting by Z^t on each atom in M_0, we get another monomer, which is naturally labeled by M_t. Apart from giving practical way to count the atoms, this conclusion shows that complete information on the entire polymer is given by the group of generalized translations and chemical and geometrical structure of a single monomer. In other words, physical properties of polymer are determined by properties of a single monomer and symmetry of the arrangement of the monomers.

2.1.2 Axial Point Groups: Intrinsic Monomer Symmetry

Single monomer possesses its own symmetry. As monomer is considered to be a finite molecule or ion, its symmetry transformations cannot include translations, and their form is $X = (X|0)$. Such transformations are gathered into *point groups* [2]. Here we analyze which of these transformations contribute to the symmetry of the compound and classify all the relevant intrinsic symmetry groups.

Let $\boldsymbol{P}_{\mathrm{M}}$ be symmetry group of an arbitrary monomer M_0. Any transformation P from $\boldsymbol{P}_{\mathrm{M}}$ leaves M_0 invariant. However, to be a symmetry of the whole polymer, this transformation in addition has to map any monomer M_t into itself or into another monomer $\mathrm{M}_{t'}$. Generally, this cannot be fulfilled unless P leaves z-axis invariant. Hence, only the maximal subgroup $\boldsymbol{P}$ of $\boldsymbol{P}_{\mathrm{M}}$ leaving z-axis invariant may contribute

[1] Capital A denotes the translational period of the commensurate helical group, only. The period of the total group is denoted as a.

to the symmetry of the polymer. As the maximal orthogonal group which preserves the z-axis is the $\boldsymbol{D}_{\infty\mathrm{h}}$, we conclude that the point group $\boldsymbol{P} = \boldsymbol{P}_{\mathrm{M}} \cap \boldsymbol{D}_{\infty\mathrm{h}}$ is the maximal subgroup of the monomer symmetry group relevant for the polymer.

The orthogonal transformations leaving z-axis invariant are rotations C_n^s for angle $\phi = 2\pi s/n$ around z-axis, rotations U for π around an axis perpendicular onto z-axis, horizontal and vertical mirror planes (σ_{h} and σ_{v}, and their combinations, such as roto-reflectional plane $S_{2n} = C_{2n}\sigma_{\mathrm{h}}$). It turns out that with such operations one can build seven infinite families of the axial point groups (Table 2.1, Fig. 2.1). The groups within a family differ by the order $n = 1, 2, \dots$ of the principal axis. In fact, each axial point group has a subgroup $\boldsymbol{C}_n$ of index $n_F = |\boldsymbol{P}_n|/n$, with $n_F = 1$ for family 1, $n_F = 2$ for families 2, 3, 4, and 5, and $n_F = 4$ for families 6 and 7. This allows decomposition onto cosets:

$$\boldsymbol{P}_n = \sum_{i=1}^{n_F} p_i^{(F)} \boldsymbol{C}_n, \tag{2.1}$$

with $p_1^{(F)} = e$, and the remaining coset representatives listed in Table 2.1.

The transformations of these groups may reverse the direction of z-axis; such elements will be called *negative* to differ from the *positive elements* preserving the z-axis direction. As the product of negative elements is positive, set of positive elements is either the whole group (*positive group* $\boldsymbol{P}_n^+$) or a halving subgroup (which is also axial point group as $\boldsymbol{C}_n$ is positive), when the group is called *negative*:

$$\boldsymbol{P}_n^- = \boldsymbol{P}_n^+ + p^- \boldsymbol{P}_n^+. \tag{2.2}$$

The first two families, $\boldsymbol{C}_n$ and $\boldsymbol{S}_{2n}$, are cyclic groups. All other families are factorized into products of cyclic groups. The first three families, $\boldsymbol{C}_n$, $\boldsymbol{S}_{2n}$, and $\boldsymbol{C}_{n\mathrm{h}}$, are the abelian groups.

Table 2.1 Axial point groups. For each group $\boldsymbol{P}_n$ ($n = 1, 2, \dots$), its order $|\boldsymbol{P}_n|$, factorization F, generators g, limiting continual groups for $n = \infty$, positive subgroup $\boldsymbol{P}_n^+$ (for positive groups $\boldsymbol{P}_n^+ = \boldsymbol{P}_n$), and the coset representative(s) $p^{(F)}$ of $\boldsymbol{C}_n$ are given. For $\boldsymbol{D}_{n\mathrm{d}}$ and $\boldsymbol{D}_{n\mathrm{h}}$ (when $|\boldsymbol{P}_n| = 4n$), the first coset representative satisfies $\boldsymbol{P}_n^+ = \boldsymbol{C}_n + p^{(1)}\boldsymbol{C}_n$. In the last two rows the extension $\boldsymbol{P}_n^{\mathcal{I}} = \boldsymbol{P}_n + \mathcal{I}\boldsymbol{P}_n$ of $\boldsymbol{P}_n$ by spatial inversion is given for n odd and even

$\boldsymbol{P}_n$	$\boldsymbol{C}_n$	$\boldsymbol{S}_{2n}$	$\boldsymbol{C}_{n\mathrm{h}}$	$\boldsymbol{D}_n$	$\boldsymbol{C}_{n\mathrm{v}}$	$\boldsymbol{D}_{n\mathrm{d}}$	$\boldsymbol{D}_{n\mathrm{h}}$
$\|\boldsymbol{P}_n\|$	n	$2n$	$2n$	$2n$	$2n$	$4n$	$4n$
F	$\boldsymbol{C}_n$	$\boldsymbol{S}_{2n}$	$\boldsymbol{C}_n\boldsymbol{C}_{1\mathrm{h}}$	$\boldsymbol{C}_n\boldsymbol{D}_1$	$\boldsymbol{C}_n\boldsymbol{C}_{1\mathrm{v}}$	$\boldsymbol{S}_{2n}\boldsymbol{C}_{1\mathrm{v}}$	$\boldsymbol{D}_n\boldsymbol{C}_{1\mathrm{v}}$
g	C_n	$C_{2n}\sigma_{\mathrm{h}}$	C_n, σ_{h}	C_n, U	C_n, σ_{v}	$C_{2n}\sigma_{\mathrm{h}}, U_{\mathrm{d}}$	$C_n, U, \sigma_{\mathrm{v}}$
$\boldsymbol{P}_\infty$	$\boldsymbol{C}_\infty$	$\boldsymbol{C}_{\infty\mathrm{h}}$	$\boldsymbol{C}_{\infty\mathrm{h}}$	$\boldsymbol{D}_\infty$	$\boldsymbol{C}_{\infty\mathrm{v}}$	$\boldsymbol{D}_{\infty\mathrm{h}}$	$\boldsymbol{D}_{\infty\mathrm{h}}$
$\boldsymbol{P}_n^+$	$\boldsymbol{C}_n$	$\boldsymbol{C}_n$	$\boldsymbol{C}_n$	$\boldsymbol{C}_n$	$\boldsymbol{C}_{n\mathrm{v}}$	$\boldsymbol{C}_{n\mathrm{v}}$	$\boldsymbol{C}_{n\mathrm{v}}$
$p^{(F)}$		$C_{2n}\sigma_{\mathrm{h}}$	σ_{h}	U	σ_{v}	$\sigma_{\mathrm{v}}, C_{2n}\sigma_{\mathrm{h}}, U_{\mathrm{d}}$	$\sigma_{\mathrm{v}}, \sigma_{\mathrm{h}}, U$
$\boldsymbol{P}_{n=2i+1}^{\mathcal{I}}$	$\boldsymbol{S}_{2n}$	$\boldsymbol{S}_{2n}$	$\boldsymbol{C}_{2n\mathrm{h}}$	$\boldsymbol{D}_{n\mathrm{d}}$	$\boldsymbol{D}_{n\mathrm{d}}$	$\boldsymbol{D}_{n\mathrm{d}}$	$\boldsymbol{D}_{2n\mathrm{h}}$
$\boldsymbol{P}_{n=2i}^{\mathcal{I}}$	$\boldsymbol{C}_{n\mathrm{h}}$	$\boldsymbol{C}_{2n\mathrm{h}}$	$\boldsymbol{C}_{n\mathrm{h}}$	$\boldsymbol{D}_{n\mathrm{h}}$	$\boldsymbol{D}_{n\mathrm{h}}$	$\boldsymbol{D}_{2n\mathrm{h}}$	$\boldsymbol{D}_{n\mathrm{h}}$

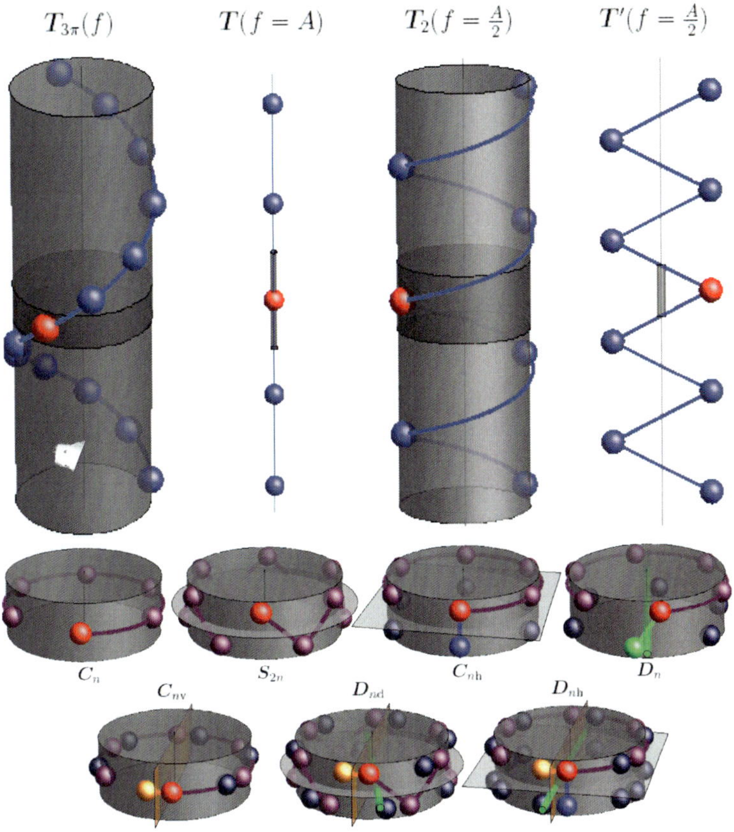

Fig. 2.1 Factors of line groups. Generalized translational groups (up, from *left* to *right*): incommensurate (and chiral) helical axis, translational group, zigzag group, glide plane; fractional translations are shown by *dark gray* cylinders on which the initial (*red*) atoms sit. Axial point groups (down): the groups with $n = 6$ from all the families are presented by the points obtained from the initial one (*red*). Those obtained by the successive action of the generators (according to Table 2.1) are in special colors and connected by the lines: *purple, blue, green*, and *orange* correspond to C_n^s or $(C_{2n}\sigma_h)^s$, σ_h, U and σ_v; accordingly, vertical mirror plane and U-axis are orange and green. *Gray square* and *circle* are horizontal mirror and roto-reflectional planes

In the limit $n = \infty$ each family gives a continuous group $\boldsymbol{P}_\infty$. These limiting groups correspond to symmetry of linear molecules and also appear as the isogonal groups of incommensurate systems (Sect. 2.1.5).

2.1.3 Compatible Intrinsic and Arrangement Symmetries

The final part in the construction of the line groups is to examine compatibility of the intrinsic monomer symmetry and of the symmetry of the arrangement. Precisely, we construct a group $\boldsymbol{L}$ with elements being the products $\ell = Z^t P$ of the generalized translations Z^t from $\boldsymbol{Z}$ (all elements of $\boldsymbol{Z}$ are powers of the generator Z) and intrinsic symmetries P from $\boldsymbol{P}$. If such a set has a group structure, then $\boldsymbol{P}$ and $\boldsymbol{Z}$ are its subgroups (for fixed $t = 0$ and $P = I$, respectively). In other words, we want to find all pairs $\boldsymbol{P}$ and $\boldsymbol{Z}$ giving a group $\boldsymbol{L} = \boldsymbol{PZ}$.

It is text book knowledge [3] that product of two subgroups is a group if and only if these subgroups commute,[2] i.e., $\boldsymbol{PZ}$ is group if and only if $\boldsymbol{PZ} = \boldsymbol{ZP}$. Thus, we have to check the compatibility of each axial point group with each group of generalized translations. Here we only briefly sketch these results, in order to get the classification of the line groups.

At first, we combine rotational groups $\boldsymbol{C}_n$ with screw-axes. The rotations around z-axis and translations along it commute, implying that all the helical operations $(C_Q|f)^t$ commute with pure rotations around z-axis. Therefore any group $\boldsymbol{C}_n$ is compatible with any $\boldsymbol{T}_Q(f)$. Their products $\boldsymbol{T}_Q(f)\boldsymbol{C}_n$ comprise the first family line groups, the simplest line groups which are subgroups of all the other line groups. More subtle details on these groups studied in Sect. 2.2 are necessary to construct and analyze the structure of other groups (Sect. 2.3). Quite analogously, all $\boldsymbol{T}_Q(f)$ are compatible with the groups $\boldsymbol{D}_n$, since $U(C_Q|f)^t = (C_Q|f)^{-t}U$.

However, mirror planes (of the intrinsic symmetry) can be retained in the helical arrangement only in the very special cases of screw-axes. For example, if in the point factor besides $\boldsymbol{C}_n$ there is a horizontal mirror plane, then the compatibility condition $\sigma_{\mathrm{h}}(C_Q|f)^t = (C_Q|f)^{-t}\sigma_{\mathrm{h}}C_Q^{2t}$ becomes $C_Q^{2t} = C_n^s$. In other words, for each t there must be s such that $2t/Q = s/n$, which is fulfilled only if n is multiple of $Q/2$, i.e., $Q = 2n/j$ for some $j = 1, 2, \ldots$. At first, this means that Q is rational, i.e., only commensurate helical groups allow mirror plane. Moreover, generator $(C_Q|f) = (C_{2n}^j|f)$ becomes $Z'C_n^i$, where C_n^i is rotation from $\boldsymbol{C}_n$, while $Z' = (I|f)$ for j even and $Z' = (C_{2n}|f)C_n^i$ for j odd; therefore, the group $\boldsymbol{T}_Q(f)\boldsymbol{C}_n$ is equal to the group $\boldsymbol{T}(f)\boldsymbol{C}_n$ for j even and to $\boldsymbol{T}_{2n}^1(f)\boldsymbol{C}_n$ for j odd.[3] Quite similar argumentation can be performed for all other axial point groups, giving the same result: besides $\boldsymbol{C}_n$ and $\boldsymbol{D}_n$ which are compatible with any helical group, the other axial point groups are compatible only with special helical groups $\boldsymbol{T}(f = a)$ and $\boldsymbol{T}_{2n}^1(f = a/2)$, where n is order of the principle axis of the point group.

On the other hand, glide plane group is compatible with all the axial point groups. The only requirement when the point group contains mirror planes or U-axes is that glide plane either coincides with them or bisects them, and different groups are obtained in these two cases. As glide plane is commensurate, such are all the resulting groups.

This way we obtain [1] all the products of the point factor $\boldsymbol{P}$ and generalized translation group $\boldsymbol{Z}$:

$$\boldsymbol{L} = \boldsymbol{ZP}. \tag{2.3}$$

---

[2] This does not mean that the elements of $\boldsymbol{P}$ and $\boldsymbol{Z}$ commute, but only that for each P and t there is a choice of P' and t' such that $PZ^t = Z^{t'}P'$.

[3] Essentially, we exploited here nonuniqueness of the helical subgroup of the first family line groups, which will be discussed in Sect. 2.2.2.

Table 2.2 Line groups. For each family F different factorizations, roto-helical subgroup $\boldsymbol{L}^{(1)}$, generators, and the isogonal point group $\boldsymbol{P}_{\mathrm{I}}$, are given in the first line. Below follow $N_F = |\boldsymbol{L}^{(F)}|/|\boldsymbol{L}^{(1)}|$, international symbol (of commensurate groups only), family F^+ of the positive subgroup $\boldsymbol{L}^+$ (for positive groups $F = F^+$), and the coset representatives $\ell_i^{(F)}$ for $i > 1$ of (2.29) (when $n_F = 4$, the first one gives $\boldsymbol{L}^+$). $\boldsymbol{T}'_{\mathrm{d}}$ and U_{d} are glide plane and horizontal axes bisecting vertical mirror planes, while $S_{2n} = C_{2n}\sigma_{\mathrm{h}}$. For families 1 and 5, the order q of the isogonal principle axis is given by $Q = q/r$ (according to (2.12)) for commensurate groups, while $q = \infty$ otherwise

F / N_F	n even	Factorizations / International	n odd	$\boldsymbol{L}^{(1)}$ / F^+	Generators / $\ell_i^{(F)}$	$\boldsymbol{P}_{\mathrm{I}}$
1		$\boldsymbol{T}_Q \otimes \boldsymbol{C}_n$		$\boldsymbol{T}_Q \otimes \boldsymbol{C}_n$	$(C_Q\vert f), C_n$	$\boldsymbol{C}_q$
1		$\boldsymbol{L}q_p$		1		
2		$\boldsymbol{T} \wedge \boldsymbol{S}_{2n}$		$\boldsymbol{T} \otimes \boldsymbol{C}_n$	$(I\vert a), S_{2n}$	$\boldsymbol{S}_{2n}$
2	$\boldsymbol{L}\overline{2n}$		$\boldsymbol{L}\bar{n}$	1	S_{2n}	
3		$\boldsymbol{T} \wedge \boldsymbol{C}_{n\mathrm{h}}$		$\boldsymbol{T} \otimes \boldsymbol{C}_n$	$(I\vert a), C_n, \sigma_{\mathrm{h}}$	$\boldsymbol{C}_{n\mathrm{h}}$
2	$\boldsymbol{L}n/m$		$\boldsymbol{L}\overline{2n}$	1	σ_{h}	
4		$\boldsymbol{T}^1_{2n}\boldsymbol{C}_{n\mathrm{h}} = \boldsymbol{T}^1_{2n}\boldsymbol{S}_{2n}$		$\boldsymbol{T}^1_{2n}\boldsymbol{C}_n$	$(C_{2n}\vert a/2), C_n, \sigma_{\mathrm{h}}$	$\boldsymbol{C}_{2n\mathrm{h}}$
2		$\boldsymbol{L}2n_n/m$		1	σ_{h}	
5		$\boldsymbol{T}_Q \wedge \boldsymbol{D}_n$		$\boldsymbol{T}_Q \otimes \boldsymbol{C}_n$	$(C_Q\vert f), C_n, U$	$\boldsymbol{D}_q$
2	$\boldsymbol{L}q_p22$		$\boldsymbol{L}q_p2$	1	U	
6		$\boldsymbol{T} \otimes \boldsymbol{C}_{n\mathrm{v}} = \boldsymbol{C}_{n\mathrm{v}} \wedge \boldsymbol{T}'$		$\boldsymbol{T} \otimes \boldsymbol{C}_n$	$(I\vert a), C_n, \sigma_{\mathrm{v}}$	$\boldsymbol{C}_{n\mathrm{v}}$
2	$\boldsymbol{L}nmm$		$\boldsymbol{L}nm$	6	σ_{v}	
7		$\boldsymbol{C}_n \wedge \boldsymbol{T}'$		7 $\boldsymbol{T} \otimes \boldsymbol{C}_n$	$(\sigma_{\mathrm{v}}\vert a/2), C_n$	$\boldsymbol{C}_{n\mathrm{v}}$
2	$\boldsymbol{L}ncc$		$\boldsymbol{L}nc$	7	$(\sigma_{\mathrm{v}}\vert a/2)$	
8		$\boldsymbol{C}_{n\mathrm{v}} \wedge \boldsymbol{T}^1_{2n} = \boldsymbol{C}_{n\mathrm{v}} \wedge \boldsymbol{T}'_{\mathrm{d}}$		$\boldsymbol{T}^1_{2n} \otimes \boldsymbol{C}_n$	$(C_{2n}\vert a/2), C_n, \sigma_{\mathrm{v}}$	$\boldsymbol{C}_{2n\mathrm{v}}$
2		$\boldsymbol{L}2n_nmc$		8	σ_{v}	
9		$\boldsymbol{T} \wedge \boldsymbol{D}_{n\mathrm{d}} = \boldsymbol{T}' \wedge \boldsymbol{D}_{n\mathrm{d}}$		$\boldsymbol{T} \otimes \boldsymbol{C}_n$	$(I\vert a), C_n, U_{\mathrm{d}}, \sigma_{\mathrm{v}}$	$\boldsymbol{D}_{n\mathrm{d}}$
4	$\boldsymbol{L}\overline{2n}2m$		$\boldsymbol{L}\bar{n}m$	6	$\sigma_{\mathrm{v}}, U_{\mathrm{d}}, S_{2n}$	
10		$\boldsymbol{T}'\boldsymbol{S}_{2n} = \boldsymbol{T}'_{\mathrm{d}}\boldsymbol{D}_n$		$\boldsymbol{T} \otimes \boldsymbol{C}_n$	$(\sigma_{\mathrm{v}}\vert a/2), S_{2n}$	$\boldsymbol{D}_{n\mathrm{d}}$
4	$\boldsymbol{L}\overline{2n}2c$		$\boldsymbol{L}\bar{n}c$	7	$(\sigma_{\mathrm{v}}\vert a/2), S_{2n}, (U_{\mathrm{d}}\vert a/2)$	
11		$\boldsymbol{T} \wedge \boldsymbol{D}_{n\mathrm{h}} = \boldsymbol{T}'\boldsymbol{D}_{n\mathrm{h}}$		$\boldsymbol{T} \otimes \boldsymbol{C}_n$	$(I\vert a), C_n, U, \sigma_{\mathrm{v}}$	$\boldsymbol{D}_{n\mathrm{h}}$
4	$\boldsymbol{L}n/mmm$		$\boldsymbol{L}\overline{2n}2m$	7	$\sigma_{\mathrm{v}}, U, \sigma_{\mathrm{h}}$	
12		$\boldsymbol{T}'\boldsymbol{C}_{n\mathrm{h}} = \boldsymbol{T}'\boldsymbol{D}_n$		$\boldsymbol{T} \otimes \boldsymbol{C}_n$	$(\sigma_{\mathrm{v}}\vert a/2), C_n, \sigma_{\mathrm{h}}$	$\boldsymbol{D}_{n\mathrm{h}}$
4	$\boldsymbol{L}n/mcc$		$\boldsymbol{L}\overline{2n}2c$	7	$(\sigma_{\mathrm{v}}\vert a/2), U, (S_{2n}\vert a/2)$	
13		$\boldsymbol{T}^1_{2n}\boldsymbol{D}_{n\mathrm{h}} = \boldsymbol{T}^1_{2n}\boldsymbol{D}_{n\mathrm{d}} = \boldsymbol{T}'_{\mathrm{d}}\boldsymbol{D}_{n\mathrm{h}} = \boldsymbol{T}'_{\mathrm{d}}\boldsymbol{D}_{n\mathrm{d}}$		$\boldsymbol{T}^1_{2n} \otimes \boldsymbol{C}_n$	$(C_{2n}\vert a/2), C_n, U, \sigma_{\mathrm{v}}$	$\boldsymbol{D}_{2n\mathrm{h}}$
4		$\boldsymbol{L}2n_n/mcm$		8	$\sigma_{\mathrm{v}}, U, \sigma_{\mathrm{h}}$	

Although with different factors, some of these products are equal, giving different factorization of the same line group. Taking this into account, we get 13 infinite families of the line groups in the factorized form (Table 2.2). Each family includes all groups (with various parameters Q, f, n) with fixed type of $\boldsymbol{Z}$ and $\boldsymbol{P}$. The incommensurate line groups are either from the first or from the fifth family (commensurate ones from these families are singled out by the condition $Q = q/r$), while in all other families generalized translational group is either glide plane, pure translational group $\boldsymbol{T}$, or zigzag group $\boldsymbol{T}_{2n}^{1}$. In Appendix B all of the 75 groups satisfying crystallographic conditions from all families are singled out (i.e., subgroups of space groups); these groups are known as rod groups.

As the intersection of $\boldsymbol{P}$ and $\boldsymbol{Z}$ is the identity element only (point operations are without translational part, and the identity Z^0 is the only such element in $\boldsymbol{Z}$), the products are weak direct ones. In some cases one or both subgroups are invariant, and this product becomes semi-direct ($\wedge$, with the first factor invariant) and direct ($\otimes$), respectively.

2.1.4 Monomer, Elementary Cell, Symcell

After clarifying the structure of the line groups, we are able to define precisely the basic structural ingredients of the systems with line group symmetry.

At first, by *monomer* we denote the minimal part of the system sufficient to generate the whole system by the action of generalized translations $\boldsymbol{Z}$ only. For commensurate systems we can also introduce the *elementary cell*, which generates the polymer by the translations only. Finally, with the help of the full symmetry we generate the system with only a part of monomer which we call *symcell* or *symmetry cell*.

None of these parts of the system is uniquely defined, but the number of atoms it contains is. In fact, the symcell is any set of orbit representatives (orbits of line groups will be discussed in details in Sect. 3.1). Further, as in commensurate case translational group is a subgroup of $\boldsymbol{Z}$, elementary cell contains $|\boldsymbol{Z}|/|\boldsymbol{T}|$ monomers.

2.1.5 Isogonal Groups

For some physical applications, like study of the excitations in the external fields (e.g., first order Raman spectra), the translational parts of the symmetry transformations are not important. Therefore, for such purposes only the set of all the orthogonal parts X from the line group elements $(X|f)$ is relevant. This set is obviously an axial point group $\boldsymbol{P}_{\mathrm{I}}$, called *isogonal point group*. As it includes all elements of the point factor, $\boldsymbol{P}$ is a subgroup of $\boldsymbol{P}_{\mathrm{I}}$. Note that the elements of $\boldsymbol{P}_{\mathrm{I}}$ being not in $\boldsymbol{P}$ are not the symmetries of the considered system.

For an incommensurate line group, the isogonal group is infinite. In order to prove this, we assume the opposite, i.e., that the set of rotations R_ϕ included in the elements $(R_\phi|f)$ (where f may be zero) is finite. As the line group is infinite, this means that the same rotation, R_ϕ, in the line group must be accompanied by (infinitely many) different translational parts. Thus, there are elements $(R_\phi|f)$ and $(R_\phi|f')$ in $\boldsymbol{L}$. Then, there is also a pure translation $(R_\phi|f)(R_\phi|f')^{-1} = (I|f-f')$, which contradicts the assumption of incommensurability. Consequently, the isogonal groups of the incommensurate first and fifth family line groups are continuous groups $\boldsymbol{C}_\infty$ and $\boldsymbol{D}_\infty$, respectively.[4]

For the commensurate line groups, the translational group $\boldsymbol{T}$ is an invariant subgroup, and the isogonal group is the factor group $\boldsymbol{P}_\mathrm{I} = \boldsymbol{L}/\boldsymbol{T}$. This means that the line group may be partitioned into disjoint cosets[5]:

$$\boldsymbol{L} = \sum_{X \in \boldsymbol{P}_\mathrm{I}} (X|f_X)\boldsymbol{T}. \qquad (2.4)$$

Obviously, only when $\boldsymbol{Z} = \boldsymbol{T}$ all of the elements $(X|f_X)$ may be taken from $\boldsymbol{P}$ (having $f_X = 0$), showing that then $\boldsymbol{P}_\mathrm{I} = \boldsymbol{P}$. These line groups are called symmorphic. Beside some special line groups of the families 1 and 5, symmorphic are the whole families 6, 9, and 11.

For a commensurate line group of the first family, the isogonal point group contains products of the rotations C_q^{rt} (different among them are for $t = 0, \ldots, q-1$) from the helical group, and C_n^s from the point factor. As q and r are co-primes, the set C_q^{rt} is same as the set C_q^t. Thus, we have the products $C_q^t C_n^s$ of the elements of two cyclic groups $\boldsymbol{C}_q$ and $\boldsymbol{C}_n$. This group is $\boldsymbol{C}_q\boldsymbol{C}_n = \boldsymbol{C}_{\mathrm{LCM}(q,n)}$, since the intersection of factors is $\boldsymbol{C}_{\mathrm{GCD}(q,n)}$. Consequently, $\boldsymbol{C}_{\mathrm{LCM}(q,n)}$ is the isogonal group of the commensurate line groups of the first family. Analogously, for the fifth family one gets $\boldsymbol{D}_{\mathrm{LCM}(q,n)}$. For other families the isogonal group is easily found (Table 2.2). Later on (Sect. 2.2.2) we will show that in the commensurate groups q may be taken to be a multiple of n. Thus, within this convention, q is the order of the principle axis of the isogonal group.

2.1.6 Spatial Inversion and Chirality

Many molecules and polymers have left and right form, i.e., two different conformations being mapped one into another by the spatial inversion $\mathcal{I}$. Although chemically the same, and with the same properties when isolated, the two forms may be distinguished during interaction with some "oriented" external probe, like polarized light, where they rotate the polarization of the transmitted (and also reflected) beam in the opposite way, which is known as *dichroism* or *birefringence*. Such systems are called *chiral*, while the *achiral* ones have the same left and right form. Chirality is important property of quasi-one-dimensional systems. In technology it has applications in optical devices and spectroscopical methods. In addition chirality is

---

[4] Strictly, number of elements R_ϕ is countable; however, the obtained angles ϕ are dense in the interval $[0, 2\pi)$, and for physical applications, when the involved quantities are continuous functions on ϕ, there is no difference between such countable groups and mentioned continuous ones.

[5] This type of coset decomposition is applied in crystallography for derivation of the space groups, and it is reflected in the international notation of these groups (see Sect. 2.2.3). The (complicated) mathematical construction used for this purpose is known as extension from the translational subgroup by the isogonal group. It gives [4] classification of the commensurate line groups only.

correlated with many structural properties of symmetry groups and through this with a wide scale of physical properties of system. So, we are going to define criteria to distinguish chiral and achiral system.

Let us start with a bit simplified approach, which is also of physical interest in some quantum mechanical analysis involving selection rules for transitions in various external fields. Obviously, if symmetry group contains spatial inversion $\mathcal{I}$, then the system is invariant under $\mathcal{I}$ and automatically achiral. Hence line groups containing spatial inversion are to be singled out. Having in mind that $\mathcal{I}$ may be written as a product of a mirror plane and the rotation for π around the perpendicular axis to it, e.g., $\mathcal{I} = C_2\sigma_{\rm h} = U\sigma_{\rm v}^{\perp}$ (mirror plane perpendicular to U axis), it is easy to find such groups: spatial inversion appears in all the groups of the families 4 and 13; families 2, 9, and 10 contain $\mathcal{I}$ for n odd, and the families 3, 11, and 12 for n even. In fact, in these cases the isogonal group contains spatial inversion (Table 2.2).

However, the question whether the initial and the inverted systems are indistinguishable is more subtle. Indistinguishability means that the inversion does not change the relative positions of atoms: the atoms of the left form are mutually related in the same way as the atoms of the right form. In order to clarify this, we begin with a mono-orbit system. Let $\boldsymbol{L}_{\rm R}$ be the symmetry of its right-handed form. Then all the atoms are mutually connected by the operations of $\boldsymbol{L}_{\rm R}$: any two atoms positioned at $\boldsymbol{r}$ and $\boldsymbol{r}'$ there is a symmetry transformation $\ell_{\rm R} \in \boldsymbol{L}_{\rm R}$ such that $\boldsymbol{r}' = \ell_{\rm R}\boldsymbol{r}$. In the inverted system, i.e., in the left form, we get $\mathcal{I}\boldsymbol{r}' = \mathcal{I}\ell_{\rm R}\boldsymbol{r} = (\mathcal{I}\ell_{\rm R}\mathcal{I}^{-1})\mathcal{I}\boldsymbol{r}$. Thus, the inverted positions are related by the elements $\ell_{\rm L} = \mathcal{I}\ell_{\rm R}\mathcal{I}$ (obviously $\mathcal{I} = \mathcal{I}^{-1}$) of the left symmetry group $\boldsymbol{L}_{\rm L} = \mathcal{I}\boldsymbol{L}_{\rm R}\mathcal{I}$. The relative positions of the atoms are same if for each $\ell_{\rm R}$ there is $\ell_{\rm L} = (R|\boldsymbol{v})\ell_{\rm R}(R|\boldsymbol{v})^{-1}$. Here R and $\boldsymbol{v}$ are arbitrary rotation and translation (but same for all $\ell_{\rm R}$), enabling to move the system as a rigid body; essentially, they interrelate various right frames. Even more, if the elements $\ell_{\rm L}$ are not equal to the corresponding element $(R|\boldsymbol{v})\ell_{\rm R}(R|\boldsymbol{v})^{-1}$, but they are still elements of the set $(R|\boldsymbol{v})\boldsymbol{L}_{\rm R}(R|\boldsymbol{v})^{-1}$, we would get indistinguishable system, as this means that at the position $\mathcal{I}\boldsymbol{r}'$ is an equivalent atom to that being at $\boldsymbol{r}'$ in the right form. So, the criterion of (a)chirality of a mono-orbit system is related exclusively to the symmetry groups of the right and left conformations: the system is achiral if the left and right groups are *geometrically conjugated* (note that they are always conjugated by $\mathcal{I}$ and thus isomorphic):

$$\boldsymbol{L} = (R|\boldsymbol{v})^{-1}\mathcal{I}\boldsymbol{L}\mathcal{I}(R|\boldsymbol{v}). \tag{2.5}$$

Otherwise it is chiral. Still, in the case of multi-orbit systems, with symcell atoms at $\boldsymbol{r}_1, \ldots, \boldsymbol{r}_S$, the above analysis provides achirality criterion for each orbit separately, but the relative positions of the spatially reversed symcell atoms (being at $\mathcal{I}\boldsymbol{r}_1, \ldots, \mathcal{I}\boldsymbol{r}_S$) are to be analyzed independently. Obviously, this is reduced to comparison of the sets of vectors $\boldsymbol{r}_{ij} = \boldsymbol{r}_i - \boldsymbol{r}_j$ and $(R|\boldsymbol{v})\mathcal{I}\boldsymbol{r}_{ij}$ for each pair of the symcell atoms. Achirality appears only if simultaneously with condition (2.5) these two sets are equivalent, in the sense that they are the same up to the possible permutation of the chemically same symcell atoms.

Obviously, if spatial inversion belongs to $\boldsymbol{L}$, left and right groups are the same and the system is achiral.

All orthogonal transformations (pure rotations, including U-axes, mirror planes, and roto-reflections) commute with $\mathcal{I}$ (i.e., $\mathcal{I}\ell\mathcal{I} = \ell$ for such an element ℓ), so these symmetries cannot provide chirality. Also, as $\mathcal{I}(\sigma_{\text{v}}|a/2)^t\mathcal{I} = (\sigma_{\text{v}}|a/2)^{-t}$, spatially inverted glide plane $\boldsymbol{T}'(a/2)$ remains the same (only its elements are reordered). Therefore, chirality is completely determined by helical axis solely, i.e., by the first family subgroup of the line group. As we will show in the Sect. 2.2.5, there are only two *achiral groups* of the first family: *translational* or *symmorphic* $\boldsymbol{T}(a)\boldsymbol{C}_n$ and *zigzag* or *non-symmorphic* $\boldsymbol{T}^1_{2n}(a/2)\boldsymbol{C}_n$. Thus, all the line groups with such $\boldsymbol{L}^{(1)}$ are achiral.

However, vast majority of the line group families are achiral, namely, chiral groups are only in the two (out of 13) families, 1 and 5. Besides, not all the groups from these families are chiral: two infinite series ($Q = n$ and $Q = 2n$) are achiral line groups. In other words, mirror and glide planes, as well as roto-reflectional axes are compatible only with achiral roto-translations.

2.2 First Family Line Groups

The first family line groups are used in the construction of all the other line groups and their irreducible representations. Here we derive the basic properties of the first family groups. Generally, they are direct products of a helical and rotational group, $\boldsymbol{T}_Q(f) \otimes \boldsymbol{C}_n$, being thus abelian and non-symmorphic.

2.2.1 Helix Generated by the Helical Group

Let us chose an arbitrary point $\boldsymbol{r}_0 = (\rho, \varphi_0, z_0)$ (cylindrical coordinates) which is not at the z-axis (i.e., $\rho > 0$). Transformations of the screw-axis $\boldsymbol{T}_Q(f)$ map $\boldsymbol{r}_0$ into the infinite set of points:

$$\boldsymbol{r}_t = (C_Q|f)^t \boldsymbol{r}_0 = \left(\rho, \varphi_0 + t\frac{2\pi}{Q}, z_0 + tf\right). \tag{2.6}$$

All these points are at a cylinder of radius ρ, and at a helix given by (2.6) when t is continuously changed (Fig. 2.2). The inclination of the helix is $\chi = \arctan\frac{fQ}{2\pi\rho}$, while the increase of z in a single turn is *helix step* $h = fQ$. Among all right-handed arcs (right-handed rotations are accompanied by increase of z) connecting successive points r_t, the minimal in length is just the one along the helix. Also, one can draw another left-handed minimal helix connecting the points, involving left rotations for $2\pi/Q'$, with $Q' = Q/(Q-1)$, i.e., generated by $(C^{-1}_{Q'}|f)$. The inclinations of the two helices are related by $1/\tan\chi + 1/\tan\chi' = 2\pi\rho/f$.

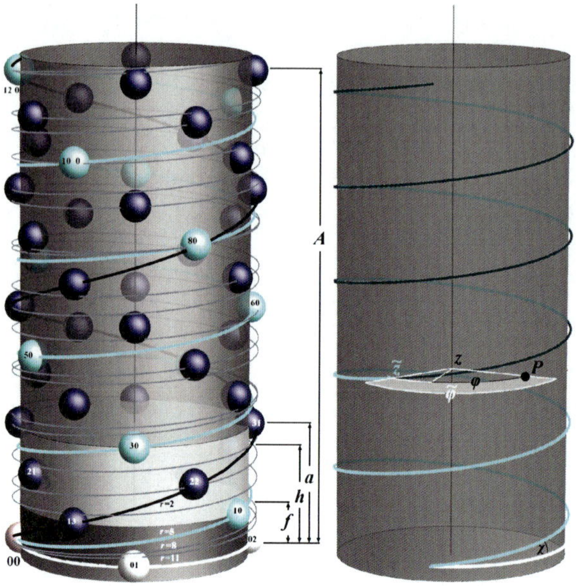

Fig. 2.2 Helical coordinates. Some of the points r_{ts} obtained by $\boldsymbol{L} = \boldsymbol{T}_{12}^{5}(f)\boldsymbol{C}_4$ from $\boldsymbol{r}_{00}$ in accordance with (2.10) are on the *left* denoted by pairs ts; among them, $t0$ are on the standard (convention C1) helix ($r = 5$, *thick line*), while other helices correspond to the alternative helical groups $\boldsymbol{T}_{12}^{r_0=2}(f)$ (convention C0, *black*), $\boldsymbol{T}_{12}^{r_2=8}(f)$ and $\boldsymbol{T}_{12}^{r_3=11}(f)$. Fractional translation f, line group period a and period A of $\boldsymbol{T}_{12}^{5}(f)$ are indicated as *dark, light*, and *medium* parts of cylinder. Standard helix is used (*right*) to define coordinates $\tilde{\varphi}$ and $\tilde{z}$ of an arbitrary point p. As $r = 5$, helix step h is $12f/5$

We use a helix generated by a screw-axis in order to define the coordinate system in which action of the helical generator changes a single coordinate. Given an arbitrary point $\boldsymbol{r} = (\rho, \varphi, z)$ of the cylinder, starting from the point $x = \rho$, $y = z = 0$, we move along the helix (2.6) with $\varphi_0 = z_0 = 0$ until we reach the height z; the corresponding length of the helix is $\tilde{z}$. To reach $\boldsymbol{r}$ it remains to move along the horizontal circle for (right-handed) angle $\tilde{\varphi}$. Obviously, the *helical coordinates* $(\rho, \tilde{\varphi}, \tilde{z})$ uniquely define $\boldsymbol{r}$. They are related to the cylindrical ones as:

$$\rho = \rho, \quad z = \frac{h}{\tilde{h}}\tilde{z} = \tilde{z}\sin\chi, \quad \varphi = \tilde{\varphi} + \frac{2\pi}{\tilde{h}}\tilde{z} = \tilde{\varphi} + \frac{\cos\chi}{\rho}\tilde{z}, \tag{2.7}$$

where $\tilde{h} = \sqrt{4\pi^2\rho^2 + h^2} = h/\sin\chi$ is the increase of $\tilde{z}$ per turn. Note that h is determined by the group parameters only, while χ and $\tilde{h}$ depend additionally on the radial coordinate ρ.

Momenta $\tilde{p}$ and $\tilde{l}_z$ conjugated to the helical coordinates $\tilde{z}$ and $\tilde{\varphi}$ are combinations of the z components of linear momentum p_z (conjugated to z) and angular momentum l_z (conjugated to φ):

$$\tilde{p} = \frac{h}{\tilde{h}} p_z + \frac{2\pi}{\tilde{h}} l_z = \sin\chi p_z + \frac{\cos\chi}{\rho} l_z, \quad \tilde{l}_z = l_z. \tag{2.8}$$

The last equality manifests that the coordinates φ and $\tilde{\varphi}$ have the same physical sense: as the corresponding coordinate lines are the same circles, with different starting points only, the angular momenta l_z and $\tilde{l}_z$ are equal. On the contrary, helical momentum $\tilde{p}$ combines linear and angular momenta in the ρ-dependent way, as its z and circumferential components are proportional, respectively, to the sine and cosine of the inclination angle, which decreases with ρ (the step h is fixed).

2.2.2 Different Factorizations and Conventions

To get the points generated by the first family line group, it remains to apply group $\boldsymbol{C}_n$ to the set of points (2.6). In other words, the elements of the group are of the roto-helical form:

$$\ell_{ts} = (C_Q|f)^t C_n^s. \tag{2.9}$$

which manifests the factorization of the line group onto the cyclic helical $\boldsymbol{T}_Q$ and rotational $\boldsymbol{C}_n$ subgroups. Obviously, $\ell_{ts} = \ell_{10}^t \ell_{01}^s$. Alternatively, we shall use $\tilde{\ell}(t,s)$ instead of ℓ_{ts}.

The elements ℓ_{ts} map $\boldsymbol{r}_0 = \boldsymbol{r}_{00}$ into the points

$$\boldsymbol{r}_{ts} = (C_Q|f)^t C_n^s \boldsymbol{r}_0 = \left(\rho, \varphi_0 + 2\pi\left(\frac{t}{Q} + \frac{s}{n}\right), z_0 + tf\right). \tag{2.10}$$

For the fixed s set of the points $\boldsymbol{r}_{ts}$ lay on the helix $\boldsymbol{T}_Q(f)$. However, n points $\boldsymbol{r}_{1s}$ (at the height $z_0 + f$) may also be connected with $\boldsymbol{r}_0$, giving the arcs of n different helixes counted by s (Fig. 2.2). Note that non-minimal helixes, making more than one turn before passing through $\boldsymbol{r}_{1s}$ are not taken into account as $Q \geq 1$ is assumed. All of them may be defined with the help of rotational angles $\phi = 2\pi/Q$ and $\phi_s = \phi + s\frac{2\pi}{n} = 2\pi/Q_s$ ($s = 0, \pm 1 \ldots$), i.e., these helixes correspond to the transformations $Z^{(s)} = (C_Q|f)C_n^s = (C_{Q_s}|f)$, with $Q_s = \frac{Qn}{n+sQ}$. Therefore, although each $Z^{(s)}$ generates different helical group $\boldsymbol{T}_{Q_s}(f)$, all the products $\boldsymbol{T}_{Q_s}\boldsymbol{C}_n$ are the same and equal to $\boldsymbol{T}_Q(f)\boldsymbol{C}_n$ (obtained for $s = 0$). Thus, to classify the first family line groups we have to resolve this non-uniqueness by a convention.

To this end we consider arbitrary factorization $\boldsymbol{T}_{Q'}(f)\boldsymbol{C}_n$ and determine all equivalent $Q'_s = \frac{nQ'}{n+sQ'}$. Then, by the convention which will be referred to as C0, we assign to Q the maximal finite Q'_s. This means that we take s which provides the minimal positive denominator $n + sQ'$. It is easy to find that

$$\text{Convention C0:} \quad Q = \begin{cases} Q' & \text{if } n \le Q' \text{ (for } s = 0), \\ \frac{nQ'}{n+Q'+Q'\left[-\frac{n}{Q'}\right]} & \text{if } n \ge Q' \left(\text{for } s = 1 + \left[-\frac{n}{Q'}\right]\right) \end{cases} \tag{2.11}$$

($[x]$ is integer part of x). Consistently, for $n = Q'$ the both expressions give $Q = Q'$. Geometrically, the introduced convention means that the helix with the maximal inclination is chosen.[6] It is easy to show that C0 always give $Q \geq n$.

2.2.3 Commensurability

Now we analyze commensurability of the first family groups. If exist, pure translations form a cyclic group generated by the minimal among them $(I|a)$, i.e., a denotes the *translational period*. Therefore, arbitrary translation is of the form $(I|tf)$, and according to (2.9), commensurability requires the existence of t and s such that $(I|tf) = (C_Q^t|tf)C_n^s$. This gives equation $\frac{2\pi}{Q}t + \frac{2\pi}{n}s \stackrel{2\pi}{=} 0$, and it is solvable if and only if $\frac{t}{Q} + \frac{s}{n}$ is an integer. Obviously, when the helical factor is commensurate, the whole group $\boldsymbol{T}_Q(f)\boldsymbol{C}_n$ is commensurate, since additional symmetry $\boldsymbol{C}_n$ cannot break translational periodicity. On the contrary, in the case of the incommensurate helical group the rational number s/n cannot cancel the irrational $1/Q$, and the condition cannot be satisfied. Thus, the first family line groups are commensurate if and only if the helical factor is commensurate.

As for commensurate groups, the convention on the helical factor can be more specified. Namely, for $Q' = q'/r'$, with positive co-primes q' and r' (all the other cases may be reduced to this one) (2.11) becomes

$$Q = \frac{q}{r}, \text{ with } q = \text{LCM}(q', n) = n\tilde{q}, \; r = \begin{cases} \frac{q}{n}\left\{\frac{r'n}{q'}\right\} & \text{if } \frac{q}{n}\left\{\frac{r'n}{q'}\right\} \neq 0, \\ 1 & \text{otherwise.} \end{cases} \tag{2.12}$$

Recall (Sect. 2.1.5) that here q is the order of the principle axis of the isogonal group $\boldsymbol{C}_q$. Also, we introduced quantity $\tilde{q} = q/n$, which proves to be useful in further discussions. It is easy to show that r and $\tilde{q} = q/n = q'/\text{GCD}(n, q')$ are co-primes. However, r may have common factors with n and thus with q. If these factors are canceled, and Q written in the simplest form as assumed by the convention C0, the convenient form of q will be lost. To remedy this, we slightly change the convention. In fact, various helical generators may now be written in the form $(C_q^{r_s}|f) = (C_q^{r_0}|f)C_n^s$, $r_s = r_0 + s\tilde{q}$, with $r = r_0$. Each r_s is co-prime with $\tilde{q}$, but at least one of them is also co-prime with q. So, by the convention C1 we choose r as the minimal r_s which is co-prime with q, retaining thus the more useful form of q.

[6] Also, one can find the minimal Q_s not less than one; in the above-mentioned cases, it is $Q_{\min} = nQ'/(n + nQ' - Q')$ (for $n \leq Q'$) and $Q_{\min} = nQ'/(n + Q'[(nQ' - n)/Q'])$ (for $n \geq Q'$).

Note that the translational period of the helical factor $\boldsymbol{T}_q^{r_s}(f)$ is $qf/\mathrm{GCD}(n, r_s)$ and that within convention C1 it gets simple form: qf.

Although $\boldsymbol{C}_n$ cannot break the commensurability of the helical factor, the translational period of $\boldsymbol{T}_q^r(f)\boldsymbol{C}_n$ is in general decreased with respect to that of the helical factor. In order to find this period, we determine t solving the commensurability condition, $\frac{t}{Q} + \frac{s}{n}$ being integer. Knowing that $Q = \tilde{q}n/r$, this condition transforms to $rt + s\tilde{q} = iq$, $i = 0, \pm 1, \ldots$, equation of the type (C.7). As $\mathrm{GCD}(r, \tilde{q}) = 1$, it has always solutions[7] (C.8). The translational period of the whole group is defined as tf for the minimal positive solution in t. This is $t = \tilde{q}$ (then $s = -r$), obtained for $i = 0$. Thus, the minimal pure translation is $(C_q^r|f)^{\tilde{q}}C_n^{-r} = (I|\tilde{q}f)$, and the translational period of $\boldsymbol{T}_q^r(f)\boldsymbol{C}_n$ is

$$a = \tilde{q}f. \tag{2.13}$$

As the translational period of the helical factor alone is $qf/\mathrm{GCD}(r, n)$, it turns out that the rotational group $\boldsymbol{C}_n$ decreases period, i.e., increases translation symmetry of $\boldsymbol{T}_q^r(f)$ by $n/\mathrm{GCD}(r, n)$. Particularly, within the convention C1 the period is decreased n times. It immediately follows that in this case the step of the generated helix is $h = an/r$. Apart from n and a defining periodicity of the system along coordinates φ and z, r is necessary to define the inclination, and therefore we call it *helicity parameter*.

Summarizing all these, the commensurate line groups of the first family are uniquely given by the integers r, n, $\tilde{q}$, and real f, in the form (according to the conventions (a) and (b)):

$$\boldsymbol{L} = \boldsymbol{T}_{\tilde{q}n}^r(f) \otimes \boldsymbol{C}_n, \quad n = 1, 2, \ldots, \quad \begin{cases} \mathrm{C0}: r = 0, 1, \ldots \tilde{q}, & \mathrm{GCD}(\tilde{q}, r) = 1, \\ \mathrm{C1}: r = 0, 1, \ldots \tilde{q}n, & \mathrm{GCD}(\tilde{q}n, r) = 1. \end{cases} \tag{2.14}$$

Factorized form of the general element of these groups is

$$\ell_{ts} = \left(C_q^{rt}|t\frac{a}{\tilde{q}}\right)C_n^s, \quad s = 0, \ldots, n-1; \; t = 0, \pm 1, \ldots \tag{2.15}$$

The achiral groups have within the both conventions the same factorization: $\boldsymbol{T}_n^1(f = a)\boldsymbol{C}_n$ and $\boldsymbol{T}_{2n}^1(f = a/2)\boldsymbol{C}_n$, translational and zigzag groups, respectively. Note that $\boldsymbol{T}(a)$ instead of the alternative for notation $\boldsymbol{T}_n^0(a)$ (i.e., $r = 0$, with the helix reduced to the vertical line) can be used.

Finally, let us mention that when it is convenient we shall unify notation for commensurate and incommensurate groups, writing helical factor either as $\boldsymbol{T}_Q(f)$ (assuming $Q = q/r$ for commensurate groups) or as $\boldsymbol{T}_q^r(f)$ (assuming irrational $q = Q$ and $r = 1$ for incommensurate groups).

[7] The solvability, i.e., commensurability, has been already provided by rationality of Q.

2.2.3.1 Translational Factorization and International Notation

As discussed in Sect. 2.1.5, when $\boldsymbol{L}$ is commensurate, pure translations form its invariant subgroup. For the commensurate first family line group the factor group (isogonal point group) is $\boldsymbol{C}_q$, and $\boldsymbol{L}$ can be decomposed into q cosets of the translational group. The coset representatives may be taken in the form $(C_q^j|f_j)$ ($j = 0, \ldots, q-1$), where f_j are the corresponding fractional translations (less than a). To find f_j we write coset representatives in the factorized form (2.15): $(C_q^j|f_j) = (C_q^{rt_j}|t_j\frac{a}{\tilde{q}})C_n^{s_j}$; obviously all t_j are less than $\tilde{q}$. For $j = 1$ the rotational part gives the condition $C_q = C_q^{rt_1}C_n^{s_1}$, i.e., $rt_1 + s_1\tilde{q} = 1$. This is Diophantine equation (C.7) in t_1 and s_1, with co-primes r and $\tilde{q}$ (independently of the convention used for r) having solutions (C.8). The only solution in t_1 which is less than $\tilde{q}$ is $t_1 = \tilde{p} = r^{\mathrm{Eu}(\tilde{q})-1} = r^{-1}_{(\tilde{q})}$ (inverse of r modulo $\tilde{q}$; see Appendix C). Thus, the first coset representative is $(C_q|a\frac{\tilde{p}}{\tilde{q}}) = (C_q|a\frac{p}{q})$, where $p = n\tilde{p}$ is integer less than q. Clearly, $\tilde{p}$ indicates the monomer at the minimal height (equal to $z = \tilde{p}f$) with the atoms rotated for $2\pi/q$ (the minimal rotation involved with respect to the initial monomer[8] (at $z = 0$)). The other coset representatives can be the powers of the first one; as we want that f_j is less than a, we subtract appropriate number of periods from jpa/q and take them in the form $(C_q^j|\{\frac{jp}{q}\}a)$, getting the coset decomposition:

$$\boldsymbol{L} = \sum_{j=1}^{q} \left(C_q^j| \left\{\frac{jp}{q}\right\} a\right) \boldsymbol{T}(a). \tag{2.16}$$

This gives another factorization of the elements of the line group:

$$\ell(t, j) = \left(C_q^j|ta + \left\{\frac{jp}{q}\right\} a\right), \quad j = 0, \ldots, q-1; \; t = 0, \pm 1, \ldots. \tag{2.17}$$

Obviously, the translational subgroup is generated by $\ell(1, 0)$ and contains all the elements $\ell(t, 0)$. The element $\ell(0, 1) = (C_q|a\frac{p}{q})$ generates the helix with step $h = pa$ and inclination $\tan\chi = pa/2\pi\rho$. The coset representatives $\ell(0, j)$ do not form a subgroup in $\boldsymbol{L}$ unless $p = 0$, as their translational part is fractional translation $\{jp/q\}a$ instead of jpa/q of $\ell^j(0, 1)$. Accordingly, we call the factorization (2.17) *translational form*. Only in the case of symmorphic groups ($p = 0$), using the choice $r = 0$ for the helical factor (i.e., $\boldsymbol{T}(a) = \boldsymbol{T}_n^0(a)$), the two factorization become the same: $\ell_{ts} = \ell(t, s)$.

Starting from q, r, and n, we have found p. Oppositely, if q and p, are given, then as we have seen $n = \mathrm{GCD}(p, q)$, while r is modular (with respect to q/n) inverse of $\tilde{p}$. Therefore, the first family line group is given by q, f, and either p or

---

[8] Note that r and p, respectively, correspond to the minimal (possibly not pure) rotation and translation involved: r is (convention C0) chosen such that C_q^r is the minimal rotation mapping initial monomer to the monomer at this minimal height f.

the pair (r, n). The transition rules are

$$n = \mathrm{GCD}(n, p), \quad r = \tilde{p}^{-1}_{(\tilde{q})} + l\tilde{q}, \quad p = nr^{-1}_{(\tilde{q})}; \tag{2.18}$$

here the inverse modular modulo $\tilde{q}$ is used (see C.5). In the second expression, l is zero within the convention C0, while in C1 it must be calculated. Note again that p is independent on the convention used for r. As q, p, and period a completely determine line group, they are used in the so-called *international notation* (Table 2.2): a line group of the first family is denoted as $\boldsymbol{L}q_p(a)$ ($q = 1, 2, \ldots$; $p = 0, 1, \ldots, q-1$). It is now easily seen that for any positive integer q there are q different commensurate first family line groups $\boldsymbol{L}q_p$ with the isogonal group $\boldsymbol{C}_q$. Among them, only for $p = 0$ the symmorphic line group $\boldsymbol{T}(a)\boldsymbol{C}_{q=n}$ is obtained.

2.2.3.2 Transition Between the Two Factorizations

As both factorization (2.15) and (2.17) have some advantages, we will use both. Here we interrelate them to enable direct switch from one to another. For this purpose it is convenient to use notation $\tilde{\ell}(t, s) = \ell_{ts}$.

To get the translational form of the roto-helical generators we find the corresponding t and j of (2.17). From $\tilde{\ell}(0, 1) = \ell(t, j) = (C_q^j|(t + \{\frac{jp}{q}\})a)$ we get $t + \{\frac{jp}{q}\} = 0$, i.e., $t = 0$ and $\{\frac{j\tilde{p}}{\tilde{q}}\} = 0$. As $\tilde{p}$ and $\tilde{q}$ are co-primes, this is the case if and only if j is multiple of $\tilde{q}$, yielding $\tilde{\ell}(0, 1) = \ell(0, \tilde{q})$. From $\tilde{\ell}(1, 0) = \ell(t, j)$, equality of the translational parts $t + \{\frac{j\tilde{p}}{\tilde{q}}\} = 1/\tilde{q}$ gives $t = 0$, while the fractional parts are equal to $1/\tilde{q}$ for $j = r$, i.e., $\tilde{\ell}(1, 0) = \ell(0, r)$. Analogously, for the roto-helical form of the translational generators, we solve (2.15) in t and s for each of them. From $\ell(0, 1) = \tilde{\ell}(t, s) = (C_q^{rt+\tilde{q}s}|ta/\tilde{q})$, the translational part immediately gives $t = \tilde{p}$, while s must satisfy $s\tilde{q} \stackrel{q}{=} 1 - r\tilde{p}$, i.e., $s = n\{\frac{1-r\tilde{p}}{q}\}$. Thus, $\ell(0, 1) = \tilde{\ell}(\tilde{p}, n\{\frac{1-r\tilde{p}}{q}\})$. Finally, solving $\ell(1, 0) = \tilde{\ell}(t, s)$ we get $\ell(1, 0) = \tilde{\ell}(\tilde{q}, -n\{\frac{r}{n}\})$. Altogether, the two set of generators of the commensurate groups are related as follows:

$$\tilde{\ell}(1, 0) = \left(C_q^r|\frac{a}{\tilde{q}}\right) = \ell(0, r), \qquad \tilde{\ell}(0, 1) = C_n = \ell(0, \tilde{q}), \tag{2.19a}$$

$$\ell(1, 0) = (I|a) = \tilde{\ell}\left(\frac{q}{n}, -n\left\{\frac{r}{n}\right\}\right), \quad \ell(0, 1) = \left(C_q|\frac{p}{q}a\right) = \tilde{\ell}\left(\frac{p}{q}, n\left\{\frac{1 - r\tilde{p}}{q}\right\}\right). \tag{2.19b}$$

2.2.4 *Isomorphisms and Physical Equivalence*

The first family line groups are determined by the group parameters Q, f, and n. As Q and f can take any real value, the group parameters make a continuous set. In principle, each of these parameters may be experimentally measured, and therefore these groups are physically different.

However, all the helical groups $\boldsymbol{T}_Q(f)$ are infinite cyclic groups, and therefore they are mutually isomorphic. Hence, nonisomorphic groups are distinguished by the natural parameter n only.

In many experiments, only the isogonal group can be manifested. If it is finite, then the line group is commensurate. This opens the question which are different commensurate groups with the same q. Nonisomorphic groups are defined by different possible values of n, i.e., by different divisors of q. Despite continual parameter f, it remains to determine different values of r. To count them we use convention C0, when r is co-prime with $\tilde{q}$. As each such co-prime defines different group, the number of groups with fixed q and n equals the number of co-primes with $\tilde{q}$ less than $\tilde{q}$, i.e., to the Euler function $\mathrm{Eu}(\tilde{q})$. Hence, summing over all divisors of q, we get $\sum_n \mathrm{Eu}(\frac{q}{n}) = q$ (which is a number theory theorem). Indeed, as shown at the end of Sect. 2.2.3.1, these groups are counted by the parameter $p = 0, \dots, q-1$.

2.2.5 Chirality

To discuss chirality of the first family group $\boldsymbol{L}_{\mathrm{R}} = \boldsymbol{T}_{Q_{\mathrm{R}}}(f)\boldsymbol{C}_n$, we have to determine its spatially inverted group $\boldsymbol{L}_{\mathrm{L}} = \mathcal{I}\boldsymbol{L}_{\mathrm{R}}^{(1)}\mathcal{I}$. Since $\mathcal{I}(C_Q|f)^t C_n^s \mathcal{I} = (C_Q|-f)^t C_n^s$, we see that the rotational factor $\boldsymbol{C}_n$ is the same. Further, to get positive fractional translation, instead of $(C_Q|-f)$ we use its inverse $(C_Q^{-1}|f)$ generating the same helical factor of the "left" group. Therefore, also f is the same (i.e., R and L indices are not necessary for n and f), only in the "left" helical factor it is coupled with $C_Q^{-1} = C_{Q'_{\mathrm{L}}}$; as $2\pi/Q'_{\mathrm{L}} = 2\pi - 2\pi/Q_{\mathrm{R}}$, we find $Q'_{\mathrm{L}} = Q_{\mathrm{R}}/(Q_{\mathrm{R}}-1)$. To find "left" helical group according to the convention C0 we apply (2.11), for $Q'_{\mathrm{L}} \le n$ (this is fulfilled as in C0 convention $Q_{\mathrm{R}} \ge n \ge 1$). This gives

$$Q_{\mathrm{L}} = \frac{nQ_{\mathrm{R}}}{nQ_{\mathrm{R}} - n + Q_{\mathrm{R}} + Q_{\mathrm{R}}\left[-n + \frac{n}{Q_{\mathrm{R}}}\right]} = \begin{cases} n, & \text{if } Q_{\mathrm{R}} = n, \\ \frac{nQ_{\mathrm{R}}}{Q_{\mathrm{R}}-n}, & \text{otherwise.} \end{cases} \tag{2.20}$$

The equation $Q_{\mathrm{L}} = Q_{\mathrm{R}} = Q$ has solutions $Q = n, 2n$, i.e., the only *achiral* first family groups are *translational* (*symmorphic*) $\boldsymbol{T}_n^1\boldsymbol{C}_n$ and *zigzag* (*non-symmorphic*) $\boldsymbol{T}_{2n}^1\boldsymbol{C}_n$, both commensurate with $\tilde{q} = 1, 2$, respectively, and $r = 1$. For other commensurate groups, substituting $Q_{\mathrm{R}} = q_{\mathrm{R}}/r_{\mathrm{R}}$ general solution (2.20) gives $q_{\mathrm{L}}/r_{\mathrm{L}} = q_{\mathrm{R}}/(q_{\mathrm{R}}/n - r_{\mathrm{R}})$, i.e.,

$$q_{\mathrm{R}} = q_{\mathrm{L}} = q, \quad r_{\mathrm{L}} = \tilde{q} - r_{\mathrm{R}} \quad \text{(convention C0).} \tag{2.21}$$

2.3 Other Families

Here we briefly consider the most important structural properties of the other line group families. Although these groups are more complicated, their properties may be derived using their maximal first family subgroup.

2.3.1 Elements

The factorization (2.3) of the line groups enables to generalize the form (2.15) and to factorize elements of any line group as follows $\ell_{tp} = Z^t P$. Here Z generates the generalized translations subgroup, and P is the axial point group element. Hence P can be further factorized to the rotation C_n^s around the system axis and additional one or two *parities* P_1, P_2 (generators of the second order, i.e., $P_i^2 = I$), so that the general form of the line group element is

$$\ell_{tsp_1p_2} = z^t C_n^s P_1^{p_1} P_2^{p_2}, \quad t = 0, \pm 1, \ldots; \ s = 0, \ldots, n-1; \ p_1, p_2 = 0, 1. \tag{2.22}$$

Only for second family line groups the last generator S_{2n} is not a parity, being of order $2n$. However, as $S_{2n}^{2p} = C_n^p$, the general form is $\ell_{tsp} = (I|a)^t C_n^s S_{2n}^p = (I|a)^t S_{2n}^{2s+p}$ for $p = 0, 1$, and (2.22) is valid.[9]

The action of all the line group transformations ℓ in the Euclidean space leaves the radial coordinate ρ invariant. Therefore, it is effectively reduced to the cylinder and can be explicated both in the cylindrical and helical coordinates. Using the general form (2.22), from the following results we can generate this action for any line group element:

$$(C_Q|f)(\varphi, z) = \left(\varphi + \frac{2\pi}{Q}, z + f\right), \quad (C_Q|f)(\tilde{\varphi}, \tilde{z}) = \left(\tilde{\varphi}, \tilde{z} + \frac{\tilde{h}}{Q}\right), \tag{2.23}$$

$$C_n(\varphi, z) = \left(\varphi + \frac{2\pi}{n}, z\right), \quad C_n(\rho, \tilde{\varphi}, \tilde{z}) = \left(\tilde{\varphi} + \frac{2\pi}{n}, \tilde{z}\right), \tag{2.24}$$

$$(I|f)(\varphi, z) = (\varphi, z + f), \quad (I|f)(\tilde{\varphi}, \tilde{z}) = \left(\tilde{\varphi} - \frac{2\pi}{h} f, \tilde{z} + \frac{\tilde{h}}{h} f\right), \tag{2.25}$$

$$U(\varphi, z) = (-\varphi, -z), \quad U(\tilde{\varphi}, \tilde{z}) = (-\tilde{\varphi}, -\tilde{z}), \tag{2.26}$$

[9] For the families 2 and 9 S_{2n} can be used instead of C_n to get the general forms $\ell_{ts} = (I|a)^t S_{2n}^s$ and $\ell_{tsp} = (I|a)^t S_{2n}^s \sigma_v^p$, respectively (this reduces the number of generators to 2 and 3).

$$C_n\sigma_\mathrm{h}(\varphi,z)=\left(\varphi+\frac{2\pi}{n},-z\right),\quad C_n\sigma_\mathrm{h}(\tilde\varphi,\tilde z)=\left(\tilde\varphi+\frac{4\pi}{\tilde h}\tilde z,-\tilde z\right),\tag{2.27}$$

$$C_n\sigma_\mathrm{v}(\varphi,z)=\left(-\varphi+\frac{2\pi}{n},z\right),\quad C_n\sigma_\mathrm{v}(\tilde\varphi,\tilde z)=\left(-\tilde\varphi-\frac{4\pi}{\tilde h}\tilde z+\frac{2\pi}{n},\tilde z\right).\tag{2.28}$$

The action of the mirror planes is obtained from the last two equations for $n=0$. Note that $(C_Q|f)$ and C_n change only a single helical coordinate, $\tilde z$ and $\tilde\varphi$, respectively.

2.3.2 First Family Subgroup

The set of all roto-helical transformations forms a subgroup $\boldsymbol{L}^{(1)}$ of any line group. In the case of the first family groups, $\boldsymbol{L}^{(1)}$ is the group itself, i.e., the trivial subgroup. For the families $F=2,\ldots,8$ it is a halving subgroup, $N_F=|\boldsymbol{L}^{(F)}|/|\boldsymbol{L}^{(1)}|=2$, and for the remaining families $F=9,\ldots,13$, the first family subgroup is a subgroup of index four: $N_F=4$. Assuming $\ell_1^{(F)}=(I|0)$, the line group decomposition is analogous to the one of the point factor (2.1):

$$\boldsymbol{L}^{(F)}=\sum_{i=1}^{N_F}\ell_i^{(F)}\boldsymbol{L}^{(1)}.\tag{2.29}$$

For the families 9–13, with $\boldsymbol{L}^{(1)}$ and only one of the three cosets one gets one of the groups from families $2,\ldots,8$, as a halving subgroup. The coset representatives $\ell^{(F)}$ are listed in Table 2.2.

Let us emphasize once again that the first family subgroup in all the families except the first and fifth ones are achiral (either translational or zigzag) and also commensurate. Thus, only the first and fifth family groups can be chiral (when $Q\neq n,2n$) and incommensurate (for irrational Q), while the other families are achiral and commensurate.

2.3.3 Subgroups Preserving *z*-Axis

Leaving z-axis invariant, the line group transformations may reverse its orientation. Hence, there are two classes of such elements: *positive* ones preserve the direction of z-axis, while *negative* elements reverse it. As products of the positive elements are always positive they form a subgroup $\boldsymbol{L}^{+}$ of any line group. On the other hand, if there is a negative element, ℓ^{-}, then its product with any positive element is negative, meaning that the negative elements are coset of $\boldsymbol{L}^{+}$. Therefore, there are positive and negative line groups. *Positive groups* contain only positive elements, i.e., $\boldsymbol{L}=\boldsymbol{L}^{+}$. As neither of the generalized translations reverses z-axis, the type of the line group is effectively determined by the type of the point factor: *positive* are the families 1, 6, 7, and 8. The remaining families pertain to the *negative groups*, with positive elements forming a halving subgroup (which is a line group itself, Table 2.2):

$$\boldsymbol{L} = \boldsymbol{L}^{+} + \ell^{-}\boldsymbol{L}^{+}. \qquad (2.30)$$

Obviously, the first family subgroup is always contained in $\boldsymbol{L}^{+}$.

2.3.4 International Notation

According to the standard crystallographic prescription, the commensurate line groups have the international symbols $\boldsymbol{L}P$ given in Table 2.2 . Here, when $\boldsymbol{L}$ is symmorphic P is the international symbol P_I of the isogonal point group $\boldsymbol{P}_\mathrm{I}$. For non-symmorphic groups, P is obtained by modifying symbol of P_I in order to indicate the non-symmorphic elements: the letter c is used instead of the letter m to indicate the glide plane and also q_p instead of q denotes the corresponding screw-axis.

The international notation [5, 4] gives the isogonal point group, slightly modified in the cases of the non-symmorphic line groups. It is based on the crystallographic conventions, covering only commensurate groups; thus it cannot be generalized to the full range of the line groups. Factorized notation explicitly manifests structure of the line groups, being convenient for various group-theoretical constructions and derivation of the properties related to the groups structure. These are the main reasons that throughout the text the factorized notation is used.

On the other hand, $\boldsymbol{Z}$ and $\boldsymbol{P}$ (including the parameter Q or the pair q and r) in the factorized notation are not unique, and additional conventions are necessary to fix them. Thus, when line groups (and later on magnetic line groups) are themselves derived, it is important to give their international symbols, in order to avoid discussions on the equivalence of the groups with different factors and parameters. Finally, some conclusions are derived using factorized notation within the most suitable convention; in such situations, expressing the result in terms of international notation (e.g., besides q, r, and n, also p is found) automatically provides the convention independent consideration (e.g., derivation of the magnetic line groups).

References

1. M. Damnjanović, M. Vujičić, Phys. Rev. B **25**, 6987 (1982) 3
2. T. Janssen, *Crystallographic groups* (North-Holland, Amsterdam, 1973)
3. L. Jansen, M. Boon, *Theory of Finite Groups. Applications in Physics* (North-Holland, Amsterdam, 1967)
4. M. Vujičić, I. Božović, F. Herbut, J. Phys. A **10**, 1271 (1977)
5. V. Kopsky, D. Litvin, *Subperiodic Groups, International Tables for Crystallography*, vol. E (Kluwer, Dordrecht, 2003)

Chapter 3
Symmetrical Compounds

Abstract Acting by a whole line group to a single atom one obtains orbit, i.e., an elementary system invariant under that line group. There are altogether 15 infinite classes of such elementary systems, and their various combinations give any system periodic along one direction, exhausting all line group symmetries. The most efficient way to completely determine configuration of a system is to define coordinates for one atom from each orbit.

3.1 Orbits of the Line Groups

In this section we list and depict all the systems having line group symmetry. As there are infinitely many line groups, the number of such systems is infinite and we give their general classification [1] and illustrate results for one group from each family.

3.1.1 Orbit, Stabilizer, and Transversal

Recall that the invariance of some polymer under the line group $\boldsymbol{L}$ means that any transformation ℓ of the line group maps any particular atom x into another one (of the same chemical type necessarily), say $x' = \ell' x$. We define *orbit* S_x of the atom x as the set of atoms obtained by the action of $\boldsymbol{L}$ on this atom. In other words, S_x is the set of atoms x' for which there is at least one element ℓ' of $\boldsymbol{L}$ such that $x' = \ell' x$, i.e., $S_x = \boldsymbol{L}x$. Note that the transformation ℓ'^{-1} maps x' into x. Consequently, if ℓ'' maps x into another member $x'' = \ell'' x$ of S_x, then also x' is mapped into x'' by $\ell''\ell'^{-1}$. This shows that the orbit of x is the same as the orbit of any other member of S_x, and that two orbits S_x and $S_{x'}$ are the same if and only if x is from $S_{x'}$ (than x' is from S_x, as well). We conclude that the whole orbit is completely determined by the position of any of its atoms. On the contrary, if x' is not in the orbit of x, then S_x and $S_{x'}$ are disjoint. Thus, choosing an atom x of the considered system, we can find its orbit S_x. If there are other atoms, we take arbitrary one x' being not in S_x, and acting on it by $\boldsymbol{L}$ we get $S_{x'}$. The procedure is repeated, until the whole system is partitioned into the disjoint orbits, each of them being invariant under $\boldsymbol{L}$.

Damnjanović, M., Milošević, I.: *Symmetrical Compounds*. Lect. Notes Phys. **801**, 29–46 (2010)
DOI 10.1007/978-3-642-11172-3_3

Obviously, the orbits are the simplest systems invariant under $\boldsymbol{L}$. They are the building blocks of any invariant system and gather the atoms being physically equivalent regarding both chemical properties and interaction with the neighbors (this is the symmetry-based definition of the *physical equivalence* of the atoms). Thus, each system is uniquely decomposed into a number of disjoint orbits, and the classification of the symmetrical compounds is reduced to determination of all the possible orbits of the line groups. However, orbits may be mutually quite different. For example, if x is an atom on the z-axis, then this is the case for all atoms of the orbit of x. As the operations of $\boldsymbol{C}_n$ leave x fixed, the number of atoms in such orbit is at least n times less than in the orbits with atoms out of z-axis. Therefore, the different types of orbits are to be found. This is performed with help of the *stabilizer* or *little group* $\boldsymbol{L}_x$ of atom x, which is in a sense complementary notion to the orbit S_x. It is the set of the elements of $\boldsymbol{L}$ for which x is a *fixed point*: $\ell_x x = x$. As it is easily verified, $\boldsymbol{L}_x$ is a subgroup of $\boldsymbol{L}$, and the partition of $\boldsymbol{L}$ into the cosets $c_i \boldsymbol{L}_x$ yields the factorization of the line group elements in the form $\ell = c_i \ell_x$. The whole coset maps the atom x into the same atom of the orbit S_x, meaning that the different atoms of the orbit are determined by the coset representatives c_i. Consequently, the orbit S_x is completely determined by the set of the coset representatives or *transversal* $\boldsymbol{Y}_x$. Further, it is straightforward to show that the atom $x_i = c_i x$ (from the orbit of x) has conjugated (and accordingly isomorphic) stabilizer $\boldsymbol{L}_{x_i} = c_i \boldsymbol{L}_x c_i^{-1}$. This is used to define *orbit type* as the set of the orbits with the conjugated stabilizers [2]. Considering the symmetry-based properties, the orbits of the same type are equivalent, making classification of the orbit types important for applications in physics. Hence, to find different orbits, we have to determine non-conjugated stabilizers of the line groups.

3.1.2 Construction of the Orbit Types of the Line Groups

Factorization of the line groups (2.3) is the starting point in the construction of the orbit types. In fact, it enables to employ already known [3] analogous classification of the point group orbits to solve this problem. As $\boldsymbol{Z}$ is an infinite cyclic group, generated by $Z = (R|f)$ without fixed points, it has no common elements with $\boldsymbol{L}_x$, and $\boldsymbol{L}_x$ must be finite. Therefore, the expansion of $\boldsymbol{L}_x$ into cosets of the point *stabilizer* subgroup $\boldsymbol{P}_x = \boldsymbol{P} \cap \boldsymbol{L}_x$ has the following form [4]: $\boldsymbol{L}_x = \sum_{i=0}^{K} Z^{t_i} p_i \boldsymbol{P}_x$, where Z^{t_i} and p_i are from $\boldsymbol{Z}$ and $\boldsymbol{P}$, respectively, and $Z^{t_0} = Z^0 = p_0 = (I|0)$. Besides the trivial situation, $K = 0$, when $\boldsymbol{L}_x = \boldsymbol{P}_x$, there are cases when there are cosets of $\boldsymbol{P}_x$ in $\boldsymbol{L}_x$. Then p_i must reverse z-axis for $i > 0$, because otherwise $Z^{t_i} p_i$ cannot have fixed points and cannot be in the stabilizer; thus the stabilizer form is $\boldsymbol{L}_x = \boldsymbol{P}_x + \sum_{i=1}^{K} Z^{t_i} p^- p_i^+ \boldsymbol{P}_x$, where p_i^+ are from $\boldsymbol{P}^+$ (see (2.2)), and $\boldsymbol{P}_x$ is a subgroup of $\boldsymbol{P}^+$, i.e., $\boldsymbol{P}_x = \boldsymbol{P}_x^+ = \boldsymbol{L}_x \cap \boldsymbol{P}^+$. For any two coset representatives $Z^{t_i} p^- p_i^+$ and $Z^{t_j} p^- p_j^+$ $(i, j \neq 0)$, the product $(Z^{t_i} p^- p_i^+)^{-1} Z^{t_j} p^- p_j^+$ must be in $\boldsymbol{L}_x$. However, this is possible only for $i = j$, since a simple calculation shows that this composite transformation diminishes z-coordinate of x for $f(t_j - t_i)$, and otherwise cannot be in the stabilizer. Accordingly $p_i^+ = p_j^+$, meaning that there is at most one coset: $\boldsymbol{L}_x = \boldsymbol{P}_x^+ + Z^t p^- p^+ \boldsymbol{P}_x^+$. Moreover, by conjugating with Z^s we verify that for t even $\boldsymbol{L}_x$ is conjugated to a purely point stabilizer (i.e., the mentioned trivial case is achieved again), and otherwise to a subgroup with $t = 1$. To summarize, there are two types of stabilizers:

$$\text{(a)} \quad \boldsymbol{L}_x = \boldsymbol{P}_x, \qquad \text{(b)} \quad \boldsymbol{L}_x = \boldsymbol{P}_x^+ + Zp^- \boldsymbol{P}_x^+. \tag{3.1}$$

Stabilizers of the type (a) are the stabilizers of the axial point group $\boldsymbol{P}$, which are known [3], while the second case can arise within nine families of the negative line groups (Sect. 2.3.3). Even in this case, all the subgroups $\boldsymbol{P}_x^+$ are the stabilizers of the point group $\boldsymbol{P}^+$, and it remains only to find various combinations Zp^-. In other words, for given $\boldsymbol{P}_x^+$ one should find possible elements p^- such that $\boldsymbol{P}_x^+ + Zp^- \boldsymbol{P}_x^+$ is a group, i.e., such that simultaneously satisfy conditions: (i) $(Zp^-)^2 \in \boldsymbol{P}_x^+$ and (ii) $p^- \boldsymbol{P}_x^+ = \boldsymbol{P}_x^+ p^-$ (when $\boldsymbol{Z}$ is an invariant subgroup (i) simplifies to $p^{-2} \in \boldsymbol{P}_x^+$).

Thus we have obtained a straightforward algorithm for classifying orbit types of line groups. If $\boldsymbol{P} = \boldsymbol{P}^+$, each orbit type of $\boldsymbol{P}$ generates one orbit type of $\boldsymbol{L}$, giving the complete set of $\boldsymbol{L}$-orbits. It turns out that this correspondence is bijective, except in the case of the groups $\boldsymbol{L}2n_n mc$ for n even, when the point group orbits b and c give the same orbit type of the line group (two stabilizers are conjugated by Z). In the case $\boldsymbol{P} = \boldsymbol{P}^-$, orbit types of $\boldsymbol{P}$ with the stabilizers reversing z-axis bijectively correspond to a part of $\boldsymbol{L}$-orbit types. However, $\boldsymbol{P}$-orbit type with the stabilizer from $\boldsymbol{P}^+$ (only orthogonal transformations are considered) is in general split when $\boldsymbol{Z}$ is introduced: (a) there are points in $\mathbb{R}^3$ for which $\boldsymbol{P}_x^+$ remains the stabilizer in $\boldsymbol{L}$, and they give one $\boldsymbol{L}$-orbit type (again, the correspondence is bijective, except that for the groups $\boldsymbol{L}2n_n/mcm$, where point group orbits b and c generate the same orbit type of the line group) and (b) for other points the stabilizer in $\boldsymbol{Z}$ is twice greater. In this case all the subgroups of the type (b) are to be found (using the conditions (i) and (ii)), and among them only those for which $\boldsymbol{P}_x^+$ and the coset have the same fixed points are retained. Finally, non-conjugated subgroups obtained in this manner generate bijectively the rest of the $\boldsymbol{L}$-orbit types.

3.1.3 Monomers and Orbit Orders

After this discussion on the stabilizers, a brief consideration of the corresponding orbits will be made. The whole polymer must be disjoint union of the orbits of $\boldsymbol{Z}$. Each $\boldsymbol{Z}$ orbit has $|\ \boldsymbol{Z}\ |$ points, since generalized translations have no fixed points. This enables to define *monomer* M, as the set containing one atom from each orbit of $\boldsymbol{Z}$. Then the whole system S is generated from M by the action of $\boldsymbol{Z}$ in the form of disjoint union $S = \sum_{t=-\infty}^{\infty} Z^t M$, and formally it contains $|\ S\ |=|\ M\ ||\ \boldsymbol{Z}\ |$ atoms.

If S_x is an $\boldsymbol{L}$-orbit of the type (a) there is no p in $\boldsymbol{P}$ and Z^t in $\boldsymbol{Z}$ such that $px = Z^t x$, otherwise $Z^{-t} p$ belongs to $\boldsymbol{L}_x$. In other words, S_x is factorized into orbits of $\boldsymbol{P}$ and orbits of $\boldsymbol{Z}$, giving the simple form of the monomer: $M = \boldsymbol{P}x = \{px | p \in \boldsymbol{P}\}$, i.e., one monomer is the orbit of $\boldsymbol{P}$ with x and $|\ S_x\ |=|\ \boldsymbol{Z}\ |\ \frac{|\boldsymbol{P}|}{|\boldsymbol{P}_x|}$. When S_x is of the type (b), then $Z^{-1}x = p^{-}x$ and clearly $\boldsymbol{P}x = \boldsymbol{P}^{+}x + Z\boldsymbol{P}^{+}x$ (to show this in detail note that $Z\boldsymbol{P}^{+} = \boldsymbol{P}^{+}Z$ and $p^{-}\boldsymbol{P}^{+} = \boldsymbol{P}^{+}p^{-}$). Hence the role of monomer is taken by $M = \boldsymbol{P}^{+}x$, and the order of the orbit is $| S_x |=| \boldsymbol{Z} | \frac{|\boldsymbol{P}|}{2|\boldsymbol{P}_x|}$.

The symmetry group $\boldsymbol{P}_{\mathrm{M}}$ of the monomer has been defined as the maximal axial point group $\boldsymbol{P}_{\mathrm{M}}$ leaving the monomer invariant. Still, when in various applications the line group of the system is fixed it is useful to introduce symmetry group of the monomer as the subgroup $\boldsymbol{M}$ containing only the transformations from $\boldsymbol{L}$, i.e., $\boldsymbol{M} = \boldsymbol{P}_{\mathrm{M}} \cap \boldsymbol{L}$. This group is isomorphic to the point factor $\boldsymbol{P}$, but some of its (negative) elements can have fractional translations.

3.1.4 Results

The result of this procedure applied to the factorization first listed in Table 2.2 is given in Table 3.1. For each line group $\boldsymbol{L} = \boldsymbol{Z}\boldsymbol{P}$, its orbit types are in the intersection of the row $\boldsymbol{P}$ and the column $\boldsymbol{Z}$ (there is "Not group" when $\boldsymbol{Z}$ and $\boldsymbol{P}$ do not combine into a group, or another factorization of the same group if it is used). For the orbits with the stabilizer $\boldsymbol{P}_x + Zp^{-}\boldsymbol{P}_x$ in the fourth column is $\boldsymbol{P}_x$, and under corresponding $\boldsymbol{Z}$ is p^{-} or "None" if the stabilizer is $\boldsymbol{P}_x$ only. For $\boldsymbol{Z}$ equal to $\boldsymbol{T}(a)$, $\boldsymbol{T}^{1}_{2n}(a/2)$, $\boldsymbol{T}_Q(f)$, and $\boldsymbol{T}'(a/2)$, the generators Z are $(E|1)$, $(C_{2n}|\frac{1}{2})$, $(C_Q|f)$, and $(\sigma_{\mathrm{v}}|\frac{1}{2})$, respectively. Symbols $a, b, \ldots$ for orbits are as in [3], with indices distinguishing between different $\boldsymbol{L}$-orbit types generated from the same $\boldsymbol{P}$-orbit type. The order of monomer is in the third column. As usual, σ_{h} is xy mirror plane, while σ_{v} and U are along x-axis, except in $\boldsymbol{D}_{n\mathrm{d}}$ where only σ_{v} is along x-axis. Finally, the symmetry group $\boldsymbol{M}$ of monomer is in the fifth column. Note that for the families 1 and 5, the cases when $\boldsymbol{Z}$ is $\boldsymbol{T}$ and $\boldsymbol{T}^{1}_{2n}$ are separately considered in the table for the purposes of the following sections.

To visualize the obtained results, for one group from each of the 13 line group families the different orbits are illustrated in Fig. 3.1 and explained in detail in Table 3.2. The coordinate system is adapted to the positions of characteristic symmetry elements (mirror planes, horizontal axes, which are also indicated in the figures):

- horizontal mirror or roto-reflectional plane is xy-plane (depicted in gray, roto-reflectional planes are round);
- vertical mirror or glide plane matches xz-plane (orange, glide plane distinguished by zigzag edges);
- U-axis (green lines) coincides with the x-axis, except in the ninth family, where U_{d} bisects vertical mirror planes (along x-axis).

The representative atom is connected to the atoms obtained by the action of the group generators: blue for Z, purple for C_n or S_{2n}, green for z-reversing elements, and orange for vertical mirror/glide planes. Each monomer consists of the atoms within dark gray part of the cylinder (with z between $-\frac{f}{2}$ and fh). The subgroup $\boldsymbol{M} = \boldsymbol{P}_{\mathrm{M}} \cap \boldsymbol{L}$ leaving the monomer invariant is explicitly given in the third row

Table 3.1 Classification of the orbits of the line groups (see explanation in the text)

$\mathbf{P}$	Condition	Orbit	$\lvert M\rvert$	$\mathbf{P}_x$	$\mathbf{P}_{\mathrm{M}}$	$\mathbf{T}$	$\mathbf{T}^1_{2n}$	$\mathbf{T}^r_q$	$\mathbf{T}'$
$\mathbf{C}_n$		a_1	n	$\mathbf{C}_1$	$\mathbf{D}_{n\mathrm{h}}$	None	None	None	None
		b_1	1	$\mathbf{C}_n$	$\mathbf{D}_{\infty\mathrm{h}}$	None	None	None	None
$\mathbf{S}_{2n}$		a_1	$2n$	$\mathbf{C}_1$	$\mathbf{D}_{n\mathrm{d}}$	None	$\mathbf{T}^1_{2n}\mathbf{C}_{n\mathrm{h}}$	Not	None
		a_2	n			\		group	S_{2n}^{-1}
	n even	a_3	n			\			S_{2n}
		b_1	2	$\mathbf{C}_n$	$\mathbf{D}_{\infty\mathrm{h}}$	None			None
		b_2	1			$C_{2n}\sigma_{\mathrm{h}}$			S_{2n}
		c_1	1	$\mathbf{S}_{2n}$	$\mathbf{D}_{\infty\mathrm{h}}$	None			None
$\mathbf{C}_{n\mathrm{h}}$		a_1	$2n$	$\mathbf{C}_1$	$\mathbf{D}_{n\mathrm{h}}$	None	None	Not	None
		a_2	n			σ_{h}	\	group	σ_{h}
	n even	a_3	n			\	\		$\sigma_{\mathrm{h}}C_n$
		b_1	n	$\mathbf{C}_{1\mathrm{h}}$	$\mathbf{D}_{n\mathrm{h}}$	None	None		None
		c_1	2	$\mathbf{C}_n$	$\mathbf{D}_{\infty\mathrm{h}}$	None	None		None
		c_2	1		$\mathbf{D}_{\infty\mathrm{h}}$	σ_{h}	σ_{h}		σ_{h}
		d_1	1	$\mathbf{C}_{n\mathrm{h}}$	$\mathbf{D}_{\infty\mathrm{h}}$	None	None		None
$\mathbf{C}_{n\mathrm{v}}$		a_1	$2n$	$\mathbf{C}_1$	$\mathbf{D}_{n\mathrm{h}}$	None	None	Not	$\mathbf{T}^1_{2n}$
		b_1	n	$\mathbf{C}_{1v}$	$\mathbf{D}_{n\mathrm{h}}$	None	None	group	$\mathbf{T}\mathbf{C}_{n\mathrm{v}}$
	n even	c_1	n	$\mathbf{C}_{1v}$	$\mathbf{D}_{n\mathrm{h}}$	None	\		
		d_1	1	$\mathbf{C}_{n\mathrm{v}}$	$\mathbf{D}_{\infty\mathrm{h}}$	None	None		
$\mathbf{D}_n$		a_1	$2n$	$\mathbf{C}_1$	$\mathbf{D}_n$	None	None	None	$\mathbf{T}'\mathbf{S}_{2n}$
		a_2	n			U	U	U	$\mathbf{T}'\mathbf{C}_{n\mathrm{h}}$
	n even	a_3	n			UC_n	UC_n	UC_n	
		b_1	n	$\mathbf{D}_1$	$\mathbf{D}_{n\mathrm{h}}$	None	None	None	
	n even	c_1	n	$\mathbf{D}_1$	$\mathbf{D}_{n\mathrm{h}}$	None	None	None	
		d_1	2	$\mathbf{C}_n$	$\mathbf{D}_{\infty\mathrm{h}}$	None	None	None	
		d_2	1			U	U	U	
		e_1	1	$\mathbf{D}_n$	$\mathbf{D}_{\infty\mathrm{h}}$	None	None	None	
$\mathbf{D}_{n\mathrm{d}}$		a_1	$4n$	$\mathbf{C}_1$	$\mathbf{D}_{n\mathrm{d}}$	None	$\mathbf{T}^1_{2n}\mathbf{D}_{n\mathrm{h}}$	Not	$\mathbf{T}\mathbf{D}_{n\mathrm{d}}$
		a_2	$2n$			U		group	
		b_1	$2n$	$\mathbf{C}_{1v}$	$\mathbf{D}_{n\mathrm{d}}$	None			
		c_1	$2n$	$\mathbf{D}_1$	$\mathbf{D}_{2n\mathrm{h}}$	None			
		d_1	2	$\mathbf{C}_{n\mathrm{v}}$	$\mathbf{D}_{\infty\mathrm{h}}$	None			
		d_2	1			U			
		e_1	1	$\mathbf{D}_{n\mathrm{d}}$	$\mathbf{D}_{\infty\mathrm{h}}$	None			
$\mathbf{D}_{n\mathrm{h}}$		a_1	$4n$	$\mathbf{C}_1$	$\mathbf{D}_{n\mathrm{h}}$	None	None	Not	$\mathbf{T}\mathbf{D}_{n\mathrm{h}}$
		a_2	$2n$			σ_{h}	U	group	$\mathbf{T}^1_{2n}\mathbf{D}_{n\mathrm{h}}$
		b_1	$2n$	$\mathbf{C}_{1v}$	$\mathbf{D}_{n\mathrm{h}}$	None	None		$\mathbf{T}'\mathbf{D}_{n\mathrm{d}}$
		b_2	n			σ_{h}	\		
	n even	c_1	$2n$	$\mathbf{C}_{1v}$	$\mathbf{D}_{n\mathrm{h}}$	None	\		
	n even	c_2	n			σ_{h}	\		
		d_1	$2n$	$\mathbf{C}_{1\mathrm{h}}$	$\mathbf{D}_{n\mathrm{h}}$	None	None		
		e_1	n	$\mathbf{D}_{1\mathrm{h}}$	$\mathbf{D}_{n\mathrm{h}}$	None	None		
	n even	f_1	n	$\mathbf{D}_{1\mathrm{h}}$	$\mathbf{D}_{n\mathrm{h}}$	None	None		
		g_1	2	$\mathbf{C}_{n\mathrm{v}}$	$\mathbf{D}_{\infty\mathrm{h}}$	None	None		
		g_2	1			σ_{h}	σ_{h}		
		h_1	1	$\mathbf{D}_{n\mathrm{h}}$	$\mathbf{D}_{\infty\mathrm{h}}$	None	None		

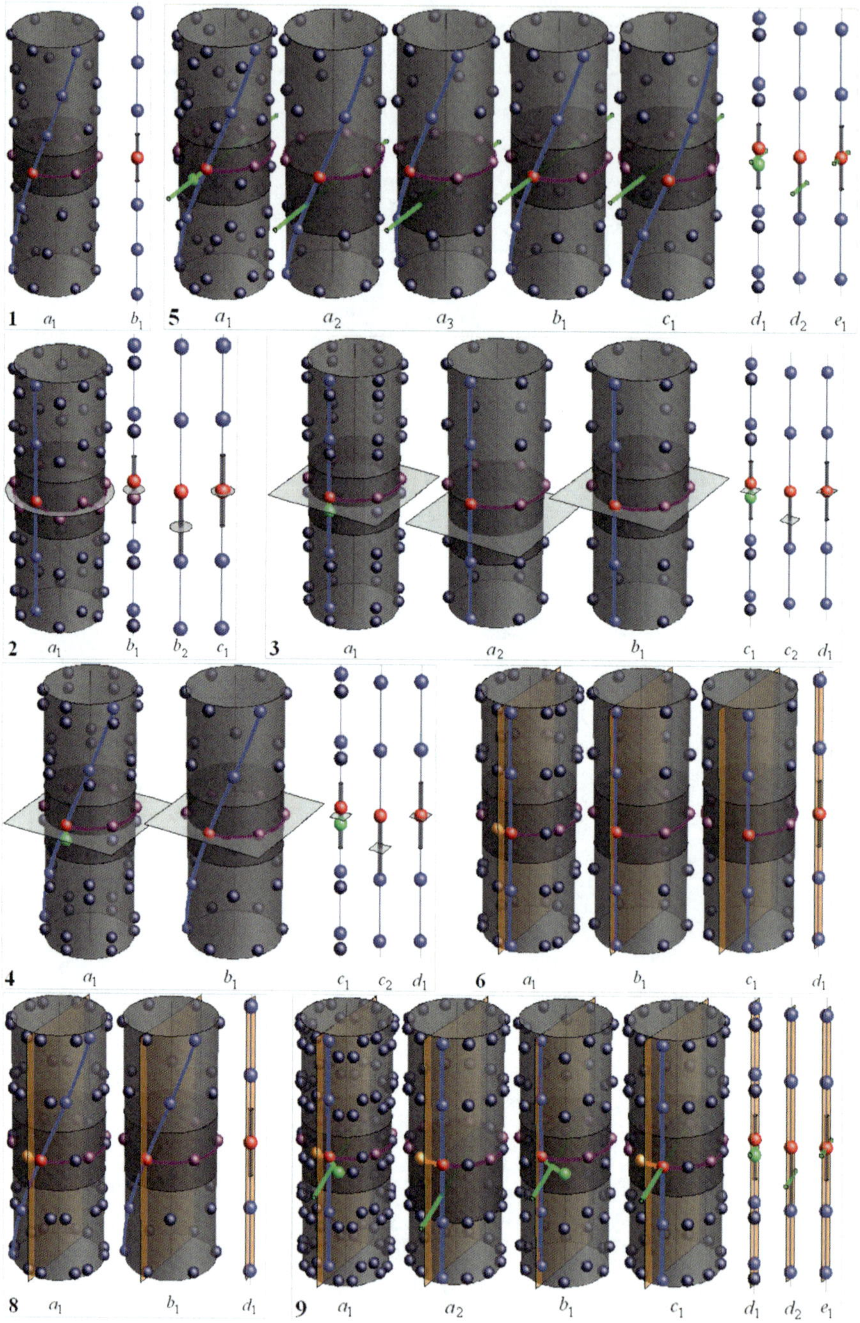

Fig. 3.1 (continued)

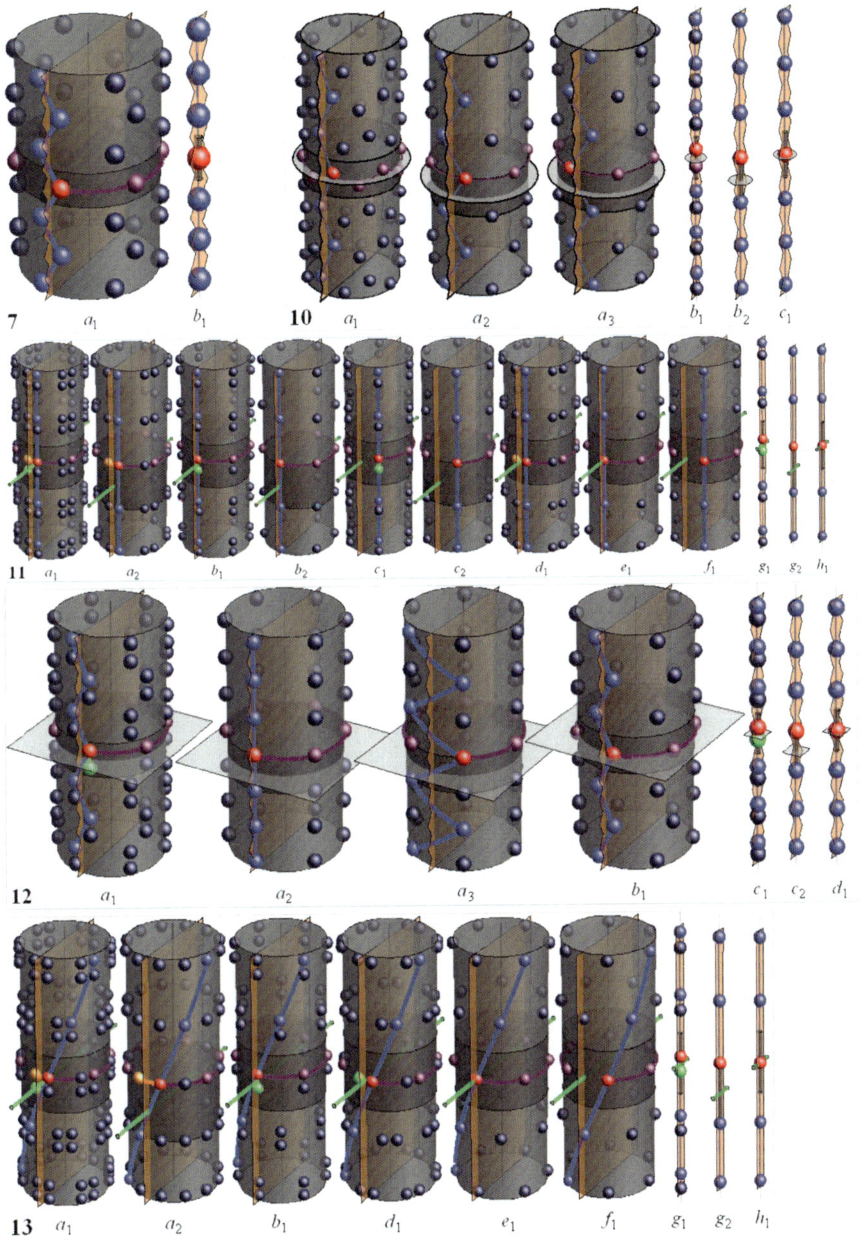

Fig. 3.1 Orbits of the line groups. Each box shows orbits of the group from family F (given in the *down right corner*) with $n = 6$; also, $Q = 18$ and $Q = 12$ for the families 1 and 5, respectively

of the table, and the stabilizer in the fourth row. Finally, in the last row a possible transversal $\boldsymbol{Y}$ is suggested. It is not unique, and the choice is restricted by the convention that the transversal is a subgroup of $\boldsymbol{L}$. However, this is not possible in general, and the transversals of the orbits b_1 of the groups with even n of the families 2 and 10 are not groups. In some other cases the chosen subgroups are not unique, and some other (listed in the caption) may be taken instead. Note that some orbit types exist only for special values of n. First, there are orbit types appearing only for even n, as in for n odd they become another orbit type specified in the square bracket. Second, for $n = 1$ some linear orbit types disappear, becoming another orbit type (given in the bracket) with reduced stabilizer.

3.2 Conformation Classes and Their Symmetry

Inspection of Table 3.2 shows that for different line groups there are orbits generated by the same transversal. The conformations of such orbits are the same, and only their stabilizers may differ in accordance with group. In this sense, we find 15 *conformation classes* corresponding to different transversals (Fig. 3.2). Thirteen of them are the generic orbit types a_1 of all the line group families; their transversals are the groups $\boldsymbol{Y}^{(F)} = \boldsymbol{L}^{(F)}$. The rest are the two types of linear (arranged along z-axis) orbits. The first of them is generated by the transversals being generalized translations $\boldsymbol{Y} = \boldsymbol{Z}(f)$ (along z-axis all of them act as pure translations, $\boldsymbol{T}(f)$). For the second one the transversal contains additional z-reversing element, thus having the form $\boldsymbol{Y} = \boldsymbol{Z} + p^- \boldsymbol{Z}$; however, all such elements along z-axis act as σ_{h}, making that all these transversals (namely $\boldsymbol{T}_Q\boldsymbol{D}_1$, $\boldsymbol{T}_Q\boldsymbol{C}_{1\mathrm{h}}$, $\boldsymbol{T}_Q\boldsymbol{D}_{1\mathrm{d}}$, $\boldsymbol{T}'\boldsymbol{C}_{1\mathrm{h}}$, $\boldsymbol{T} + S_{2n}\boldsymbol{T}$, and $\boldsymbol{T}' + S_{2n}\boldsymbol{T}'$) act effectively as $\boldsymbol{T}(f)\boldsymbol{C}_{1\mathrm{h}}$. It is natural to denote the classes by their transversals: the first 13 conformation classes correspond to the line group families $\boldsymbol{L}^{(F)}$, while for the linear conformations, despite the transversal is not unique, we use $\boldsymbol{Y}^{(14)} = \boldsymbol{T}(f)$ and $\boldsymbol{Y}^{(15)} = \boldsymbol{T}(f)\boldsymbol{C}_{1\mathrm{h}}$. However, this is not completely unique, as the transversals themselves are not unique (alternative transversals are listed in Table 3.2).

When the transversal is a group, then the whole group is obviously a product of the transversal and stabilizer, $\boldsymbol{L} = \boldsymbol{Y}_x\boldsymbol{L}_x$ (one of the factorizations of $\boldsymbol{L}$). This has a consequence that such transversals are simultaneously left and right: $\boldsymbol{L} = \boldsymbol{Y}_x\boldsymbol{L}_x = \boldsymbol{L}_x\boldsymbol{Y}_x$. All the transversals of the line groups are again the line groups (with $n = 1$ for linear orbits), except $\boldsymbol{T} + S_{2n}\boldsymbol{T}$ and $\boldsymbol{T}' + S_{2n}\boldsymbol{T}'$. Even for these exceptional transversals it is easy to show directly that they are both sided, making this property universal for the line group orbits.

It follows from the definition that each orbit of $\boldsymbol{L}$ is invariant under $\boldsymbol{L}$, but most of the orbits have additional symmetries. Actually, the full or *covering symmetry* group $\tilde{\boldsymbol{Y}}$ of an orbit is related only to its conformation class, i.e., to the transversal $\boldsymbol{Y}$. For example, all isolated orbits a_1 of the line group $\boldsymbol{L} = \boldsymbol{T}(a)\boldsymbol{C}_n$ are invariant under the group $\tilde{\boldsymbol{Y}} = \boldsymbol{T}(a)\boldsymbol{D}_{n\mathrm{h}}$ (these are orbits c_1, b_2, c_2, or e_1 of $\boldsymbol{L}^{(13)}$) and $\boldsymbol{L}$ is only a subgroup of the symmetry of such orbits. Also, independent of the line group, the linear orbits have symmetry $\boldsymbol{T}\boldsymbol{D}_{\infty\mathrm{h}}$. Altogether, the symmetry of the orbits is one of the groups $\boldsymbol{T}_Q\boldsymbol{D}_n$, $\boldsymbol{T}\boldsymbol{D}_{n\mathrm{d}}$, $\boldsymbol{T}'S2n$, $\boldsymbol{T}\boldsymbol{D}_{n\mathrm{h}}$, $\boldsymbol{T}'\boldsymbol{C}_{n\mathrm{h}}$, $\boldsymbol{T}^1_{2n}\boldsymbol{D}_{n\mathrm{h}}$, and $\boldsymbol{T}\boldsymbol{D}_{\infty\mathrm{h}}$. In Table 3.3

Table 3.2 Characteristics of the orbits of the line groups. For each line group family F orbits (OL) are listed in the decreasing generality order: allowed coordinates (x, y, z) (column C) of the orbit representative (Wyckoff positions) are arbitrary but cannot take the values of the coordinates of orbits given below; in short we use $c_X = \cos\frac{\pi}{X}$, $s_X = \sin\frac{\pi}{X}$, $c_{n,Q} = \cos(\frac{\pi}{Q} - \frac{\pi}{n})$, and $s_{n,Q} = \sin(\frac{\pi}{Q} - \frac{\pi}{n})$. Then the monomer symmetry group $\boldsymbol{M}$ and stabilizer (site symmetry) group $\boldsymbol{S}$ follow. If possible transversal $\boldsymbol{Y}$ is chosen as a subgroup; this is unique except when alternative choices are given in $\boldsymbol{Y}'$: 1 $\boldsymbol{T}\boldsymbol{S}_2$ (for n odd only), 2 $\boldsymbol{T}'\boldsymbol{S}_2$ (for n odd only), 3 $\boldsymbol{T}\boldsymbol{D}_n$, 4 $\boldsymbol{T}\boldsymbol{D}_1$, 5 $\boldsymbol{T}^1_{2n}\boldsymbol{C}_{nh}$, 6 $\boldsymbol{T}^1_{2n}\boldsymbol{D}_n$, and 7 $\boldsymbol{T}^1_{2n}\boldsymbol{D}_1$. Among the orbit types belonging to the same conformation class (CC), bold-faced class denotes one used as a representative of the class in the text

F	OL	A	$\boldsymbol{M}$	$\boldsymbol{S}$	Y	Y'	CC
1	a_1	(x, y, z)	$\boldsymbol{C}_n$	$\boldsymbol{C}_1$	$\boldsymbol{L}$		**1**
	$b_1\ (a_1)$	$(0, 0, z)$	$\boldsymbol{C}_n$	$\boldsymbol{C}_n$	$\boldsymbol{T}_Q(f)$		**14**
2	a_1	(x, y, z)	$\boldsymbol{S}_{2n}$	$\boldsymbol{C}_1$	$\boldsymbol{L}$		**2**
	$b_1\ (a_1)$	$(0, 0, z)$	$\boldsymbol{S}_{2n}$	$\boldsymbol{C}_n$	$\boldsymbol{T} + \boldsymbol{S}_{2n}\boldsymbol{T}$	1	**15**
	b_2	$(0, 0, \frac{a}{2})$	$\boldsymbol{C}_n + (S_{2n}\|a)\boldsymbol{C}_n$	$\boldsymbol{C}_n + (S_{2n}\|a)\boldsymbol{C}_n$	$\boldsymbol{T}$		14
	c_1	$(0, 0, 0)$	$\boldsymbol{S}_{2n}$	$\boldsymbol{S}_{2n}$	$\boldsymbol{T}$		14
3	a_1	(x, y, z)	$\boldsymbol{C}_{nh}$	$\boldsymbol{C}_1$	$\boldsymbol{L}$		**3**
	a_2	$(x, y, \frac{a}{2})$	$\boldsymbol{C}_n + (\sigma_h\|a)\boldsymbol{C}_n$	$\{e, (\sigma_h\|a)\}$	$\boldsymbol{T}\boldsymbol{C}_n$		1_1
	b_1	$(x, y, 0)$	$\boldsymbol{C}_{nh}$	$\boldsymbol{C}_{1h}$	$\boldsymbol{T}\boldsymbol{C}_n$		1_1
	$c_1\ (a_1)$	$(0, 0, z)$	$\boldsymbol{C}_{nh}$	$\boldsymbol{C}_n$	$\boldsymbol{T}\boldsymbol{C}_{1h}$		15
	$c_2\ (a_2)$	$(0, 0, \frac{a}{2})$	$\boldsymbol{C}_n + (\sigma_h\|a)\boldsymbol{C}_n$	$\boldsymbol{C}_n + (\sigma_h\|a)\boldsymbol{C}_n$	$\boldsymbol{T}$		15
	$d_1\ (b_1)$	$(0, 0, 0)$	$\boldsymbol{C}_{nh}$	$\boldsymbol{C}_{nh}$	$\boldsymbol{T}$		14
4	a_1	(x, y, z)	$\boldsymbol{C}_{nh}$	$\boldsymbol{C}_1$	$\boldsymbol{L}$		**4**
	b_1	$(x, y, 0)$	$\boldsymbol{C}_{nh}$	$\boldsymbol{C}_{1h}$	$\boldsymbol{T}^1_{2n}\boldsymbol{C}_n$		1_2
	$c_1\ (a_1)$	$(0, 0, z)$	$\boldsymbol{C}_{nh}$	$\boldsymbol{C}_n$	$\boldsymbol{T}^1_{2n}\boldsymbol{C}_{1h}$		15
	c_2	$(0, 0, \frac{a}{4})$	$\boldsymbol{C}_n + (S_{2n}\|\frac{a}{2})\boldsymbol{C}_n$	$\boldsymbol{C}_n + (S_{2n}\|\frac{a}{2})\boldsymbol{C}_n$	$\boldsymbol{T}^1_{2n}$		14
	$d_1\ (b_1)$	$(0, 0, 0)$	$\boldsymbol{C}_{nh}$	$\boldsymbol{C}_{nh}$	$\boldsymbol{T}^1_{2n}$		14
5	a_1	(x, y, z)	$\boldsymbol{D}_n$	$\boldsymbol{C}_1$	$\boldsymbol{L}$		**5**
	a_2	$(\rho c_Q, \rho s_Q, \frac{f}{2})$	$\boldsymbol{C}_n + (C_Q\|f)U\boldsymbol{C}_n$	$\{e, (C_Q\|f)U\}$	$\boldsymbol{T}_Q(f)\boldsymbol{C}_n$		1
	$a_3\ [a_2]$	$(\rho c_{n,Q}, \rho s_{n,Q}, \frac{f}{2})$	$\boldsymbol{C}_n + (C_Q\|f)U\boldsymbol{C}_n$	$\{e, (C_Q\|f)C_n^{-1}U\}$	$\boldsymbol{T}_Q(f)\boldsymbol{C}_n$		1
	b_1	$(x, 0, 0)$	$\boldsymbol{D}_n$	$\boldsymbol{D}_1$	$\boldsymbol{T}_Q(f)\boldsymbol{C}_n$		1
	$c_1\ [a_2]$	$(\rho c_n, \rho s_n, 0)$	$\boldsymbol{D}_n$	$\{e, C_nU\}$	$\boldsymbol{T}_Q(f)\boldsymbol{C}_n$		1
	$d_1\ (a_1)$	$(0, 0, z)$	$\boldsymbol{D}_n$	$\boldsymbol{C}_n$	$\boldsymbol{T}_Q(f)\boldsymbol{D}_1$		15
	$d_2\ (a_2)$	$(0, 0, \frac{f}{2})$	$\boldsymbol{C}_n + (C_QU\|f)\boldsymbol{C}_n$	$\boldsymbol{C}_n + (C_QU\|f)\boldsymbol{C}_n$	$\boldsymbol{T}_Q(f)$		14
	$e_1\ (b_1)$	$(0, 0, 0)$	$\boldsymbol{D}_n$	$\boldsymbol{D}_n$	$\boldsymbol{T}_Q(f)$		14
6	a_1	(x, y, z)	$\boldsymbol{C}_{nv}$	$\boldsymbol{C}_1$	$\boldsymbol{L}$		**6**
	b_1	$(x, 0, z)$	$\boldsymbol{C}_{nv}$	$\boldsymbol{C}_{1v}$	$\boldsymbol{T}\boldsymbol{C}_n$		1_1
	$c_1\ [b_1]$	$(\rho c_n, \rho s_n, z)$	$\boldsymbol{C}_{nv}$	$\{e, C_n\sigma_v\}$	$\boldsymbol{T}\boldsymbol{C}_n$		1_1
	$d_1\ (b_1)$	$(0, 0, z)$	$\boldsymbol{C}_{nv}$	$\boldsymbol{C}_{nv}$	$\boldsymbol{T}$		14
7	a_1	(x, y, z)	$\boldsymbol{C}_n$	$\boldsymbol{C}_1$	$\boldsymbol{L}$		**7**
	$b_1\ (a_1)$	$(0, 0, z)$	$\boldsymbol{C}_n$	$\boldsymbol{C}_n$	$\boldsymbol{T}'$		14
8	a_1	(x, y, z)	$\boldsymbol{C}_{nv}$	$\boldsymbol{C}_1$	$\boldsymbol{L}$		**8**
	b_1	$(x, 0, z)$	$\boldsymbol{C}_{nv}$	$\boldsymbol{C}_{1v}$	$\boldsymbol{T}^1_{2n}\boldsymbol{C}_n$		1_2
	$d_1\ (b_1)$	$(0, 0, z)$	$\boldsymbol{C}_{nv}$	$\boldsymbol{C}_{nv}$	$\boldsymbol{T}^1_{2n}$		14
9	a_1	(x, y, z)	$\boldsymbol{D}_{nd}$	$\boldsymbol{C}_1$	$\boldsymbol{L}$		**9**
	a_2	$(\rho c_{2n}, \rho s_{2n}, \frac{a}{2})$	$\boldsymbol{C}_{nv} + (U_d\|a)\boldsymbol{C}_{nv}$	$\{e, (U_d\|a)\}$	$\boldsymbol{T}\boldsymbol{C}_{nv}$		1_1
	b_1	$(x, 0, z)$	$\boldsymbol{D}_{nd}$	$\boldsymbol{C}_{1v}$	$\boldsymbol{T}\boldsymbol{D}_n$		2
	c_1	$(\rho c_{2n}, \rho s_{2n}, 0)$	$\boldsymbol{D}_{nd}$	$\boldsymbol{D}_1$	$\boldsymbol{T}\boldsymbol{C}_{nv}$		1_1

Table 3.2 (continued)

	$d_1(b_1)$	$(0, 0, z)$	$\boldsymbol{D}_{nd}$	$\boldsymbol{C}_{nv}$	$\boldsymbol{TD}_{1d}$		15
	d_2	$(0, 0, \frac{a}{2})$	$\boldsymbol{C}_{nv} + (U_d\|a)\boldsymbol{C}_{nv}$	$\boldsymbol{C}_{nv} + (U_d\|a)\boldsymbol{C}_{nv}$	$\boldsymbol{T}$		14
	e_1	$(0, 0, 0)$	$\boldsymbol{D}_{nd}$	$\boldsymbol{D}_{nd}$	$\boldsymbol{T}$		14
10	a_1	(x, y, z)	$\boldsymbol{S}_{2n}$	$\boldsymbol{C}_1$	$\boldsymbol{L}$		**10**
	a_2	$(\rho c_{2n}, \rho s_{2n}, \frac{a}{4})$	$\boldsymbol{C}_n + (\sigma_v\|\frac{a}{2})S_{2n}\boldsymbol{C}_n$	$\{e, (\sigma_v\|\frac{a}{2})S_{2n}^{-1}\}$	$\boldsymbol{T}'\boldsymbol{C}_n$		1_2
	$a_3\ [a_2]$	$(\rho c_{2n}, -\rho s_{2n}, \frac{a}{4})$	$\boldsymbol{C}_n + (\sigma_v\|\frac{a}{2})S_{2n}\boldsymbol{C}_n$	$\{e, (\sigma_v\|\frac{a}{2})S_{2n}\}$	$\boldsymbol{T}'\boldsymbol{C}_n$		1_2
	$b_1\ (a_1)$	$(0, 0, z)$	$\boldsymbol{S}_{2n}$	$\boldsymbol{C}_n$	$\boldsymbol{T}' + S_{2n}\boldsymbol{T}'$	2	15
	$b_2\ (a_2)$	$(0, 0, \frac{a}{4})$	$\boldsymbol{C}_n + (\sigma_v\|\frac{a}{2})S_{2n}\boldsymbol{C}_n$	$\boldsymbol{C}_n + (\sigma_v\|\frac{a}{2})S_{2n}\boldsymbol{C}_n$	$\boldsymbol{T}'$		14
	c_1	$(0, 0, 0)$	$\boldsymbol{S}_{2n}$	$\boldsymbol{S}_{2n}$	$\boldsymbol{T}'$		14
11	a_1	(x, y, z)	$\boldsymbol{D}_{nh}$	$\boldsymbol{C}_1$	$\boldsymbol{L}$		**11**
	a_2	$(x, y, \frac{a}{2})$	$\boldsymbol{C}_{nv} + (\sigma_h\|1)\boldsymbol{C}_{nv}$	$\{e, (\sigma_h\|a)\}$	$\boldsymbol{TC}_{nv}$	3	6
	b_1	$(x, 0, z)$	$\boldsymbol{D}_{nh}$	$\boldsymbol{C}_{1v}$	$\boldsymbol{TC}_{nh}$	3	3
	b_2	$(x, 0, \frac{a}{2})$	$\boldsymbol{C}_{nv} + (\sigma_h\|a)\boldsymbol{C}_{nv}$	$\{e, \sigma_v, (\sigma_h\|a), (U\|a)\}$	$\boldsymbol{TC}_n$		1_1
	$c_1\ [b_1]$	$(\rho c_n, \rho s_n, z)$	$\boldsymbol{D}_{nh}$	$\{e, C_n\sigma_v\}$	$\boldsymbol{TC}_{nh}$	3	3
	$c_2\ [b_2]$	$(\rho c_n, \rho s_n, \frac{a}{2})$	$\boldsymbol{C}_{nv} + (\sigma_h\|a)\boldsymbol{C}_{nv}$	$\{e, C_n\sigma_v, (\sigma_h\|a), (U\|a)\}$	$\boldsymbol{TC}_n$		1_1
	d_1	$(x, y, 0)$	$\boldsymbol{D}_{nh}$	$\boldsymbol{C}_{1h}$	$\boldsymbol{TC}_{nv}$	3	6
	e_1	$(x, 0, 0)$	$\boldsymbol{D}_{nh}$	$\boldsymbol{D}_{1h}$	$\boldsymbol{TC}_n$		1_1
	$f_1\ [e_1]$	$(\rho c_n, \rho s_n, 0)$	$\boldsymbol{D}_{nh}$	$\{e, C_n\sigma_v, \sigma_h, C_nU\}$	$\boldsymbol{TC}_n$		1_1
	$g_1\ (b_1)$	$(0, 0, z)$	$\boldsymbol{D}_{nh}$	$\boldsymbol{C}_{nv}$	$\boldsymbol{TC}_{1h}$	4	15
	$g_2\ (b_2)$	$(0, 0, \frac{a}{2})$	$\boldsymbol{C}_{nv} + (\sigma_h\|a)\boldsymbol{C}_{nv}$	$\boldsymbol{C}_{nv} + (\sigma_h\|a)\boldsymbol{C}_{nv}$	$\boldsymbol{T}$		14
	$h_1\ (e_1)$	$(0, 0, 0)$	$\boldsymbol{D}_{nh}$	$\boldsymbol{D}_{nh}$	$\boldsymbol{T}$		14
12	a_1	(x, y, z)	$\boldsymbol{C}_{nh}$	$\boldsymbol{C}_1$	$\boldsymbol{L}$		**12**
	a_2	$(x, 0, \frac{a}{4})$	$\boldsymbol{C}_n + (\sigma_v\|\frac{a}{2})\sigma_h\boldsymbol{C}_n$	$\{e, (\sigma_v\|\frac{a}{2})\sigma_h\}$	$\boldsymbol{T}'\boldsymbol{C}_n$		1_1
	$a_3\ [a_2]$	$(\rho c_n, \rho s_n, \frac{a}{4})$	$\boldsymbol{C}_n + (\sigma_v\|\frac{a}{2})\sigma_h\boldsymbol{C}_n$	$\{e, (\sigma_v\|\frac{a}{2})C_n^{-1}\sigma_h\}$	$\boldsymbol{T}'\boldsymbol{C}_n$		1_1
	b_1	$(x, y, 0)$	$\boldsymbol{C}_{nh}$	$\boldsymbol{C}_{1h}$	$\boldsymbol{T}'\boldsymbol{C}_n$		7
	$c_1\ (a_1)$	$(0, 0, z)$	$\boldsymbol{C}_{nh}$	$\boldsymbol{C}_n$	$\boldsymbol{T}'\boldsymbol{C}_{1h}$		15
	$c_2\ (a_2)$	$(0, 0, \frac{a}{4})$	$\boldsymbol{C}_n + (\sigma_v\sigma_h\|\frac{a}{2})\boldsymbol{C}_n$	$\boldsymbol{C}_n + (\sigma_v\|\frac{a}{2})\sigma_h\boldsymbol{C}_n$	$\boldsymbol{T}'$		14
	$d_1\ (b_1)$	$(0, 0, 0)$	$\boldsymbol{C}_{nh}$	$\boldsymbol{C}_{nh}$	$\boldsymbol{T}'$		14
13	a_1	(x, y, z)	$\boldsymbol{D}_{nh}$	$\boldsymbol{C}_1$	$\boldsymbol{L}$		**13**
	$a_2^{(5)}$	$(\rho c_{2n}, \rho s_{2n}, \frac{a}{4})$	$\boldsymbol{C}_{nv} + (C_{2n}\sigma_h\|\frac{a}{2})\boldsymbol{C}_{nv}$	$\{e, (C_{2n}\|\frac{a}{2})U\}$	$\boldsymbol{T}^1_{2n}\boldsymbol{C}_{nv}$	5	1_1
	$b_1^{(6)}$	$(x, 0, z)$	$\boldsymbol{D}_{nh}$	$\boldsymbol{C}_{1v}$	$\boldsymbol{T}^1_{2n}\boldsymbol{C}_{nh}$	6	4
	$d_1^{(6)}$	$(x, y, 0)$	$\boldsymbol{D}_{nh}$	$\boldsymbol{C}_{1h}$	$\boldsymbol{T}^1_{2n}\boldsymbol{C}_{nv}$	6	8
	e_1	$(x, 0, 0)$	$\boldsymbol{D}_{nh}$	$\boldsymbol{D}_{1h}$	$\boldsymbol{T}^1_{2n}\boldsymbol{C}_n$		1_2
	$f_1\ [e_1]$	$(\rho c_n, \rho c_n, 0)$	$\boldsymbol{D}_{nh}$	$\{e, \sigma_h, C_nU, C_n\sigma_v\}$	$\boldsymbol{T}^1_{2n}\boldsymbol{C}_n$		1_2
	$g_1^{(7)}\ (b_1)$	$(0, 0, z)$	$\boldsymbol{D}_{nh}$	$\boldsymbol{C}_{nv}$	$\boldsymbol{T}^1_{2n}\boldsymbol{C}_{1h}$	7	15
	g_2	$(0, 0, \frac{a}{4})$	$\boldsymbol{C}_{nv} + (C_{2n}\sigma_h\|\frac{a}{2})\boldsymbol{C}_{nv}$	$\boldsymbol{C}_{nv} + (C_{2n}\sigma_h\|\frac{a}{2})\boldsymbol{C}_{nv}$	$\boldsymbol{T}^1_{2n}$		14
	$h_1\ (e_1)$	$(0, 0, 0)$	$\boldsymbol{D}_{nh}$	$\boldsymbol{D}_{nh}$	$\boldsymbol{T}^1_{2n}$		14

we list the symmetry of all different orbits of the line groups, which is important in some applications. Some orbit types for a special position of orbit representatives have even larger symmetry (e.g., orbit type a_1 of the group $\boldsymbol{T}(a)\boldsymbol{S}_{2n}$ for $z = a/4$ has doubled symmetry $\boldsymbol{T}^1_{2n}(a/2)\boldsymbol{D}_{nh}$). Such special cases of $\boldsymbol{Y}^{(i)}$ are listed immediately below row of $\boldsymbol{Y}^{(i)}$; as the orbit type and transversal are the same, the condition selecting the special case is given instead. Also, the orbits of the achiral groups ($Q = n, 2n$) of the first family have increased symmetry with respect to the chiral ones, and for these special cases of the transversal $\boldsymbol{Y}^{(1)}$ we use notation $\boldsymbol{Y}^{(1_1)}$ and $\boldsymbol{Y}^{(1_2)}$, respectively. All the elements of symmetry are depicted in Fig. 3.2.

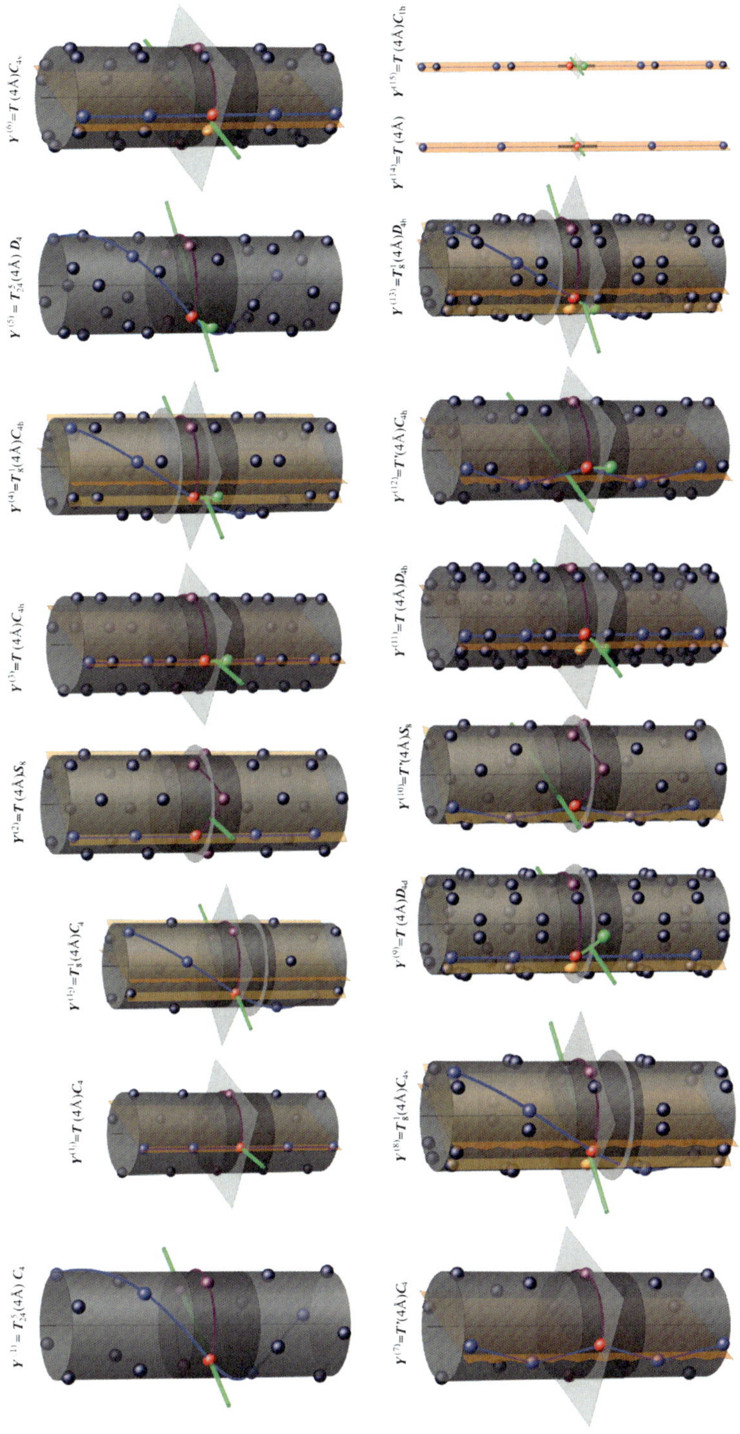

Fig. 3.2 Conformation classes with full symmetry. From each class the representative with $n = 4$ and $f = 4\text{Å}$ is depicted. For $\boldsymbol{Y}^{(1)}$ and $\boldsymbol{Y}^{(5)}$ in the chiral cases $q = 24, r = 5$. Orbit representatives (*red* atoms) are positioned at $(\rho, \varphi, z) = (3\text{Å}, \pi/18, 0.8\text{Å})$ except for the last two classes when $(\rho, \varphi, z) = (0, 0, 0.8\text{Å})$

Table 3.3 Symmetry of the orbits of the line groups. For all line group families (column F) and each orbit type (column OT) the transversal $\boldsymbol{Y}$ and the full symmetry group $\tilde{\boldsymbol{Y}}$ are given. The first appearance of a transversal is labeled (column CC) as its configuration class, $\boldsymbol{Y}^{(i)}$, and the special cases are listed below; the other appearances are denoted by i only

F	CC	OT	$\boldsymbol{Y}$	$\tilde{\boldsymbol{Y}}$
1	$\boldsymbol{Y}^{(1)}$	a_1	$\boldsymbol{T}_Q(f)\boldsymbol{C}_n$	$\boldsymbol{T}_Q(f)\boldsymbol{D}_n$
	1_1		$Q=1$	$\boldsymbol{T}(f)\boldsymbol{D}_{n\mathrm{h}}$
	1_2		$Q=2n$	$\boldsymbol{T}^1_{2n}(f)\boldsymbol{D}_{n\mathrm{h}}$
	$\boldsymbol{Y}^{(14)}$	b_1	$\boldsymbol{T}_Q(f)$	$\boldsymbol{T}(f)\boldsymbol{D}_{\infty\mathrm{h}}$
2	$\boldsymbol{Y}^{(2)}$	a_1	$\boldsymbol{T}(f)\boldsymbol{S}_{2n}$	$\boldsymbol{T}(f)\boldsymbol{D}_{n\mathrm{d}}$
	1_1		$z=\frac{f}{2}$	$\boldsymbol{T}(f)\boldsymbol{D}_{2n\mathrm{h}}$
	1_2		$z=\frac{f}{4}$	$\boldsymbol{T}^1_{2n}(\frac{f}{2})\boldsymbol{D}_{n\mathrm{h}}$
	$\boldsymbol{Y}^{(15)}$	b_1	$\boldsymbol{T}(f)\{e, S_{2n}\}$	$\boldsymbol{T}(f)\boldsymbol{D}_{\infty\mathrm{h}}$
	14		$z=\frac{f}{4}$	$\boldsymbol{T}(\frac{f}{2})\boldsymbol{D}_{\infty\mathrm{h}}$
	14	b_2 c_1	$\boldsymbol{T}(f)$	$\boldsymbol{T}(f)\boldsymbol{D}_{\infty\mathrm{h}}$
3	$\boldsymbol{Y}^{(3)}$	a_1	$\boldsymbol{T}(f)\boldsymbol{C}_{n\mathrm{h}}$	$\boldsymbol{T}(f)\boldsymbol{D}_{n\mathrm{h}}$
	1_1		$z=\frac{f}{4}$	$\boldsymbol{T}(\frac{f}{2})\boldsymbol{D}_{n\mathrm{h}}$
	1_1	a_2 b_1	$\boldsymbol{T}(f)\boldsymbol{C}_n$	$\boldsymbol{T}(f)\boldsymbol{D}_{n\mathrm{h}}$
	15	c_1	$\boldsymbol{T}(f)\boldsymbol{C}_{1\mathrm{h}}$	$\boldsymbol{T}(f)\boldsymbol{D}_{\infty\mathrm{h}}$
	14	c_2 d_1	$\boldsymbol{T}(f)$	$\boldsymbol{T}(f)\boldsymbol{D}_{\infty\mathrm{h}}$
4	$\boldsymbol{Y}^{(4)}$	a_1	$\boldsymbol{T}^1_{2n}(f)\boldsymbol{C}_{n\mathrm{h}}$	$\boldsymbol{T}^1_{2n}(f)\boldsymbol{D}_{n\mathrm{h}}$
	1_1		$z=\frac{f}{2}$	$\boldsymbol{T}(f)\boldsymbol{D}_{2n\mathrm{h}}$
	1_2	b_1	$\boldsymbol{T}^1_{2n}(f)\boldsymbol{C}_n$	$\boldsymbol{T}^1_{2n}(f)\boldsymbol{D}_{n\mathrm{h}}$
	15	c_1	$\boldsymbol{T}^1_{2n}(f)\boldsymbol{C}_{1\mathrm{h}}$	$\boldsymbol{T}(f)\boldsymbol{D}_{\infty\mathrm{h}}$
	14	c_2 d_1	$\boldsymbol{T}^1_{2n}(f)$	$\boldsymbol{T}(f)\boldsymbol{D}_{\infty\mathrm{h}}$
5	$\boldsymbol{Y}^{(5)}$	a_1	$\boldsymbol{T}_Q(f)\boldsymbol{D}_n$	$\boldsymbol{T}_Q(f)\boldsymbol{D}_n$
	2		$Q=1, \varphi=\frac{\pi}{2n}$	$\boldsymbol{T}(f)\boldsymbol{D}_{n\mathrm{d}}$
	3		$Q=1, \varphi=0$	$\boldsymbol{T}(f)\boldsymbol{D}_{n\mathrm{h}}$
	4		$Q=2n, \varphi=0, \frac{\pi}{2n}$	$\boldsymbol{T}^1_{2n}(f)\boldsymbol{D}_{n\mathrm{h}}$
	1	a_2,a_3 b_1,c_1	$\boldsymbol{T}_Q(f)\boldsymbol{C}_n$	$\boldsymbol{T}_Q(f)\boldsymbol{D}_n$
	15	d_1	$\boldsymbol{T}(f)\boldsymbol{D}_1$	$\boldsymbol{T}(f)\boldsymbol{D}_{\infty\mathrm{h}}$
	14	d_2 e_1	$\boldsymbol{T}(f)$	$\boldsymbol{T}(f)\boldsymbol{D}_{\infty\mathrm{h}}$
6	$\boldsymbol{Y}^{(6)}$	a_1	$\boldsymbol{T}(f)\boldsymbol{C}_{n\mathrm{v}}$	$\boldsymbol{T}(f)\boldsymbol{D}_{n\mathrm{h}}$
	1_1		$\varphi=\frac{\pi}{2n}$	$\boldsymbol{T}(f)\boldsymbol{D}_{2n\mathrm{h}}$
	1_1	b_1 c_1	$\boldsymbol{T}(f)\boldsymbol{C}_n$	$\boldsymbol{T}(f)\boldsymbol{D}_{n\mathrm{h}}$
	14	d_1	$\boldsymbol{T}(f)$	$\boldsymbol{T}(f)\boldsymbol{D}_{\infty\mathrm{h}}$
7	$\boldsymbol{Y}^{(7)}$	a_1	$\boldsymbol{T}'(f)\boldsymbol{C}_n$	$\boldsymbol{T}'(f)\boldsymbol{C}_{n\mathrm{h}}$
	1_1		$\varphi=0, \frac{\pi}{n}$	$\boldsymbol{T}(f)\boldsymbol{D}_{n\mathrm{h}}$
	1_2		$\varphi=\frac{\pi}{2n}$	$\boldsymbol{T}^1_{2n}(f)\boldsymbol{D}_{n\mathrm{h}}$
	14	b_1	$\boldsymbol{T}'(f)$	$\boldsymbol{T}(f)\boldsymbol{D}_{\infty\mathrm{h}}$
8	$\boldsymbol{Y}^{(8)}$	a_1	$\boldsymbol{T}^1_{2n}(f)\boldsymbol{C}_{n\mathrm{v}}$	$\boldsymbol{T}^1_{2n}(f)\boldsymbol{D}_{n\mathrm{h}}$
	1_1		$\varphi=\frac{\pi}{2n}$	$\boldsymbol{T}(f)\boldsymbol{D}_{2n\mathrm{h}}$
	1_2	b_1	$\boldsymbol{T}^1_{2n}(f)\boldsymbol{C}_n$	$\boldsymbol{T}^1_{2n}(f)\boldsymbol{D}_{n\mathrm{h}}$
	14	d_1	$\boldsymbol{T}^1_{2n}(f)$	$\boldsymbol{T}(f)\boldsymbol{D}_{\infty\mathrm{h}}$
9	$\boldsymbol{Y}^{(9)}$	a_1	$\boldsymbol{T}(f)\boldsymbol{D}_{n\mathrm{d}}$	$\boldsymbol{T}(f)\boldsymbol{D}_{n\mathrm{d}}$
	8		$z=\frac{f}{4}$	$\boldsymbol{T}^1_{2n}(\frac{f}{2})\boldsymbol{D}_{n\mathrm{h}}$
	1_2		$z=\frac{f}{4}$	$\boldsymbol{T}^1_{2n}(\frac{f}{2})\boldsymbol{D}_{n\mathrm{h}}$
	1_1	a_2 c_1	$\boldsymbol{T}(f)\boldsymbol{C}_{n\mathrm{v}}$	$\boldsymbol{T}(f)\boldsymbol{D}_{2n\mathrm{h}}$
	2	b_1	$\boldsymbol{T}(f)\boldsymbol{D}_n$	$\boldsymbol{T}(f)\boldsymbol{D}_{n\mathrm{d}}$
	15	d_1	$\boldsymbol{T}(f)\boldsymbol{D}_{1\mathrm{d}}$	$\boldsymbol{T}(f)\boldsymbol{D}_{\infty\mathrm{h}}$
	14	d_2 e_1	$\boldsymbol{T}(f)$	$\boldsymbol{T}(f)\boldsymbol{D}_{\infty\mathrm{h}}$
10	$\boldsymbol{Y}^{(10)}$	a_1	$\boldsymbol{T}'(f)\boldsymbol{S}_{2n}$	$\boldsymbol{T}'(f)\boldsymbol{S}_{2n}$
	1_1		$\varphi=0$ $z=0$	$\boldsymbol{T}(f)\boldsymbol{D}_{2n\mathrm{h}}$
	1_2	a_2 a_3	$\boldsymbol{T}'(f)\boldsymbol{C}_n$	$\boldsymbol{T}^1_{2n}(f)\boldsymbol{D}_{n\mathrm{h}}$
	15	b_1	$\boldsymbol{T}'(f)\{e, S_{2n}\}$	$\boldsymbol{T}(a)\boldsymbol{D}_{\infty\mathrm{h}}$
	14	b_2 c_1	$\boldsymbol{T}'(f)$	$\boldsymbol{T}(a)\boldsymbol{D}_{\infty\mathrm{h}}$
11	$\boldsymbol{Y}^{(11)}$	a_1	$\boldsymbol{T}(f)\boldsymbol{D}_{n\mathrm{h}}$	$\boldsymbol{T}(f)\boldsymbol{D}_{n\mathrm{h}}$
	6		$z=\frac{f}{4}$	$\boldsymbol{T}(\frac{f}{2})\boldsymbol{D}_{n\mathrm{h}}$
	3		$\varphi=\frac{\pi}{2n}$	$\boldsymbol{T}(f)\boldsymbol{D}_{2n\mathrm{h}}$
	6	a_2 d_1	$\boldsymbol{T}(f)\boldsymbol{C}_{n\mathrm{v}}$	$\boldsymbol{T}(f)\boldsymbol{D}_{n\mathrm{h}}$
	3	b_1 c_1	$\boldsymbol{T}(f)\boldsymbol{C}_{n\mathrm{h}}$	$\boldsymbol{T}(f)\boldsymbol{D}_{n\mathrm{h}}$
	$\boldsymbol{Y}^{(1_1)}$	b_2,c_2 e_1,f_1	$\boldsymbol{T}(a)\boldsymbol{C}_n$	$\boldsymbol{T}(a)\boldsymbol{D}_{n\mathrm{h}}$
	15	g_1	$\boldsymbol{T}(f)\boldsymbol{C}_{1\mathrm{h}}$	$\boldsymbol{T}(f)\boldsymbol{D}_{\infty\mathrm{h}}$
	14	g_2 h_1	$\boldsymbol{T}(f)$	$\boldsymbol{T}(f)\boldsymbol{D}_{\infty\mathrm{h}}$
12	$\boldsymbol{Y}^{(12)}$	a_1	$\boldsymbol{T}'(f)\boldsymbol{C}_{n\mathrm{h}}$	$\boldsymbol{T}'(f)\boldsymbol{C}_{n\mathrm{h}}$
	3		$\varphi=0, \frac{\pi}{n}$	$\boldsymbol{T}(f)\boldsymbol{D}_{n\mathrm{h}}$
	4		$\varphi=\frac{\pi}{2n}$	$\boldsymbol{T}^1_{2n}(f)\boldsymbol{D}_{n\mathrm{h}}$
	1_1	a_2 a_3	$\boldsymbol{T}'(f)\boldsymbol{C}_n$	$\boldsymbol{T}(f)\boldsymbol{C}_{n\mathrm{h}}$
	7	b_1	$\boldsymbol{T}'(f)\boldsymbol{C}_n$	$\boldsymbol{T}(f)\boldsymbol{C}_{n\mathrm{h}}$
	15	c_1	$\boldsymbol{T}'(f)\boldsymbol{C}_{1\mathrm{h}}$	$\boldsymbol{T}(f)\boldsymbol{D}_{\infty\mathrm{h}}$
	14	c_2 d_1	$\boldsymbol{T}'(f)$	$\boldsymbol{T}(f)\boldsymbol{D}_{\infty\mathrm{h}}$
13	$\boldsymbol{Y}^{(13)}$	a_1	$\boldsymbol{T}^1_{2n}(f)\boldsymbol{D}_{n\mathrm{h}}$	$\boldsymbol{T}^1_{2n}(f)\boldsymbol{D}_{n\mathrm{h}}$
	4		$\varphi=\frac{\pi}{2n}$	$\boldsymbol{T}(f)\boldsymbol{D}_{2n\mathrm{h}}$
	1_1		$\varphi=\frac{\pi}{2n}, z=\frac{f}{4}$	$\boldsymbol{T}(\frac{f}{2})\boldsymbol{D}_{2n\mathrm{h}}$
	1_1	a_2	$\boldsymbol{T}^1_{2n}(f)\boldsymbol{C}_{n\mathrm{v}}$	$\boldsymbol{T}(f)\boldsymbol{D}_{2n\mathrm{h}}$
	4	b_1	$\boldsymbol{T}^1_{2n}(f)\boldsymbol{C}_{n\mathrm{h}}$	$\boldsymbol{T}^1_{2n}(f)\boldsymbol{D}_{n\mathrm{h}}$
	8	d_1	$\boldsymbol{T}^1_{2n}(f)\boldsymbol{C}_{n\mathrm{v}}$	$\boldsymbol{T}^1_{2n}(f)\boldsymbol{D}_{n\mathrm{h}}$
	$\boldsymbol{Y}^{(1_2)}$	e_1 f_1	$\boldsymbol{T}^1_{2n}(f)\boldsymbol{C}_n$	$\boldsymbol{T}^1_{2n}(\frac{a}{2})\boldsymbol{D}_{n\mathrm{h}}$
	15	g_1	$\boldsymbol{T}^1_{2n}(f)\boldsymbol{C}_{1\mathrm{h}}$	$\boldsymbol{T}(f)\boldsymbol{D}_{\infty\mathrm{h}}$
	14	g_2 h_1	$\boldsymbol{T}^1_{2n}(f)$	$\boldsymbol{T}(f)\boldsymbol{D}_{\infty\mathrm{h}}$

3.3 Symmetry Domain

As it has been defined in the Sect. 2.1.4, *symcell* is the minimal part of a system generating the whole system by the action of the translational group. However, independently of a specific system, the group action infers a partition of the whole Euclidean space. Namely, each point of this space generates an orbit of the line group, and gathering representatives of different (and automatically disjoint) orbits, one gets the set generating the whole space. It is well-known mathematical theorem on *stratification* [5] that orbit representatives may be chosen such that they form a connected *symmetry domain*. The representatives of the same orbit types form sub-domains called *strata*. The stratus of the orbit type with the least stabilizer, called *generic orbit*, almost fulfills the whole domain, while other strata, called *special*,

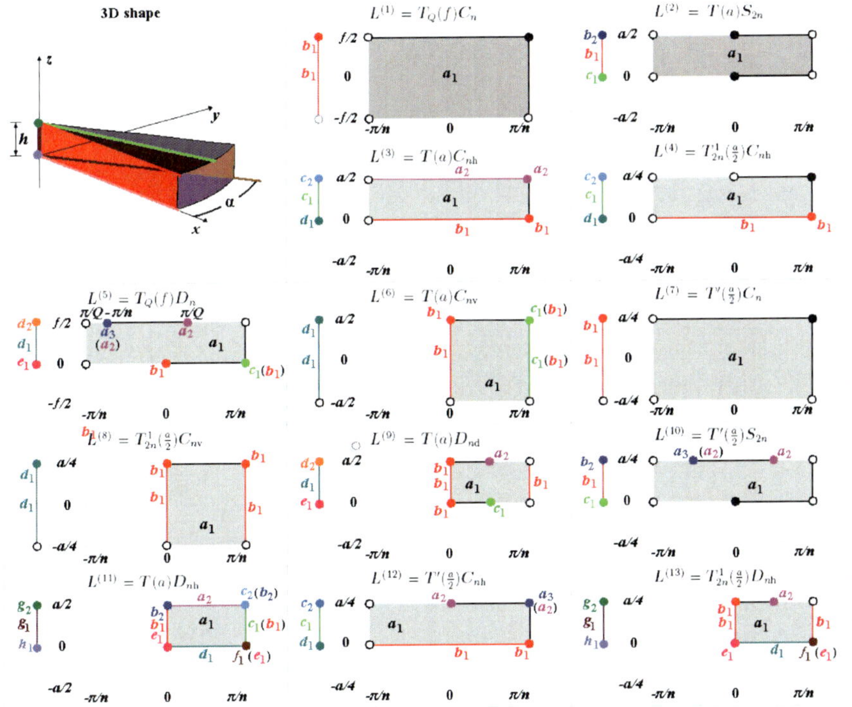

Fig. 3.3 Symmetry domains of the line groups. The *top left* panel is general shape of the domain: radial sector of the angle α of the disk of the height h. On the other panels are domains of the line group families, with specified α and h. For clarity, the part of the z-axis in the domain is given separately, while the rest of the sector is represented by its section. The generic a_1 stratus is shaded *light gray*. The special strata, surfaces and lines on the boundaries of the domain, are represented as the colored lines and points on the boundaries of the section with indicated orbit-type symbol (*braced symbols* apply for n odd). *Black lines* and *filled circles* belong to the generic strata, while the *light gray* boundaries and *empty circles* do not. For comparison between the different families, the section $-f/2 \leq z \leq f/2$ and $-\pi/n \leq \alpha \leq \pi/n$ is indicated in all cases

are on its boundary. In fact, the strata of the orbits with greater stabilizer are on the boundaries of the strata of the orbits with less stabilizers. Note that such domain is not unique; in the following we present a choice convenient for its simplicity.

As for the first family line groups, screw-axis $\boldsymbol{Z}$ generates the whole space from the disk of the infinite radius, with the height f (fractional translation of the generator $(C_Q|f)$). Further, the action of the point factor $\boldsymbol{C}_n$ generates this disk from its radial sector of the angle $2\pi/n$. Therefore, the domain is this radial sector. The whole domain is generic stratus a_1, except that its central part (on the z-axis) is the only special stratus b_1.

For other line groups, the shape of domain remains disk sector. However, for the negative line groups the disk height is the half of the translation f of the generator of $\boldsymbol{Z}$ (families 2–5 and 9–13), while the sector angle is halved in the families 6, 8, 9, 11, and 13, containing mirror planes.

The general shape and described domains for all line groups are illustrated in Fig. 3.3. It is obvious that operations from $\boldsymbol{Z}$ do not contribute to special strata. On the other hand, mirror planes produce special lines and U axes the special points. Finally, the z-axis is a special line for $\boldsymbol{C}_n$ and $\boldsymbol{S}_{2n}$, where the later has also special points.

Finally, let us mention that for any system the symcell may be chosen to be intersection of the system with the domain of its line group.

3.4 Symmetry Fixing Sets

The precise meaning of the statement that line group $\boldsymbol{L}$ is symmetry group of a system S is not only that S is invariant under all the transformations from $\boldsymbol{L}$ but also that $\boldsymbol{L}$ is the maximal group of transformations leaving S invariant. For most of the orbits of the line groups the symmetry is greater than the original line group, meaning that the symmetry of the system is in general less than the symmetry of the isolated orbits. In fact, only the operations simultaneously leaving invariant all of the orbits constitute the symmetry of the composite system. This shows that the symmetry $\boldsymbol{L}$ of a polymer S, decomposing into orbits as $S = S_1 + \ldots S_\Omega$, is the intersection $\tilde{\boldsymbol{Y}}_1 \cap \cdots \cap \tilde{\boldsymbol{Y}}_\Omega$ of the full symmetries $\tilde{\boldsymbol{Y}}_i$ of the orbits. So, combining the orbits of $\boldsymbol{L}$, we get systems with symmetry less than that of the isolated orbits, but not necessarily restricted to $\boldsymbol{L}$, i.e., some combinations give supergroups. However, if some set of orbits of $\boldsymbol{L}$ with the total symmetry being $\boldsymbol{L}$ is a subsystem in the polymer S, then the symmetry of S is $\boldsymbol{L}$, independently of other orbits.

Among the combinations of orbits having the total symmetry $\boldsymbol{L}$ the *symmetry fixing sets* or *irreducible combinations* are minimal in the sense that if any orbit from such a set is expelled, the symmetry of the remaining part of system becomes a supergroup of $\boldsymbol{L}$. Consequently, any system with symmetry $\boldsymbol{L}$ contains at least one symmetry fixing set. To determine symmetry of some concrete polymer, as well as for other physical applications (e.g., Jahn–Teller theorem discussed in Sect. 8.3.4), it is important to find out these irreducible combination of the orbits.

Table 3.4 Symmetry fixing sets of the line groups. For each line group family F, all the symmetry fixing sets are listed. Those which are symmetry fixing only for chiral cases or when n is odd or even are denoted by C, O, and E, respectively

F	Symmetry fixing sets									
1	$2a_1$	a_1+b_1								
2	$2a_1$									
3	$2a_1$	$2a_2$	$2b_1$	a_1+a_2	a_1+b_1	a_2+b_1				
4	$2a_1$	$2b_1$	a_1+b_1							
5	a_1	a_2	a_3	C: b_1	C: c_1					
6	$2a_1$	$2b_1$	$2c_1$	a_1+b_1	a_1+c_1	a_1+d_1	b_1+c_1	b_1+d_1	c_1+d_1	
7	$2a_1$									
8	$2a_1$	$2b_1$	a_1+b_1	a_1+d_1	b_1+d_1					
9	a_1	a_2	b_1	c_1						
10	E: $2a_1$	E: $2a_2$	E: a_1+a_2	O: a_1	O: a_2					
11	a_1	a_2	b_1	b_2	c_1	c_2	d_1	e_1	f_1	
12	$2a_1$	$2a_2$	$2a_3$	a_1+a_2	a_1+a_3	a_2+a_3	$2b_1$	a_1+b_1	a_2+b_1	a_3+b_1
13	a_1	a_2	b_1	d_1	e_1	f_1				

They can be found with the help of Table 3.1. In fact, the monomer of S is the union $M = M^1 + \cdots + M^\Omega$ of the monomers of the orbits contained in S. Obviously, if M is invariant under some supergroup $\boldsymbol{P}'$ of $\boldsymbol{P}$, such that $\boldsymbol{Z}\boldsymbol{P}'$ is a line group (being then a supergroup of $\boldsymbol{L} = \boldsymbol{Z}\boldsymbol{P}$, which leaves S invariant), S cannot not be a symmetry fixing set for $\boldsymbol{L}$, as it has greater symmetry. Then, the symmetry of the orbit is $\boldsymbol{P}'\boldsymbol{Z}$, where $\boldsymbol{P}'$ is the maximal subgroup of $\boldsymbol{P}_{\mathrm{M}}$ (column 5 of Table 3.1) which can be combined with $\boldsymbol{Z}$ (i.e., commutes with $\boldsymbol{Z}$). This subgroup is $\boldsymbol{P}_{\mathrm{M}}$ itself if $\boldsymbol{Z} = \boldsymbol{T}, \boldsymbol{T}', \boldsymbol{T}^1_{2n}$, because each point group can be combined with them (no "Not group" labels in the corresponding columns). Consequently, for these groups the symmetry fixing combinations of the orbits correspond to those of the point groups: if in the irreducible set for the point group (e.g., $2a + b + \ldots$) any possible combination of indices is inserted ($a_\alpha + a_\beta + b_\gamma$), the symmetry fixing set for the line group is obtained. However, chiral (including incommensurate) helical axis $\boldsymbol{T}_Q(f)$ is compatible only with $\boldsymbol{C}_n$ and $\boldsymbol{D}_n$. Further, any non-linear orbit completely defines such a helical axis. Therefore, any orbit of type a, b, or c is symmetry fixing for $\boldsymbol{T}_Q(f)\boldsymbol{D}_n$, while two orbits of $\boldsymbol{T}_Q(f)\boldsymbol{C}_n$, at least one of them non-linear (type a), are necessary to exclude U-axis. The symmetry fixing sets are given in Table 3.4.

3.5 Application: Line Group Notation for Monoperiodic Crystals

The decomposition of the system into the orbits of the line groups is used for different applications, as it will be seen in Chap. 8. In fact, this partition enables to give precise conformation of system in a very essential and efficient way: it is necessary to fix the line group and the coordinates of one orbit representative from each orbit. For many purposes only orbit type (instead of coordinates) suffices. The general form of such notation for the system with orbits $S_1, \ldots, S_\Omega$, each of them repeated N_i times, i.e., with orbit representatives at $\boldsymbol{r}_{i1} \ldots, \boldsymbol{r}_{iN_i}$, is

$$L[S_1\{Z_{11}\boldsymbol{r}_{11};\ldots;Z_{1N_1}\boldsymbol{r}_{1N_1}\},\ldots,S_\Omega\{Z_{\Omega 1},\boldsymbol{r}_{\Omega 1};\ldots;Z_{\Omega N_\Omega},\boldsymbol{r}_{\Omega N_\Omega}\}], \quad (3.2)$$

where $\boldsymbol{L}$ is symbol (international or factorized) of the particular line group and Z_{ij} is chemical symbol of the orbit representative at $\boldsymbol{r}_{ij}$. From these data, all the atomic coordinates may be found by action with the group transformations onto the orbit representative coordinates. For some purposes, when only the orbit contents is important but not the atomic coordinates, this can be reduced to $\boldsymbol{L}[N_1S_1,\ldots,N_\Omega S_\Omega]$.

For example, the symmetry formula of the *trans*-polyacetylene (Fig. 3.4) is

$$\boldsymbol{T}_2^1(f)\boldsymbol{D}_{1\mathrm{h}}[e_1\{C,(x_C,0,0);H,(x_H,0,0)\}], \quad f=\frac{a}{2}=1.22\text{Å},$$
$$x_C=0.6\,\text{Å},\ x_H=1.6\text{Å}$$

(x_C and x_H are coordinates [6] of the carbon and hydrogen orbit representatives). The short notation is $\boldsymbol{T}_2^1\boldsymbol{D}_{1\mathrm{h}}[2e_1]$. As for the other isomers we get $\boldsymbol{T}_2^1\boldsymbol{D}_{1\mathrm{h}}[2b_1]$ for *cis-transoid, trans-cisoid*, and *cis*-polyacetylene (same line group as of the *trans*-polyacetylene, also two orbits but of the different type, as the representatives are now out of the horizontal plane, i.e., z_C and z_H are not zero), and $\boldsymbol{T}\boldsymbol{D}_{1\mathrm{d}}[2b_1]$ for *trans-transoid* polyacetylene. Similarly, for the chain conformation of the stereo-regular crystalline *butadiene* [7] polymer (Fig. 3.5) we get $\boldsymbol{T}'\boldsymbol{C}_1[10a_1]$ for *cis*-1,4, $\boldsymbol{T}\boldsymbol{S}_2[5a_1]$ for *trans*-1,4, $\boldsymbol{T}_2^1\boldsymbol{D}_{1\mathrm{h}}[3a_23e_14d_1]$ for *syndiotactic* 1,2, and $\boldsymbol{T}_3^1[10a_1]$ for the *isotactic* 1,2 form.

The symmetry notation for backbone of A, B, and E types of deoxyribonucleic acid [8] is $\boldsymbol{T}_{11}^1(2.56\text{Å})\boldsymbol{D}_1[19a_1]$, $\boldsymbol{T}_{10}^1(3.38\text{Å})\boldsymbol{D}_1[19a_1]$, and $\boldsymbol{T}_{12}^1(3.61\text{Å})\boldsymbol{D}_1[19a_1]$ (the fifth family, with $n=1$), respectively. The pairs of bases arranged within double helix of backbone break both the helical and the U-axis symmetry. Only when all the pairs are same (Fig. 3.6) helical symmetry is retained. Still, geometries of different pairs are similar, and for some applications the differences may be neglected.

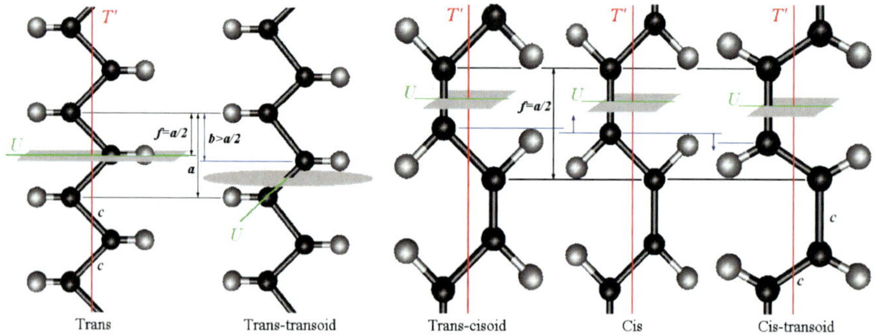

Fig. 3.4 Isomers of polyacetylene $(\mathrm{CH})_x$. Carbon atoms are *black* and hydrogen ones *gray*. The elements of the symmetry groups $L^{(9)}=\boldsymbol{T}(a)\boldsymbol{D}_{1\mathrm{d}}$ of trans-transoid and $\boldsymbol{L}(13)=\boldsymbol{T}_2^1(f)\boldsymbol{C}_1$ of other isomers are depicted as follows: glide plane $\boldsymbol{T}'$ and U-axis as *vertical* and *horizontal line*, horizontal mirror and roto-reflectional planes by *gray parallelogram* and *circle*; vertical mirror plane is xz-plane. *Arrows* emphasized deformation of *cis*-isomer transforming it to *trans-cisoid* and *cis-transoid*

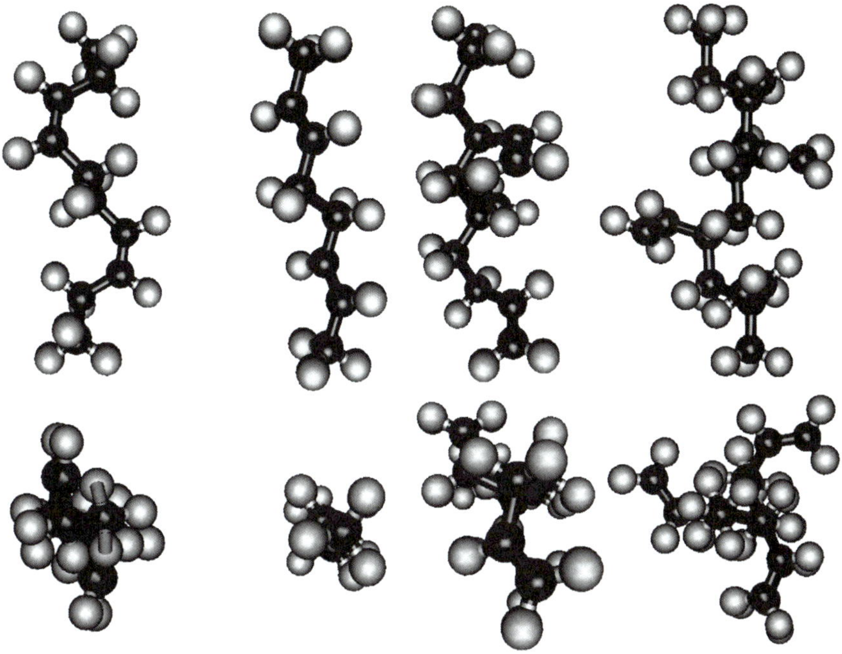

Fig. 3.5 Isomers of butadiene $(C_4H_6)_x$; from *left* to *right*: *cis*-1,4, *trans*-1,4, *syndiotactic*-1,2 and *isotactic*-1,2 (*side* and *top* view)

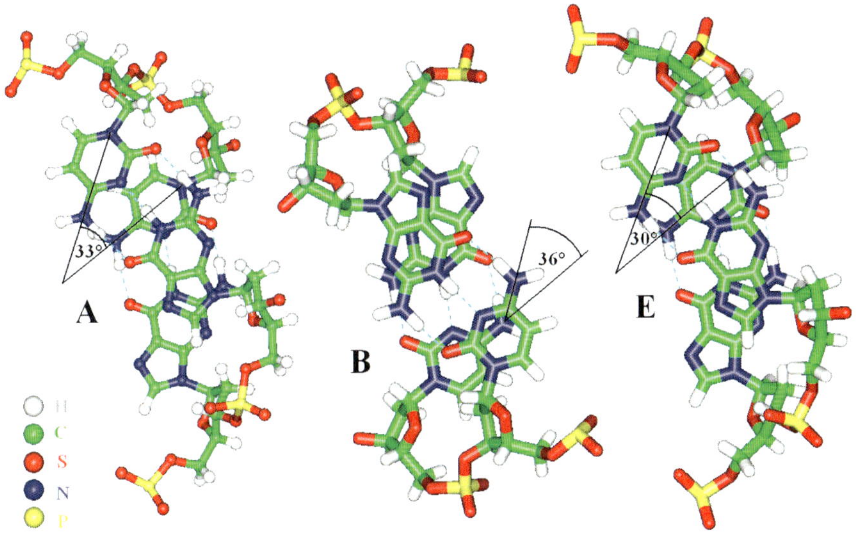

Fig. 3.6 Deoxyribonucleic acid: A, B, and E form with only cytosine–guanine pair of bases. Two successive monomers of each form are presented and the angles $2\pi/q$ of the rotation of the helical axis, with $q = 11$, 10, and 12, respectively, are indicated

Finally, for the carbon nanotubes, being single orbit systems (Sect. 9.2), we get $\boldsymbol{L}(S\{C, \boldsymbol{r}_C\}]$. For example, $\boldsymbol{T}_{28}^{11}(0.465\text{Å})\boldsymbol{D}_2[a_1\{C, (3.34, 1.3, 0.465)\}]$ fully describes the tube (8,2).

References

1. I. Milošević, M. Damnjanović, Phys. Rev. B **47**, 7805 (1993) 3
2. L. Michel, Rev. Mod. Phys **52**, 617 (1980)
3. H.A. Jahn, E. Teller, Proc. Roy. Soc. A **161**, 220 (1937)
4. M. Damnjanović, M. Vujičić, J. Phys. A **14**, 1055 (1981)
5. H. Abud, G. Sartori, Ann. Phys. **150**, 307 (1983)
6. J.C.W. Chien, *Polyacetylene* (Academic Press, New York, 1984)
7. G. Natta, Experientia Suppl. **7**, 21 (1957)
8. A. Kornberg, *DNA Synthesis* (W. H. Freeman and Company, San Francisko, 1974)

Chapter 4
Irreducible Representations

Abstract The irreducible representations of the line groups are the starting point for physical applications. Quite general Wigner's theorem [1] singles out unitary representations as the relevant ones in the quantum mechanical framework. Such representations are decomposable to the irreducible components, which are ingredients sufficient for composition of any unitary representations. Hence, in this chapter, we construct and tabulate irreducible unitary representations only, although line groups, since being not compact, have also the non-unitary representations. The construction starts with the first family groups. Then we use simple (induction) procedure to derive the representations of the families 2–8, containing the halving first family subgroup; finally, we use these representations repeating the same procedure in order to get the representations of the largest families (with the first family subgroup of index four). At the end, we make an overview of their properties and physical implications.

4.1 First Family

As the first family groups are direct product of the helical and rotational factors, their irreducible representations are found as the direct product of the irreducible representations of the factors [2]. This task is easy, since both factors are cyclic groups. In fact, the homomorphism condition $D(\ell\ell') = D(\ell)D(\ell')$ for the cyclic groups, where all the elements are powers of the single generators g, means that for given $D(g)$ all the other elements are represented as $D(g^t) = D^t(g)$. Further, Schur's Lemma [2] provides that the irreducible representations are one-dimensional. Consequently, each nonzero complex number c defines an irreducible representation by the equality $D^{(c)}(g) \stackrel{\mathrm{def}}{=} c$. Among them unitary ones are those with c of absolute value 1. Thus, the different unitary irreducible representations are $D^{(x)}(g) \stackrel{\mathrm{def}}{=} \mathrm{e}^{\mathrm{i}x}$ for $x \in (-\pi, \pi]$. For the finite groups of order n, as $g^n = (I|0)$ implying $1 = D^n(g) = \mathrm{e}^{\mathrm{i}nx}$, nx is a multiple of 2π; in other words, there are n representations given by $x = 2\pi m/n$, where m is integer from the above interval.

Damnjanović, M., Milošević, I.: *Irreducible Representations*. Lect. Notes Phys. **801**, 47–64 (2010)
DOI 10.1007/978-3-642-11172-3_4

4.1.1 Helical Quantum Numbers

Using this we find that there are infinitely many irreducible representations of the helical group $\boldsymbol{T}_Q(f)$ generated by $Z = (C_Q|f)$. They are classified with the help of parameter $\tilde{k}$:

$$_{\tilde{k}}A(Z) = \mathrm{e}^{\mathrm{i}\tilde{k}f}, \quad \tilde{k} \in \left(-\frac{\pi}{f}, \frac{\pi}{f}\right]. \tag{4.1}$$

The special form of the exponent is used in order to interpret $\hbar\tilde{k}$ as the *helical quasi-momentum*, i.e., quasi-momentum canonically conjugated to helical coordinate $\tilde{z}$ (see Sect. 2.2.1). The range $(-\frac{\pi}{f}, \frac{\pi}{f}]$ of $\tilde{k}$ is known as the *helical Brillouin zone*. Quite analogously, for the rotational subgroup $\boldsymbol{C}_n$ there are n different irreducible representations, given by the integer m:

$$A_{\tilde{m}}(C_n) = \mathrm{e}^{\mathrm{i}\tilde{m}\frac{2\pi}{n}}, \quad \tilde{m} \in \left(-\frac{n}{2}, \frac{n}{2}\right]. \tag{4.2}$$

Again, $\hbar\tilde{m}$ is z-component of *angular quasi-momentum*. Finally, we conclude that there are infinitely many irreducible representations of the first family line groups, and they are classified by the *helical quantum numbers* [3] of the helical and angular momentum $\tilde{k}$ and $\tilde{m}$:

$$_{\tilde{k}}A_{\tilde{m}}(\tilde{\ell}(1,0)) = \mathrm{e}^{\mathrm{i}\tilde{k}f}, \quad _{\tilde{k}}A_{\tilde{m}}(\tilde{\ell}(0,1)) = \mathrm{e}^{\mathrm{i}\tilde{m}\frac{2\pi}{n}} \tag{4.3}$$

(recall that $Z = \tilde{\ell}(1,0)$ and $C_n = \tilde{\ell}(0,1)$). Note that the helical quantum numbers are independently defined, which implies that the same representation is obtained for any $\tilde{m}' = \tilde{m} + \tilde{M}n$ and/or $\tilde{k}' = \tilde{k} + \tilde{K}2\pi/f$. However, the helical irreducible representations are convention dependent, as the matrices (4.1) are associated to the helical generator Z, which is fixed by the convention adopted.

4.1.2 Commensurate Groups and Linear Quantum Numbers

For the commensurate systems there is an alternative choice of quantum numbers, based on the representations of the translational subgroup $\boldsymbol{T}(a)$

$$_{k}A((I|0)) = \mathrm{e}^{\mathrm{i}ka}, \quad k \in \left(-\frac{\pi}{a}, \frac{\pi}{a}\right], \tag{4.4}$$

and of the isogonal group $\boldsymbol{C}_q$:

$$A_m(C_q) = \mathrm{e}^{\mathrm{i}m\frac{2\pi}{q}}, \quad \tilde{m} \in \left(-\frac{q}{2}, \frac{q}{2}\right]. \tag{4.5}$$

The range $(-\frac{\pi}{a}, \frac{\pi}{a}]$ of the (linear) *quasi-momentum* k is known as the *Brillouin zone*. To differ from $\tilde{m}$ which shows only the part of angular momentum which is not included in $\tilde{k}$, the quantum number m corresponds to the complete *angular quasi-momentum*. However, since $\boldsymbol{C}_q$ is not a subgroup of $\boldsymbol{L}$, these representations are not to be directly multiplied to get a representation of $\boldsymbol{L}$. Instead, due to the fractional translation for pa/q accompanying C_q in $\ell(0,1)$, the resulting representations, classified [4, 5] by the *linear quantum numbers* k and m, are

$$_kA_m(\ell(1,0)) = e^{\mathrm{i}ka}, \quad _kA_m(\ell(0,1)) = e^{\mathrm{i}k\tilde{p}\frac{a}{q}} e^{\mathrm{i}m\frac{2\pi}{q}}. \tag{4.6}$$

In contrast to the helical quantum numbers, from (4.6) it follows that $_kA_m$ is equal to $_{k'}A_{m'}$ only when the equalities $m' = m + Mn$ and $k' = k + K2\pi/a$ simultaneously hold, and the integers K and M satisfy $Mr \overset{\tilde{q}}{=} -K$. Applying (2.18) in the form $r\tilde{p} \overset{\tilde{q}}{=} 1$, we see that only the simultaneous change $k' = k + K2\pi/a$ and $m' \overset{q}{=} m - pK$ gives the same representation.

4.1.3 Transition Rules

Both sets (4.3) and (4.6) are representations of the same groups, and therefore they are biuniquely related. To find this correspondence, we use relations (2.19). First, the representations of the roto-helical and translational generators with the linear and helical quantum numbers, respectively, are easily found:

$$_kA_m(\tilde{\ell}(1,0)) = e^{\mathrm{i}k\frac{a}{q}} e^{\mathrm{i}mr\frac{2\pi}{q}}, \quad _kA_m(\tilde{\ell}(0,1)) = e^{\mathrm{i}m\frac{2\pi}{n}}; \tag{4.7a}$$

$$_{\tilde{k}}A_{\tilde{m}}(\ell(1,0)) = e^{-\mathrm{i}\tilde{m}r\frac{2\pi}{n}} e^{\mathrm{i}\tilde{k}a}, \quad _{\tilde{k}}A_{\tilde{m}}(\ell(0,1)) = e^{\mathrm{i}\tilde{m}\frac{1-r\tilde{p}}{\tilde{q}}\frac{2\pi}{n}} e^{\mathrm{i}\tilde{k}\tilde{p}\frac{a}{\tilde{q}}}. \tag{4.7b}$$

Using this we straightforwardly get the *transition rules* between the two pairs of quantum numbers of the same representation. Combining (4.3) and (4.7a) we find

$$(k,m) \to (\tilde{k}(k,m), \tilde{m}(m)) = \left(k + \frac{rm}{n}\frac{2\pi}{a} + \tilde{K}\tilde{q}\frac{2\pi}{a}, m + \tilde{M}n\right). \tag{4.8a}$$

Integers $\tilde{K}$ and $\tilde{M}$ are uniquely and independently determined by the requirement that $\tilde{k}$ and $\tilde{m}$ are from the intervals given in (4.1) and (4.2). Analogously, from (4.6) and (4.7b) it follows:

$$(\tilde{k},\tilde{m}) \to (k(\tilde{k},\tilde{m}), m(\tilde{k},\tilde{m})) = \left(\tilde{k} - \tilde{m}\frac{r}{n}\frac{2\pi}{a} - K\frac{2\pi}{a}, \tilde{m} + Kp + Mq\right). \tag{4.8b}$$

Again, K and M are integers enabling to get linear momenta from the intervals given by (4.4) and (4.5). However, as M depends on K, the value of K should be determined first.

It should be emphasized that due to the helical quantum number $\tilde{k}$, labeling the representations (4.3), the transition rules depend on the r-convention. Indeed, the choice of r determines the helicity of the momentum $\tilde{k}$: the unit change of the total momentum m leads to the same change of $\tilde{m}$ (assuming $n > 1$) and a simultaneous change of $\tilde{k}$ for $2r\pi/na$. Analogously, a change of the helical momentum by $2\pi/a$ preserves the linear momentum k and induces a jump for p of the total angular momentum m.

4.1.4 Brillouin Zones and Bands

In the both classifications the irreducible representations are grouped into the so-called bands, the property which will be inherited by all the other families, namely, for fixed $\tilde{m}$, representations differ only by $\tilde{k}$. When $\tilde{k}$ varies within the *helical Brillouin zone*, we obtain a series of representations to be referred to as a *helical* or *$\tilde{m}$-band*, physical implications of which will be given later. Analogously, for commensurate groups, fixing m we get series of the representations with k from the *Brillouin zone*, which is called *linear* or *m-bandm-*.

From (4.8a) it follows that the set of $\tilde{q}$ linear bands with m differing by multiples of n gives exactly a single $\tilde{m}$-band with $\tilde{m} \overset{n}{=} m$. As $\tilde{k}(k, m) + \frac{2\pi}{a} = \tilde{k}(k, m + p \pmod{q})$, the segment of the helical Brillouin zone of the $\tilde{m}$-band corresponding to the m-band (thus $m = \tilde{m} + Mn$ for some M), is followed by the segment corresponding to $(m + p \pmod{q})$-band; these $\tilde{q}$ different segments continuously fill up the $\tilde{m}$-band. The jumps of m occur at the points $\tilde{k} = \frac{\pi}{a}(1 + 2\tilde{m}\frac{r}{n} + 2K)$. This is illustrated in Fig. 4.1, where for the line group $\boldsymbol{T}^9_{56}\boldsymbol{C}_4 = \boldsymbol{L}56_{44}$ every $\tilde{m}$-band is divided into m-bands.

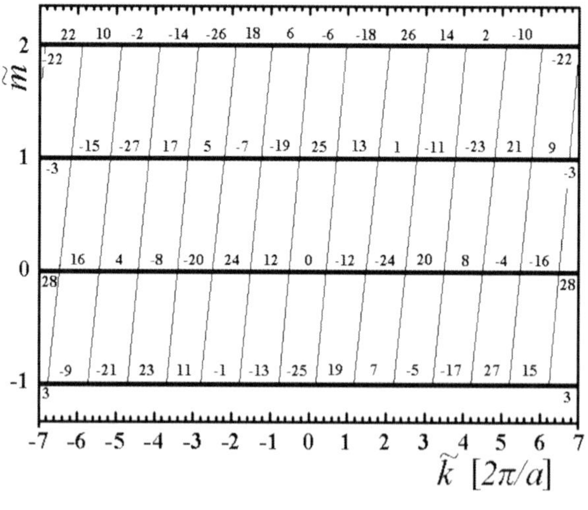

Fig. 4.1 Subdivision of $\tilde{m}$-bands into m-bands for the line group $\boldsymbol{T}^9_{56}\boldsymbol{C}_4 = \boldsymbol{L}56_{44}$. Different parts of length $2\pi/a$ (linear Brillouin zone) of $\tilde{m}$-band (of the length of helical Brillouin zone is $\tilde{q}2\pi/a$, with $\tilde{q} = 14$) are separated by the inclined lines and assigned by the corresponding quantum number m. From *left* to *right* m increases for $p = 44$, but to get standard value from the interval $(-q/2, q/2]$ (i.e., from the set $-27, \ldots, 28$) it is reduced by subtracting multiple of $q = 56$

The physical contents of the obtained quantum numbers can be seen from the irreducible representations. From the first of the equations (4.6) it follows that k is canonically conjugated to the discrete translations, i.e., k is *linear quasi-momentum*. Further, in the second equation (4.6) the same ratio p/q appearing in the generator and in the exponent means that k is associated to the fractional translation of $\ell(0, 1)$. Then the remaining m-dependent part appears due to the isogonal rotation C_q, and m is the corresponding linear *angular momentum* (component along the rotational axis of C_q). Unless $\boldsymbol{L}$ is symmorphic (when $n = q$), C_q is not element of $\boldsymbol{L}$, and m is not a conserved quantum number. As for the $\tilde{k}\tilde{m}$-numbers, from the first of equation (4.3) it follows that $\tilde{k}$ combines angular and linear momenta into the *helical momentum* conjugated to the helical coordinate $\tilde{z}$. Complementing helical momentum, $\tilde{m}$ is the "pure" angular momentum, not related to the helical momentum, as steaming from the rotations of the point subgroup $\boldsymbol{C}_n$. Both quantum numbers $\tilde{k}$ and $\tilde{m}$ are conserved. It turns out that depending on physical considerations one or the other choice of quantum numbers is more suitable.

To clarify physical meaning of the transition rules, one goes back to the relation (2.8) between the moments. Denoting by $\zeta = \sqrt{f^2 + \rho^2 4\pi^2/Q^2}$ the helical arc length between the two adjacent points connected by the helical generator $(C_Q|f)$, for the commensurate groups one gets $\zeta = \frac{\sqrt{n^2a^2+4\pi^2r^2\rho^2}}{q}$, $\sin\chi = \frac{na}{q\zeta}$, $\frac{\cos\chi}{\rho} = \frac{2\pi r}{q\zeta}$ and finally

$$\tilde{p} = \frac{nap_z + 2\pi r l_z}{q\zeta}, \quad \tilde{l}_z = l_z. \tag{4.9}$$

If $\tilde{K}$ is the quasi-momentum resulting from $\tilde{p}$ because of the periodicity along the helix (with period ζ), the irreducible representations (4.1) should be $e^{i\tilde{k}f} = e^{i\tilde{K}\zeta}$. Obviously $\tilde{k} = \tilde{K}\zeta/f = \tilde{K}/\sin\chi$, and as $\tilde{k}$ is ρ-independent, it is preferably used as the helical quantum number. According to (4.9), the corresponding momentum $\tilde{p}/\sin\chi$ is related to the linear and angular momenta as $\tilde{p}/\sin\chi = p_z + \frac{2\pi r l_z}{na}$, which clarifies the meaning of (4.8b).

4.1.5 Special Points

The center and edge point of the one-dimensional Brillouin zone (either helical or linear) have some special properties which are useful for the further constructions. As we will see later on, for the negative line groups (Sect. 2.3.3) these points are special as negative elements leave them invariant, giving rise to lower dimension of the irreducible representations. Therefore we will consider them in detail.

The representations with $\tilde{k} = 0, \tilde{q}\pi/a$ and $\tilde{m} = 0$ and (if n is even only) $\tilde{m} = n/2$ are particularly important: they are the only one-dimensional, real number representations. In particular ${}_{\tilde{k}=0}A_{\tilde{m}=0}(\ell) = 1$ for all the elements of $\boldsymbol{L}$ (the identity representation), and the remaining ones are alternating (half of the elements are represented by 1 and the other half by -1). In commensurate groups, using the transition rules (4.8b), we get special representations with $k = 0$ and $m = 0, q/2$ or $k = \pi/a$ and $m = -p/2, (q-p)/2$. While for the identity representation $\tilde{k} = \tilde{m} = 0$ we immediately get $k = m = 0$, the other representations are mutually related according to the parities of $\tilde{q}$, $\tilde{p}$, r, and A (defined by (2.18) as $r\tilde{p} = 1 + A\tilde{q}$, Appendix C):

	(k, m)			
$(\tilde{k}, \tilde{m})$	$(0, 0)$	$(0, \frac{q}{2})$	$(\frac{\pi}{a}, -\frac{p}{2})$	$(\frac{\pi}{a}, \frac{q-p}{2})$
$(0, 0)$	1–6	–	–	–
$(\tilde{q}\frac{\pi}{a}, 0)$	–	5,6	3,4	1,2
$(0, \frac{n}{2})$	–	2,4	1,6	3,5
$(\tilde{q}\frac{\pi}{a}, \frac{n}{2})$	–	1,3	2,5	4,6

$\tilde{q}\tilde{p}rA$	$\tilde{q}\tilde{p}rA$
1 oooe	4 oeeo
2 ooeo	5 eooo
3 oeoo	6 eooe

(4.10)

To use this equation, for a particular line group we first determine the parities of all four parameters and find the ordinal (between 1 and 6) of this combination in the right table ("o" and "e" stand for odd and even). Then, given a pair of linear (helical) quantum numbers, in the column (row) of the left table corresponding to this pair we fix row (column) where the found ordinal appears; the equivalent pair of helical (linear) quantum numbers is at the beginning of this row (column).

In particular, for translational groups km and $\tilde{k}\tilde{m}$ values coincide (the first column "$(\tilde{k}, \tilde{m})$" of the left table in (4.10)): in this case $q = n$, implying $\tilde{q} = 1$, $r = p = \tilde{p} = 0$, and $A = -1$, resulting in combination 4. For zigzag groups $\tilde{q} = 2$, $r = \tilde{p} = 1$, and $A = 0$, which is the combination 6. For nanotubes with hexagonal lattices, $\tilde{q} \overset{12}{=} 2$ is even, implying that $\tilde{p}$ and r are odd, and only combinations 5 and 6 can be realized, depending on the parity of A.

Note that parity of r may depend on the convention. In fact, within C0, it is co-prime with both $\tilde{q}$ and n. Thus, when n is even, which is necessary for existence of the last two helical quantum numbers, r is odd, i.e., within this convention in the last two rows of the left table the combinations 2 and 4 do not appear.

4.1.6 Zigzag Groups

Several families of the line groups are defined for $q = 2n$, and therefore it is worth to simplify the derived formula for this case. At first, as we have seen $\tilde{q} = 2$, $r = 1$, $\tilde{p} = 1$ and therefore $p = n$, meaning that

$$\boldsymbol{L} = \boldsymbol{T}^1_{2n}(a/2)\boldsymbol{C}_n = \boldsymbol{L}2n_n(a). \tag{4.11}$$

Their elements are ($t = 0, \pm 1, \dots,\ j = 0, 1, \dots, 2n-1,\ s = 0, 1, \dots, n-1$):

$$\ell(t, j) = \left(C^j_{2n} | \left(t + \left\{\frac{j}{2}\right\}\right) a\right),\ \tilde{\ell}(t, s) = \left(C^z_{2n} | z\frac{a}{2}\right) C^s_n, \tag{4.12}$$

with the transition rules between the generators

$$\tilde{\ell}(1,0) = \ell(0,1),\ \tilde{\ell}(0,1) = \tilde{\ell}(0,2),\ \ell(1,0) = \tilde{\ell}(2,-1). \tag{4.13}$$

The irreducible representations are easily seen from (4.3) and (4.6), with the transition rules:

$$(k,m) \rightarrow \left(\tilde{k} = k - \frac{m}{n}\frac{2\pi}{a} + 2\tilde{K}\frac{2\pi}{a}, \tilde{m} = m + \tilde{M}n\right), \tag{4.14a}$$

$$(\tilde{k},\tilde{m}) \rightarrow \left(k = \tilde{k} - \frac{\tilde{m}}{n}\frac{2\pi}{a} + K\frac{2\pi}{a}, m = \tilde{m} - nK + 2nM\right). \tag{4.14b}$$

Finally, the special points $(0,0)$, $(0,n/2)$, $(2\pi/a,0)$, and $(2\pi/a,n/2)$ of helical numbers correspond to the linear quantum numbers $(0,0)$, $(\pi/a,-n/2)$, $(0,n)$, and $(\pi/a,n/2)$, respectively, as follows from (4.10) for the combination 6.

4.2 Other Families

After we have found the representations of the first family groups, we use them to construct the irreducible representations of other groups. In fact, due to their structural properties explained in Sect. 2.3.2, we can apply the inductive procedure described in the Appendix D.1. The results are summarized in Tables 4.1–4.13.

Table 4.1 Irreducible representations of the line groups $\boldsymbol{T}_Q(f)\boldsymbol{C}_n$. For commensurate groups $\boldsymbol{L}q_p$ both linear and helical quantum numbers apply (with $a = f\tilde{q}$ and $Q = q/r$), while for the incommensurate $\boldsymbol{L}\infty_n$ only the latter ones. Reality: ${}_0A_0$, ${}_\pi A_0$, ${}_0A_{n/2}$, ${}_\pi A_{n/2}$, $({}_kA_m, {}_{-k}A_{-m})$

IR	(k,m)	$(C_Q\|f)$	C_n	SAB
${}_kA_m$	$k \in \left(-\frac{\pi}{a}, \frac{\pi}{a}\right]$ $m \in \left(-\frac{q}{2}, \frac{q}{2}\right]$	$e^{i\left(kf + m\frac{2\pi}{Q}\right)}$	$e^{im\frac{2\pi}{n}}$	$\|km\rangle$
${}_{\tilde{k}}A_{\tilde{m}}$	$\tilde{k} \in \left(-\frac{\pi}{f}, \frac{\pi}{f}\right]$ $\tilde{m} \in \left(-\frac{n}{2}, \frac{n}{2}\right]$	$e^{i\tilde{k}f}$	$e^{i\tilde{m}\frac{2\pi}{n}}$	$\|\tilde{k}\tilde{m}\rangle$

Table 4.2 Irreducible representations of the line groups $\boldsymbol{L}\overline{2n}$, $\boldsymbol{L}\bar{n} = \boldsymbol{T}(a)\boldsymbol{S}_{2n}$. Reality: ${}_0A_0^{\pm}$, ${}_\pi A_0^{\pm}$, ${}_kE_0$, ${}_k\bar{E}_{\frac{n}{2}}$, $({}_0A_m^{\pm}, {}_0A_{-m}^{\pm})$, $({}_0A_{n/2}^{+}, {}_0A_{n/2}^{-})$, $({}_kE_m, {}_kE_{-m})$, $({}_\pi A_m^{\pm}, {}_\pi A_{-m}^{\pm})$, $({}_\pi A_{n/2}^{+}, {}_\pi A_{n/2}^{-})$

IR	(k,m)	$(I\|a)$	$\sigma_h C_{2n}$	SAB
${}_kA_m^{\Pi_h}$	$k = 0, \frac{\pi}{a}$ $m \in \left(-\frac{n}{2}, \frac{n}{2}\right]$	e^{ika}	$\Pi_h e^{im\frac{\pi}{n}}$	$\|km\Pi_h\rangle$
${}_kE_m$	$k \in (0,\pi)$ $m \in \left(-\frac{n}{2}, \frac{n}{2}\right]$	$\begin{bmatrix} e^{ika} & 0 \\ 0 & e^{-ika} \end{bmatrix}$	$\begin{bmatrix} 0 & e^{im\frac{2\pi}{n}} \\ 1 & 0 \end{bmatrix}$	$\|km\rangle$ $\|-km\rangle$

Table 4.3 Irreducible representations of the line groups $\boldsymbol{L}n/m$, $\boldsymbol{L}\overline{2n} = \boldsymbol{T}(a)\boldsymbol{C}_{n\mathrm{h}}$. Reality: ${}_0A_0^{\pm}$, ${}_0A_{n/2}^{\pm}$, ${}_\pi A_0^{\pm}$, ${}_\pi A_{n/2}^{\pm}$, ${}_kE_0$, ${}_kE_{n/2}$, $({}_0A_m^{\pm}, {}_0A_{-m}^{\pm})$, $({}_\pi A_m^{\pm}, {}_\pi A_{-m}^{\pm})$, $({}_kE_m, {}_kE_{-m})$

IR	(k, m)	$(I\vert a)$	C_n	σ_h	SAB
${}_kA_m^{\Pi_\mathrm{h}}$	$k = 0, \frac{\pi}{a}$ $m \in (-\frac{n}{2}, \frac{n}{2}]$	$\mathrm{e}^{\mathrm{i}ka}$	$\mathrm{e}^{\mathrm{i}m\frac{2\pi}{n}}$	Π_h	$\vert km\Pi_\mathrm{h}\rangle$
${}_kE_m$	$k \in (0, \frac{\pi}{a})$ $m \in (-\frac{n}{2}, \frac{n}{2}]$	$\begin{bmatrix} \mathrm{e}^{\mathrm{i}ka} & 0 \\ 0 & \mathrm{e}^{-\mathrm{i}ka} \end{bmatrix}$	$\begin{bmatrix} \mathrm{e}^{\mathrm{i}m\frac{2\pi}{n}} & 0 \\ 0 & \mathrm{e}^{\mathrm{i}m\frac{2\pi}{n}} \end{bmatrix}$	$\begin{bmatrix} 0 & 1 \\ 1 & 0 \end{bmatrix}$	$\vert km\rangle$ $\vert -km\rangle$

Table 4.4 Irreducible representations of the line groups $\boldsymbol{L}2n_n/m = \boldsymbol{T}_{2n}^1(\frac{a}{2})\boldsymbol{C}_{n\mathrm{h}}$. Here $\tilde{k}_{\tilde{M}}(\tilde{m}) = \frac{2\pi\tilde{m}}{na} + \tilde{M}\frac{2\pi}{a}$, $\tilde{M} = 0, 1$. Reality: ${}_0A_0^{\pm}$, ${}_0A_n^{\pm}$, ${}_\pi E_0$, ${}_\pi E_{n/2}$, ${}_kE_0$, ${}_kE_n$, $({}_0A_m^{\pm}, {}_0A_{-m}^{\pm})$, $({}_kE_m, {}_kE_{-m})$, $({}_\pi E_m, {}_\pi E_{-m})$

IR	(k, m)	$(C_{2n}\vert\frac{1}{2})$	C_n	σ_h	SAB
${}_0A_m^{\Pi_\mathrm{h}}$	$k = 0$ $m \in (-n, n]$	$\mathrm{e}^{\mathrm{i}m\frac{\pi}{n}}$	$\mathrm{e}^{\mathrm{i}m\frac{2\pi}{n}}$	Π_h	$\vert km\Pi_\mathrm{h}\rangle$
${}_kE_m$	$k \in (0, \frac{\pi}{a}]$ $m \in (-n, n]$	$\begin{bmatrix} \mathrm{e}^{\mathrm{i}(m\frac{\pi}{n}+k\frac{a}{2})} & 0 \\ 0 & \mathrm{e}^{\mathrm{i}(m\frac{\pi}{n}-k\frac{a}{2})} \end{bmatrix}$	$\begin{bmatrix} \mathrm{e}^{\mathrm{i}m\frac{2\pi}{n}} & 0 \\ 0 & \mathrm{e}^{\mathrm{i}m\frac{2\pi}{n}} \end{bmatrix}$	$\begin{bmatrix} 0 & 1 \\ 1 & 0 \end{bmatrix}$	$\vert km\rangle$ $\vert -km\rangle$
${}_{\tilde{k}_{\tilde{M}}(\tilde{m})}A_{\tilde{m}}^{\Pi_\mathrm{h}}$	$\tilde{m} \in (-\frac{n}{2}, \frac{n}{2}]$	$\mathrm{e}^{\mathrm{i}\tilde{k}_{\tilde{M}}(\tilde{m})\frac{a}{2}}$	$\mathrm{e}^{\mathrm{i}\tilde{m}\frac{2\pi}{n}}$	Π_h	$\vert\tilde{k}_{\tilde{M}}(\tilde{m})m\Pi_\mathrm{h}\rangle$
${}_{\tilde{k}}E_{\tilde{m}}$	$\tilde{k} \in \left(\frac{2\pi\tilde{m}}{na}, \frac{2\pi}{a} + \frac{2\pi\tilde{m}}{na}\right)$ $\tilde{m} \in (-\frac{n}{2}, \frac{n}{2}]$	$\begin{bmatrix} \mathrm{e}^{\mathrm{i}\frac{ka}{2}} & 0 \\ 0 & \mathrm{e}^{\mathrm{i}(\tilde{m}\frac{2\pi}{n}-\tilde{k}\frac{a}{2})} \end{bmatrix}$	$\begin{bmatrix} \mathrm{e}^{\mathrm{i}\tilde{m}\frac{2\pi}{n}} & 0 \\ 0 & \mathrm{e}^{\mathrm{i}\tilde{m}\frac{2\pi}{n}} \end{bmatrix}$	$\begin{bmatrix} 0 & 1 \\ 1 & 0 \end{bmatrix}$	$\vert\tilde{k}\tilde{m}\rangle$ $\vert -\tilde{k} + \frac{4\tilde{m}\pi}{na}, \tilde{m}\rangle$

Table 4.5 Irreducible representations of line groups $\boldsymbol{T}_Q(f)\boldsymbol{D}_n$. For the commensurate groups $\boldsymbol{L}q_p22$, $\boldsymbol{L}q_p2$, both linear and helical quantum numbers apply (with $a = f\tilde{q}$ and $Q = q/r$), while for the incommensurate groups $\boldsymbol{L}\infty_n2$ there are only helical quantum numbers. Reality: all the representations of the first kind

IR	(k, m)	$(C_q^r\vert f)$	C_n	U	SAB
${}_kA_m^{\Pi_U}$	$k = 0,\ m = 0, \frac{q}{2}$ $k = \frac{\pi}{a},\ m = -\frac{p}{2}, \frac{q-p}{2}$	$\mathrm{e}^{\mathrm{i}\left(kf+m\frac{2\pi}{Q}\right)}$	$\mathrm{e}^{\mathrm{i}m\frac{2\pi}{n}}$	Π_U	$\vert km\Pi_U\rangle$
${}_kE_m$	([a])	$\begin{bmatrix} \mathrm{e}^{\mathrm{i}\left(kf+m\frac{2\pi}{Q}\right)} & 0 \\ 0 & \mathrm{e}^{-\mathrm{i}\left(kf+m\frac{2\pi}{Q}\right)} \end{bmatrix}$	$\begin{bmatrix} \mathrm{e}^{\mathrm{i}m\frac{2\pi}{n}} & 0 \\ 0 & \mathrm{e}^{-\mathrm{i}m\frac{2\pi}{n}} \end{bmatrix}$	$\begin{bmatrix} 0 & 1 \\ 1 & 0 \end{bmatrix}$	$\vert km\rangle$ $\vert -k, -m\rangle$
${}_{\tilde{k}}A_{\tilde{m}}^{\Pi_U}$	$\tilde{k} = 0, \frac{\pi}{f},\ m = 0, \frac{n}{2}$	$\mathrm{e}^{\mathrm{i}\tilde{k}f}$	$\mathrm{e}^{\mathrm{i}\tilde{m}\frac{2\pi}{n}}$	Π_U	$\vert\tilde{k}\tilde{m}\Pi_U\rangle$
${}_{\tilde{k}}E_{\tilde{m}}$	$k = 0, \frac{\pi}{f},\ m \in (0, \frac{n}{2})$ $k \in \left(0, \frac{\pi}{f}\right),\ m \in (-\frac{n}{2}, \frac{n}{2}]$	$\begin{bmatrix} \mathrm{e}^{\mathrm{i}\tilde{k}f} & 0 \\ 0 & \mathrm{e}^{-\mathrm{i}\tilde{k}f} \end{bmatrix}$	$\begin{bmatrix} \mathrm{e}^{\mathrm{i}\tilde{m}\frac{2\pi}{n}} & 0 \\ 0 & \mathrm{e}^{-\mathrm{i}\tilde{m}\frac{2\pi}{n}} \end{bmatrix}$	$\begin{bmatrix} 0 & 1 \\ 1 & 0 \end{bmatrix}$	$\vert\tilde{k}\tilde{m}\rangle$ $\vert -\tilde{k}, -\tilde{m}\rangle$

([a]) $k \in (0, \frac{\pi}{a})$ with $m \in (-\frac{q}{2}, \frac{q}{2}]$, $k = 0$ with $m \in (0, \frac{q}{2})$, and $k = \frac{\pi}{a}$ with $m \in (-\frac{p}{2}, \frac{q-p}{2})$.

Table 4.6 Irreducible representations of the line groups $\boldsymbol{L}nmm$, $\boldsymbol{L}nm = \boldsymbol{T}(a)\boldsymbol{C}_{n\mathrm{v}}$. Reality: ${}_0A_0$, ${}_0B_0$, ${}_0A_{n/2}$, ${}_0B_{n/2}$, ${}_0E_m$, ${}_\pi A_0$, ${}_\pi B_0$, ${}_\pi A_{n/2}$, ${}_\pi B_{n/2}$, ${}_\pi E_m$, $({}_kA_0, {}_{-k}A_0)$, $({}_kB_0, {}_{-k}B_0)$, $({}_kA_{n/2}, {}_{-k}A_{n/2})$, $({}_kB_{n/2}, {}_{-k}B_{n/2})$, $({}_kE_m, {}_{-k}E_m)$

IR	(k, m)	$(E\vert 1)$	C_n	σ_v	SAB
${}_kA/B_m$	$k \in (-\frac{\pi}{a}, \frac{\pi}{a}]$ $m = 0, \frac{n}{2}$	$\mathrm{e}^{\mathrm{i}ka}$	$\mathrm{e}^{\mathrm{i}m\frac{2\pi}{n}}$	Π_v	$\vert km\Pi_\mathrm{v}\rangle$
${}_kE_m$	$k \in (-\frac{\pi}{a}, \frac{\pi}{a}]$ $m \in (0, \frac{n}{2})$	$\begin{bmatrix} \mathrm{e}^{\mathrm{i}ka} & 0 \\ 0 & \mathrm{e}^{\mathrm{i}ka} \end{bmatrix}$	$\begin{bmatrix} \mathrm{e}^{\mathrm{i}m\frac{2\pi}{n}} & 0 \\ 0 & \mathrm{e}^{-\mathrm{i}m\frac{2\pi}{n}} \end{bmatrix}$	$\begin{bmatrix} 0 & 1 \\ 1 & 0 \end{bmatrix}$	$\vert km\rangle$ $\vert k, -m\rangle$

Table 4.7 Irreducible representations of the line groups ***Lncc***, ***Lnc*** $=$ $\boldsymbol{T'(\frac{a}{2})C_n}$. Reality: ${}_0A_0$, ${}_0B_0$, ${}_0A_{n/2}$, ${}_0B_{n/2}$, ${}_0E_m$, ${}_\pi\bar{E}_m$, $({}_\pi A_0, {}_\pi B_0)$, $({}_\pi A_{n/2}, {}_\pi B_{n/2})$, $({}_k A_0, {}_{-k}A_0)$, $({}_k B_0, {}_{-k}B_0)$, $({}_k A_{n/2}, {}_{-k}A_{n/2})$, $({}_k B_{n/2}, {}_{-k}B_{n/2})$, $({}_k E_m, {}_{-k}E_m)$

IR	(k, m)	$(\sigma_v \vert \frac{1}{2})$	C_n	SAB
${}_kA/B_m$	$k \in (-\frac{\pi}{a}, \frac{\pi}{a}]$ $m = 0, \frac{n}{2}$	$\Pi_v e^{i\frac{ka}{2}}$	$e^{im\frac{2\pi}{n}}$	$\vert km\Pi_v\rangle$
${}_kE_m$	$k \in (-\frac{\pi}{a}, \frac{\pi}{a}]$ $m \in (0, \frac{n}{2})$	$\begin{bmatrix} 0 & e^{i\frac{ka}{2}} \\ e^{i\frac{ka}{2}} & 0 \end{bmatrix}$	$\begin{bmatrix} e^{im\frac{2\pi}{n}} & 0 \\ 0 & e^{-im\frac{2\pi}{n}} \end{bmatrix}$	$\vert km\rangle$ $\vert k, -m\rangle$

The representations are given by the matrices representing generators of the group. This suffices to find the matrix of any other element of the group, since it is a monomial over the generators. For the irreducible representations denoted in the first column, allowed values of k and m (or $\tilde{k}$ and $\tilde{m}$) are given in the second column; when an interval corresponds to m or $\tilde{m}$, only the integral values from it are assumed. Representations with $m = n/2$ appear for n even only, while the representation does not appear at all if for given n the allowed interval of m or $\tilde{m}$ does not contain integers (in particular, several representations occur only for $n > 1$). Then the matrices of the generators follow. In the last column we indicate the corresponding symmetry-adapted basis, in terms of quantum numbers: vectors are given in the form $|km\Pi\rangle$, where Π stands for all the parities of the group; note that either all of the parities Π_U, Π_v, Π_h (this order is assumed) are present or only one of them.

The four-dimensional matrices used in the tables are ($\alpha = \frac{2\pi}{n}$):

$$A = \begin{bmatrix} e^{i(kf+m\frac{\pi}{n})} & 0 & 0 & 0 \\ 0 & e^{i(kf-m\frac{\pi}{n})} & 0 & 0 \\ 0 & 0 & e^{i(-kf+m\frac{\pi}{n})} & 0 \\ 0 & 0 & 0 & e^{-i(kf+m\frac{\pi}{n})} \end{bmatrix}, \quad B = \begin{bmatrix} 0&0&0&1 \\ 0&0&1&0 \\ 0&1&0&0 \\ 1&0&0&0 \end{bmatrix},$$

$$J = \begin{bmatrix} 0 & e^{i\frac{ka}{2}} & 0 & 0 \\ e^{i\frac{ka}{2}} & 0 & 0 & 0 \\ 0 & 0 & 0 & e^{-i\frac{ka}{2}} \\ 0 & 0 & e^{-i\frac{ka}{2}} & 0 \end{bmatrix}, \quad K = \begin{bmatrix} e^{ika} & 0 & 0 & 0 \\ 0 & e^{ika} & 0 & 0 \\ 0 & 0 & e^{-ika} & 0 \\ 0 & 0 & 0 & e^{-ika} \end{bmatrix}, \quad M(m) = \begin{bmatrix} e^{im\alpha} & 0 & 0 & 0 \\ 0 & e^{-im\alpha} & 0 & 0 \\ 0 & 0 & e^{im\alpha} & 0 \\ 0 & 0 & 0 & e^{-im\alpha} \end{bmatrix},$$

$$N = \begin{bmatrix} e^{im\alpha} & 0 & 0 & 0 \\ 0 & e^{-im\alpha} & 0 & 0 \\ 0 & 0 & e^{-im\alpha} & 0 \\ 0 & 0 & 0 & e^{im\alpha} \end{bmatrix}, \quad P = \begin{bmatrix} 0 & 1 & 0 & 0 \\ 1 & 0 & 0 & 0 \\ 0 & 0 & 0 & e^{im\alpha} \\ 0 & 0 & e^{-im\alpha} & 0 \end{bmatrix}, \quad R = \begin{bmatrix} 0&1&0&0 \\ 1&0&0&0 \\ 0&0&0&1 \\ 0&0&1&0 \end{bmatrix}, \quad S = \begin{bmatrix} 0&0&1&0 \\ 0&0&0&1 \\ 1&0&0&0 \\ 0&1&0&0 \end{bmatrix}.$$

4.3 Properties of the Representations

We finish this chapter by considering some general properties of the irreducible representations of the line groups. They will be widely used in physical applications,

Table 4.8 Irreducible representations of the line groups $\boldsymbol{L}2n_nmc = \boldsymbol{T}^1_{2n}(\frac{a}{2})\boldsymbol{C}_{n\mathrm{v}}$. Reality: ${}_0A_0$, ${}_0B_0$, ${}_0A_{n/2}$, ${}_0B_{n/2}$, ${}_0E_m$, ${}_\pi E_{n/2}$, $({}_\pi A_0, {}_\pi A_n)$, $({}_\pi B_0, {}_\pi B_n)$, $({}_kA_0, {}_{-k}A_0)$, $({}_kB_0, {}_{-k}B_0)$, $({}_kA_n, {}_{-k}A_n)$, $({}_kB_n, {}_{-k}B_n)$, $({}_kE_m, {}_{-k}E_m)$, $({}_\pi E_m, {}_\pi E_{n-m})$

IR	(k,m)	$(C_{2n}\vert\frac{1}{2})$	C_n	σ_v	SAB
${}_kA/B_m$	$k\in(-\frac{\pi}{a},\frac{\pi}{a}]$ $m=0,n$	$\mathrm{e}^{\mathrm{i}(k\frac{a}{2}+m\frac{\pi}{n})}$	1	Π_v	$\vert km\Pi_\mathrm{v}\rangle$
${}_kE_m$	$k\in(-\frac{\pi}{a},\frac{\pi}{a}]$ $m\in(0,n)$	$\begin{bmatrix}\mathrm{e}^{\mathrm{i}(k\frac{a}{2}+m\frac{\pi}{n})} & 0\\ 0 & \mathrm{e}^{\mathrm{i}(k\frac{a}{2}-m\frac{\pi}{n})}\end{bmatrix}$	$\begin{bmatrix}\mathrm{e}^{\mathrm{i}m\frac{2\pi}{n}} & 0\\ 0 & \mathrm{e}^{-\mathrm{i}m\frac{2\pi}{n}}\end{bmatrix}$	$\begin{bmatrix}0&1\\1&0\end{bmatrix}$	$\vert km\rangle$ $\vert k,-m\rangle$
${}_{\tilde{k}}A/B_0$	$\tilde{k}\in\left(-\frac{2\pi}{a},\frac{2\pi}{a}\right]$ $\tilde{m}=0$	$\mathrm{e}^{\mathrm{i}\frac{\tilde{k}a}{2}}$	1	Π_v	$\vert\tilde{k}0\Pi_\mathrm{v}\rangle$
${}_{\tilde{k}}E_{\tilde{m}}$	$\tilde{k}\in\left(-\frac{2\pi}{a},\frac{2\pi}{a}\right],\tilde{m}\in(0,\frac{n}{2})$ $\tilde{k}\in\left(0,\frac{2\pi}{a}\right],\tilde{m}=\frac{n}{2}$	$\begin{bmatrix}\mathrm{e}^{\mathrm{i}\frac{\tilde{k}a}{2}} & 0\\ 0 & \mathrm{e}^{\mathrm{i}(\tilde{k}\frac{a}{2}-\tilde{m}\frac{2\pi}{n})}\end{bmatrix}$	$\begin{bmatrix}\mathrm{e}^{\mathrm{i}\tilde{m}\frac{2\pi}{n}} & 0\\ 0 & \mathrm{e}^{-\mathrm{i}\tilde{m}\frac{2\pi}{n}}\end{bmatrix}$	$\begin{bmatrix}0&1\\1&0\end{bmatrix}$	$\vert\tilde{k}\tilde{m}0\rangle$ $\vert\tilde{k}-\frac{4\tilde{m}\pi}{na},-\tilde{m}0\rangle$

Table 4.9 Irreducible representations of the line groups $\boldsymbol{L}\overline{2n}2m$, $\boldsymbol{L}\bar{n}m = \boldsymbol{T}(a)\boldsymbol{D}_{nd}$. Reality: all the representations of the first kind, except $({}_kE^A_{n/2}, {}_kE^B_{n/2})$

IR	(k, m)	$(E\|1)$	C_n	U_d	σ_v	SAB
${}_kA/B_0^{\Pi_U}$	$k = 0, \frac{\pi}{a}$ $m = 0$	$\mathrm{e}^{\mathrm{i}ka}$	1	Π_U	Π_v	$\|k0\Pi_U\Pi_\mathrm{v}\Pi_\mathrm{h}\rangle$
${}_kE_m^{\Pi_\mathrm{h}}$	$k = 0, \frac{\pi}{a}$ $m \in (0, \frac{n}{2})$	$\mathrm{e}^{\mathrm{i}ka}\begin{bmatrix}1&0\\0&1\end{bmatrix}$	$\begin{bmatrix}\mathrm{e}^{\mathrm{i}m\frac{2\pi}{n}}&0\\0&\mathrm{e}^{-\mathrm{i}m\frac{2\pi}{n}}\end{bmatrix}$	$\Pi_\mathrm{h}\begin{bmatrix}0&\mathrm{e}^{\mathrm{i}m\frac{\pi}{n}}\\\mathrm{e}^{-\mathrm{i}m\frac{\pi}{n}}&0\end{bmatrix}$	$\begin{bmatrix}0&1\\1&0\end{bmatrix}$	$\|km00\Pi_\mathrm{h}\rangle$ $\|-k, -m00\Pi_\mathrm{h}\rangle$
${}_kE_{\frac{n}{2}}$	$k = 0, \frac{\pi}{a}$ $m = \frac{n}{2}$	$\mathrm{e}^{\mathrm{i}ka}\begin{bmatrix}1&0\\0&1\end{bmatrix}$	$\begin{bmatrix}-1&0\\0&-1\end{bmatrix}$	$\begin{bmatrix}0&1\\1&0\end{bmatrix}$	$\begin{bmatrix}1&0\\0&-1\end{bmatrix}$	$\|km0A0\rangle$ $\|km0B0\rangle$
${}_kE_m^{\Pi_\mathrm{v}}$	$k \in (0, \frac{\pi}{a})$ $m = 0, \frac{n}{2}$	$\begin{bmatrix}\mathrm{e}^{\mathrm{i}ka}&0\\0&\mathrm{e}^{-\mathrm{i}ka}\end{bmatrix}$	$\mathrm{e}^{\mathrm{i}m\frac{2\pi}{n}}\begin{bmatrix}1&0\\0&1\end{bmatrix}$	$\begin{bmatrix}0&1\\1&0\end{bmatrix}$	$\Pi_\mathrm{v}\begin{bmatrix}1&0\\0&\mathrm{e}^{\mathrm{i}m\frac{2\pi}{n}}\end{bmatrix}$	$\|km0\Pi_\mathrm{v}0\rangle$ $\|-km0, \mathrm{e}^{\mathrm{i}m\frac{2\pi}{n}}\Pi_\mathrm{v}, 0\rangle$
${}_kG_m$	$k \in (0, \frac{\pi}{a})$ $m \in (0, \frac{n}{2})$	K	N	S	P	$\|km000\rangle$ $\|k, -m000\rangle$ $\|-k, -m000\rangle$ $\|-km000\rangle$

Table 4.10 Irreducible representations of the line groups $\boldsymbol{L\overline{2n}2c}$, $\boldsymbol{L\overline{n}c} = \boldsymbol{T}'(\frac{a}{2})\boldsymbol{S}_{2n}$. Reality: ${}_0A_0^{\pm}$, ${}_0B_0^{\pm}$, ${}_0E_m^{\pm}$, ${}_0E_{n/2}$, ${}_kE_0^{\Pi_{\mathrm{v}}}$, ${}_kG_m$, ${}_\pi E_0$, $({}_kE_{n/2}^A, {}_kE_{n/2}^B)$, $({}_\pi E_m^{\pm}, {}_\pi E_m^{\mp})$

IR	(k, m)	$(\sigma_{\mathrm{v}} \vert \frac{a}{2})$	$C_{2n}\sigma_{\mathrm{h}}$	SAB
${}_kA/B_m^{\Pi_U}$	$k = 0, m = 0$ $k = \frac{\pi}{a}, m = \frac{n}{2}$	$\Pi_{\mathrm{v}} e^{ik\frac{a}{2}}$	$\Pi_{\mathrm{v}}\Pi_U e^{im\frac{\pi}{n}}$	$\vert km\Pi_U\Pi_{\mathrm{v}}\Pi_{\mathrm{h}}\rangle$
${}_kE_m$	$k = 0, m = \frac{n}{2}$ $k = \frac{\pi}{a}, m = 0$	$\begin{bmatrix} e^{ik\frac{a}{2}} & 0 \\ 0 & -e^{ik\frac{a}{2}} \end{bmatrix}$	$\begin{bmatrix} 0 & -e^{ika} \\ 1 & 0 \end{bmatrix}$	$\vert km0A0\rangle$ $\vert km0B0\rangle$
${}_kE_m^{\Pi_{\mathrm{h}}}$	$k = 0, \frac{\pi}{a}$ $m \in (0, \frac{n}{2})$	$\begin{bmatrix} 0 & e^{ik\frac{a}{2}} \\ e^{ik\frac{a}{2}} & 0 \end{bmatrix}$	$\Pi_{\mathrm{h}}\begin{bmatrix} e^{im\frac{\pi}{n}} & 0 \\ 0 & e^{i(ka - m\frac{\pi}{n})} \end{bmatrix}$	$\vert km00\Pi_{\mathrm{h}}\rangle$ $\vert k, -m00, e^{ika}\Pi_{\mathrm{h}}\rangle$
${}_kE_m^{\Pi_{\mathrm{v}}}$	$k \in (0, \frac{\pi}{a})$ $m = 0, \frac{n}{2}$	$\Pi_{\mathrm{v}}\begin{bmatrix} e^{ik\frac{a}{2}} & 0 \\ 0 & e^{i(m\frac{2\pi}{n} - k\frac{a}{2})} \end{bmatrix}$	$\begin{bmatrix} 0 & e^{im\frac{2\pi}{n}} \\ 1 & 0 \end{bmatrix}$	$\vert km0\Pi_{\mathrm{v}}0\rangle$ $\vert -km0, e^{i\frac{2\pi}{n}m}\Pi_{\mathrm{v}}0\rangle$
${}_kG_m$	$k \in (0, \frac{\pi}{a})$ $m \in (0, \frac{n}{2})$	$\begin{bmatrix} 0 & e^{ik\frac{a}{2}} & 0 & 0 \\ e^{ik\frac{a}{2}} & 0 & 0 & 0 \\ 0 & 0 & 0 & e^{-i(m\frac{2\pi}{n} + k\frac{a}{2})} \\ 0 & 0 & e^{i(m\frac{2\pi}{n} - k\frac{a}{2})} & 0 \end{bmatrix}$	$\begin{bmatrix} 0 & 0 & e^{im\frac{2\pi}{n}} & 0 \\ 0 & 0 & 0 & e^{-im\frac{2\pi}{n}} \\ 1 & 0 & 0 & 0 \\ 0 & 1 & 0 & 0 \end{bmatrix}$	$\vert km000\rangle$ $\vert k, -m000\rangle$ $\vert -km000\rangle$ $\vert -k, -m000\rangle$

Table 4.11 Irreducible representations of the line groups $\boldsymbol{TD}_{n\mathrm{h}} = \boldsymbol{Ln/mmm}, \boldsymbol{L\overline{2n}2m}$. Reality: all the representations of the first kind

IR	(k, m)	$(E\|1)$	C_n	σ_v	σ_h	SAB
${}_kA/B_m^{\Pi_\mathrm{h}}$	$k=0,\pi$; $m=0,\frac{n}{2}$	$\mathrm{e}^{\mathrm{i}ka}$	$\mathrm{e}^{\mathrm{i}m\frac{2\pi}{n}}$	Π_v	Π_h	$\|km\Pi_U\Pi_\mathrm{v}\Pi_\mathrm{h}\rangle$
${}_kE_m^{\Pi_\mathrm{h}}$	$k=0,\pi$; $m\in(0,\frac{n}{2})$	$\mathrm{e}^{\mathrm{i}ka}\begin{bmatrix}1&0\\0&1\end{bmatrix}$	$\begin{bmatrix}\mathrm{e}^{\mathrm{i}m\frac{2\pi}{n}}&0\\0&\mathrm{e}^{-\mathrm{i}m\frac{2\pi}{n}}\end{bmatrix}$	$\begin{bmatrix}0&1\\1&0\end{bmatrix}$	$\Pi_\mathrm{h}\begin{bmatrix}1&0\\0&1\end{bmatrix}$	$\|km00\Pi_\mathrm{h}\rangle$ $\|k,-m00\Pi_\mathrm{h}\rangle$
${}_kE_m^{\Pi_\mathrm{v}}$	$k\in(0,\pi)$; $m=0,\frac{n}{2}$	$\begin{bmatrix}\mathrm{e}^{\mathrm{i}ka}&0\\0&\mathrm{e}^{-\mathrm{i}ka}\end{bmatrix}$	$\mathrm{e}^{\mathrm{i}m\frac{2\pi}{n}}\begin{bmatrix}1&0\\0&1\end{bmatrix}$	$\Pi_\mathrm{v}\begin{bmatrix}1&0\\0&1\end{bmatrix}$	$\begin{bmatrix}0&1\\1&0\end{bmatrix}$	$\|km0\Pi_\mathrm{v}0\rangle$ $\|-km0\Pi_\mathrm{v}0\rangle$
${}_kG_m$	$k\in(0,\pi)$; $m\in(0,\frac{n}{2})$	K	$M(m)$	R	S	$\|km000\rangle$ $\|k,-m000\rangle$ $\|-km000\rangle$ $\|-k,-m000\rangle$

Table 4.12 Irreducible representations of the line groups $\boldsymbol{Ln/mcc}, \boldsymbol{L\overline{2n}2c} = \boldsymbol{T'}(\frac{a}{2})\boldsymbol{C}_{n\mathrm{h}}$. Reality: all the representations of the first kind except $({}_\pi E_m^{\pm}, {}_\pi E_n^{\mp})$

IR	(k, m)	$(\sigma_\mathrm{v}\|\frac{1}{2})$	C_n	σ_h	SAB
${}_0A/B_m^{\Pi_\mathrm{h}}$	$k=0$; $m=0,\frac{n}{2}$	Π_v	$\mathrm{e}^{\mathrm{i}m\frac{2\pi}{n}}$	Π_h	$\|km\Pi_U\Pi_\mathrm{v}\Pi_\mathrm{h}\rangle$
${}_kE_m^{\Pi_\mathrm{h}}$	$k=0,\pi$; $m\in(0,\frac{n}{2})$	$\mathrm{e}^{\mathrm{i}\frac{ka}{2}}\begin{bmatrix}0&1\\1&0\end{bmatrix}$	$\begin{bmatrix}\mathrm{e}^{\mathrm{i}m\frac{2\pi}{n}}&0\\0&\mathrm{e}^{-\mathrm{i}m\frac{2\pi}{n}}\end{bmatrix}$	$\Pi_\mathrm{h}\begin{bmatrix}1&0\\0&1\end{bmatrix}$	$\|km00\Pi_\mathrm{h}\rangle$ $\|k,-m00\Pi_\mathrm{h}\rangle$
${}_kE_m^{\Pi_\mathrm{v}}$	$k\in(0,\pi)$; $m=0,\frac{n}{2}$	$\Pi_\mathrm{v}\begin{bmatrix}\mathrm{e}^{\mathrm{i}\frac{ka}{2}}&0\\0&\mathrm{e}^{-\mathrm{i}\frac{ka}{2}}\end{bmatrix}$	$\mathrm{e}^{\mathrm{i}m\frac{2\pi}{n}}\begin{bmatrix}1&0\\0&1\end{bmatrix}$	$\begin{bmatrix}0&1\\1&0\end{bmatrix}$	$\|km0\Pi_\mathrm{v}0\rangle$ $\|-km0\Pi_\mathrm{v}0\rangle$
${}_\pi E_m$	$k=\pi$; $m=0,\frac{n}{2}$	$\begin{bmatrix}\mathrm{i}&0\\0&-\mathrm{i}\end{bmatrix}$	$\mathrm{e}^{\mathrm{i}m\frac{2\pi}{n}}\begin{bmatrix}1&0\\0&1\end{bmatrix}$	$\begin{bmatrix}0&1\\1&0\end{bmatrix}$	$\|\pi m0A0\rangle$ $\|\pi m0B0\rangle$
${}_kG_m$	$k\in(0,\pi)$; $m\in(0,\frac{n}{2})$	J	$M(m)$	S	$\|km000\rangle$ $\|k,-m000\rangle$ $\|-km000\rangle$ $\|-k,-m000\rangle$

because they manifest the structure of the line groups and corresponding physical systems.

4.3.1 Reduced Brillouin Zones and Bands

As discussed in Sect. 4.1.4, for fixed m irreducible representations ${}_kA_m$ of the first family line groups form continual series or bands with k running over the Brillouin zone $(-\pi/a, \pi/a]$. Analogously, in helical quantum numbers, for fixed $\tilde{m}$, there is a series ${}_{\tilde{k}}A_{\tilde{m}}$ with $\tilde{k}$ taking values the helical Brillouin zone $(-\pi/f, \pi/f]$.

Negative line groups $\boldsymbol{L}^-$ have a z-reversing coset representative. Given procedure of the construction of representations shows that in these cases the positive and negative halves of the Brillouin zone label the same representations. In fact, the parities (U or σ_h) in these groups join vectors with opposite linear momenta (from interior of the Brillouin zone) into the same irreducible representation. Therefore,

Table 4.13 Irreducible representations of $\boldsymbol{L}2n_n/mmm = \boldsymbol{T}^1_{2n}(\frac{a}{2})\boldsymbol{D}_{n\text{h}}$. Reality: all the representations of the first kind. Here, $\tilde{K}(\tilde{k},\tilde{m}) = \text{diag}[\text{e}^{\text{i}\frac{\tilde{k}a}{2}}, \text{e}^{\text{i}\frac{\tilde{k}a-2\tilde{m}\alpha}{2}}, \text{e}^{-\text{i}\frac{\tilde{k}a-2\tilde{m}\alpha}{2}}, \text{e}^{-\text{i}\frac{\tilde{k}a}{2}}]$, $\tilde{k}_{\tilde{M}}(\tilde{m}) = \frac{\alpha\tilde{m}+2\pi\tilde{M}}{a}$ $(\tilde{M}=0,1)$, $\alpha = \frac{2\pi}{n}$

IR	(k,m)	$(C^1_{2n}\vert\frac{a}{2})$	C_n	U	σ_v	SAB
${}_0A/B_m^{\Pi_\text{h}}$	$k=0$ $m=0,n$	$\text{e}^{\text{i}\frac{m\pi}{n}}$	1	Π_U	Π_v	$\vert 0m\Pi_U\Pi_\text{h}\Pi_\text{v}\rangle$
${}_0E_m^{\Pi_\text{h}}$	$k=0$ $m\in(0,n)$	$\begin{bmatrix}\text{e}^{\text{i}\frac{m\pi}{n}} & 0\\ 0 & \text{e}^{-\text{i}\frac{m\pi}{n}}\end{bmatrix}$	$\begin{bmatrix}\text{e}^{\text{i}\frac{2m\pi}{n}} & 0\\ 0 & \text{e}^{-\text{i}\frac{2m\pi}{n}}\end{bmatrix}$	$\Pi_\text{h}\begin{bmatrix}0&1\\1&0\end{bmatrix}$	$\begin{bmatrix}0&1\\1&0\end{bmatrix}$	$\vert 0m00\Pi_\text{h}\rangle$ $\vert 0,-m00\Pi_\text{h}\rangle$
${}_kE_m^{A/B}$	$k\in\left(0,\frac{\pi}{a}\right), m=0,n$ $k=\frac{\pi}{a}, m=0$	$\text{e}^{\text{i}\frac{m\pi}{n}}\begin{bmatrix}\text{e}^{\text{i}kf} & 0\\ 0 & \text{e}^{-\text{i}kf}\end{bmatrix}$	$\begin{bmatrix}0&1\\1&0\end{bmatrix}$	$\Pi_\text{v}\begin{bmatrix}0&1\\1&0\end{bmatrix}$	$\Pi_\text{v}\begin{bmatrix}1&0\\0&1\end{bmatrix}$	$\vert km0\Pi_\text{v}0\rangle$ $\vert -km0\Pi_\text{v}0\rangle$
${}_\pi E_{n/2}^{\Pi_U}$	$k=\frac{\pi}{a}$ $m=\frac{n}{2}$	$\begin{bmatrix}-1&0\\0&1\end{bmatrix}$	$-\begin{bmatrix}1&0\\0&1\end{bmatrix}$	$\Pi_U\begin{bmatrix}1&0\\0&1\end{bmatrix}$	$\begin{bmatrix}0&1\\1&0\end{bmatrix}$	$\vert\frac{\pi}{a},\frac{n}{2},\Pi_U00\rangle$ $\vert\frac{\pi}{a},-\frac{n}{2},\Pi_U00\rangle$
${}_kG_m$	$k\in\left(0,\frac{\pi}{a}\right), m\in(0,n)$ $k=\frac{\pi}{a}, m\in\left(0,\frac{n}{2}\right)$	A	$M(m)$	B	R	$\vert km000\rangle$ $\vert k,-m000\rangle$ $\vert -km000\rangle$ $\vert -k,-m000\rangle$
${}_{\tilde{k}}A/B_0^{\Pi_\text{h}}$	$\tilde{k}=0,\frac{2\pi}{a},\tilde{m}=0$	$\text{e}^{\text{i}\frac{\tilde{k}a}{2}}$	1	$\Pi_\text{h}\Pi_\text{v}$	Π_v	$\vert\tilde{k}0\Pi_\text{h}\Pi_\text{v}\rangle$
${}_{\tilde{k}_{\tilde{M}}(\tilde{m})}E_{\tilde{m}}^{\Pi_\text{h}}$	$\tilde{m}\in\left(0,\frac{n}{2}\right), \tilde{M}=0,1$ $\tilde{m}=\frac{n}{2}, \tilde{M}=0$	$(-)^{\tilde{M}}\begin{bmatrix}\text{e}^{\text{i}\frac{\tilde{m}\pi}{n}} & 0\\ 0 & \text{e}^{-\text{i}\frac{\tilde{m}\pi}{n}}\end{bmatrix}$	$\begin{bmatrix}\text{e}^{\text{i}\frac{2\tilde{m}\pi}{n}} & 0\\ 0 & \text{e}^{-\text{i}\frac{2\tilde{m}\pi}{n}}\end{bmatrix}$	$\Pi_\text{h}\begin{pmatrix}0&1\\1&0\end{pmatrix}$	$\begin{pmatrix}0&1\\1&0\end{pmatrix}$	$\vert\tilde{k}_{\tilde{M}}(\tilde{m})\tilde{m}\Pi_\text{h}\rangle$ $\vert\tilde{k}_{\tilde{M}}(\tilde{m}),-\tilde{m},\Pi_\text{h}\rangle$
${}_{\tilde{k}}E_0^{A/B}$	$\tilde{k}\in\left(0,\frac{2\pi}{a}\right),\tilde{m}=0$	$\begin{bmatrix}\text{e}^{\text{i}\tilde{k}f} & 0\\ 0 & \text{e}^{-\text{i}\tilde{k}f}\end{bmatrix}$	$\begin{pmatrix}1&0\\0&1\end{pmatrix}$	$\Pi_\text{v}\begin{pmatrix}0&1\\1&0\end{pmatrix}$	$\Pi_\text{v}\begin{pmatrix}1&0\\0&1\end{pmatrix}$	$\vert\tilde{k}0\Pi_\text{v}\rangle$ $\vert -\tilde{k},0,\Pi_\text{v}\rangle$
${}_{\frac{2\pi}{a}}E_{n/2}^{\Pi_U}$	$\tilde{k}=\frac{2\pi}{a},\tilde{m}=\frac{n}{2}$	$\begin{bmatrix}-1&0\\0&1\end{bmatrix}$	$\Pi_U\begin{pmatrix}1&0\\0&1\end{pmatrix}$	$\begin{pmatrix}0&1\\1&0\end{pmatrix}$	$\vert\frac{2\pi}{a},\frac{n}{2},\Pi_U\rangle$ $\vert 0,\frac{n}{2},\Pi_U\rangle$	
${}_{\tilde{k}}G_{\tilde{m}}$	$\tilde{k}\in(\frac{\tilde{m}\alpha}{a},\frac{2\pi}{a}+\frac{\tilde{m}\alpha}{a}), m\in\left(0,\frac{n}{2}\right)$ $\tilde{k}\in\left(0,\frac{\pi}{a}\right),\tilde{m}=\frac{n}{2}$	$\tilde{K}(\tilde{k},\tilde{m})$	$M(\tilde{m})$	B	R	$\vert\tilde{k}\tilde{m}\rangle$ $\vert\tilde{k}-\frac{2\tilde{m}\alpha}{a},-\tilde{m}\rangle$ $\vert -\tilde{k}+\frac{2\tilde{m}\alpha}{a},\tilde{m}\rangle$ $\vert -\tilde{k},-\tilde{m}\rangle$

to count all nonequivalent irreducible representations, a half of the zone suffices, and it is called *reduced Brillouin zone* or *irreducible domain* [6]. Thus, the reduced zone is the whole zone for positive line groups, i.e., for the families 1, 6, 7, and 8, while for the other ones it is the positive part $[0, \pi/a]$. Using irreducible domain, we generalize the concept of representation bands: irreducible representations are grouped into *representation bands* over the interior of the reduced Brillouin zone; these band representations extend also to the edges of the zone for the positive line groups, while for the negative groups some isolated representations of lower dimension appear in the special points (Sect. 4.3.3).

For helical quantum numbers the irreducible domain is somewhat different. For all the positive families (1, 6, 7 and 8, with helical factors $\boldsymbol{T}_Q, \boldsymbol{T}, \boldsymbol{T}$ and $\boldsymbol{T}_{2n}$, respectively) irreducible domain coincides with helical zone $(-\pi/f, \pi/f]$. For the families 2, 3, 9, 10, 11 and 12, the helical factor is translational group, and helical and linear quantum numbers coincide. Only for the family 5, the irreducible domain is $[0, \pi/f]$, analogously to the linear quantum numbers. However, for the remaining zigzag negative families 4 and 13, the irreducible domain cannot be defined in the usual sense.[1] Precisely, the $\tilde{m}$-bands are defined over intervals $(2\pi\tilde{m}/na, 2\pi/a + 2\pi\tilde{m}/na)$. In addition, ${}_{\tilde{k}}G_{n/2}$ band of the family 13 spreads over the interval $(0, \pi/a)$, i.e., over only a quarter of the Brillouin zone.

4.3.2 Symmetry-Adapted Basis

The space of the irreducible representation λ will be denoted as $\mathcal{H}^{(\lambda)}$, and its dimension, equal to the dimension of the representation matrices, by $|\lambda|$. The *symmetry-adapted basis* (SAB) $|\lambda l\rangle$ $(l = 1, \ldots, |\lambda|)$ in this space is defined as the orthonormal basis, such that each element ℓ of the line group $\boldsymbol{L}$ transforms basis vectors in the following way:

$$D^{(\lambda)}(\ell)\,|\lambda l\rangle = \sum_{l'} D^{(\lambda)}_{l'l}(\ell)\,|\lambda l'\rangle, \tag{4.15}$$

where $D^{(\lambda)}_{l'l}(\ell)$ are the elements of the matrix representing ℓ. The basis vectors are unique up to the common phase.

For the first family line groups, the irreducible representations are one-dimensional. Thus, counter l is superfluous, while the pair (k, m) or $(\tilde{k}, \tilde{m})$ of the momenta quantum numbers take the role of the label λ which defines the representation. Therefore, for the helical (linear) quantum numbers, symmetry-adapted basis of the representation ${}_{\tilde{k}}A_{\tilde{m}}$ $({}_kA_m)$ is the single vector $|\tilde{k}\tilde{m}\rangle$ $(|km\rangle)$, satisfying (4.15):

---

[1] Basically, this is because $\boldsymbol{T}^1_{2n}(a/2)$ is not invariant subgroup of these groups, in contrast to $\boldsymbol{T}(a)$.

$$
{}_{\tilde{k}}A_{\tilde{m}}\left(\tilde{\ell}_{ts}\right)|\tilde{k}\tilde{m}\rangle = \mathrm{e}^{\mathrm{i}\left(t\tilde{k}f+s\tilde{m}\frac{2\pi}{n}\right)}|\tilde{k}\tilde{m}\rangle, \tag{4.16a}
$$

$$
{}_{k}A_{m}\left(\ell_{tj}\right)|km\rangle = \mathrm{e}^{\mathrm{i}\left(\left(t+\{\frac{jp}{q}\}\right)ka+sm\frac{2\pi}{q}\right)}|km\rangle. \tag{4.16b}
$$

Note that the last equation shows that SAB vectors are common eigenbasis for all the roto-helical transformations, which is a consequence of mutual commutativity of these transformations.

For the other families, besides the momenta invoked by the first family subgroup, the parity quantum numbers are introduced. However, the parities do not commute with the roto-helical transformations, which causes appearance of multidimensional representations. For example, horizontal U-axis reverse all the momenta, k, m, $\tilde{k}$, and $\tilde{m}$. Therefore, it maps $|\tilde{k}\tilde{m}\rangle$ into $|-\tilde{k},-\tilde{m}\rangle$, and these two vectors become SAB of the irreducible representation ${}_{\tilde{k}}E_{\tilde{m}}$ of the fifth family line groups. Only in the special points $\tilde{k}=0,\pi/f$ for the special values $\tilde{m}=0,n/2$ the two vectors have the same quantum numbers, i.e., $|\tilde{k}\tilde{m}\rangle$ is an eigenvector for U. As U^2 is identity, the eigenvalues can be only 1 (vector even with respect to U) and -1 (odd vector). Thus, for these quantum numbers we obtain eight one-dimensional irreducible representations, ${}_{\tilde{k}}A_{\tilde{m}}^{\Pi_U}$, with SAB $|\tilde{k}\tilde{m}\Pi_U\rangle$, where $\Pi_U=\pm 1$ is *U-parity*. Alternatively, with the linear quantum numbers, the vectors $|km\rangle$ and $|-k,-m\rangle$ are SAB of the two-dimensional representations ${}_kE_m$. Only for $k=0$ and $m=0,q/2$ or $k=\pi/a$, and $m=-p/2,(q-p)/2$ (these values by (4.10) correspond to $\tilde{k}=0,\tilde{q}\pi/a$, and $m=0,n/2$), one-dimensional representations ${}_kA_m^{\Pi_U}$ are obtained, with single vector $|km\Pi_U\rangle$ in the SAB.

Analogously, for horizontal and vertical mirror planes the corresponding parities are denoted by Π_h and Π_v. This way SABs of all the irreducible representations are uniquely denoted by the quantum numbers of momenta and parities. It is common to denote values of Π_U and Π_h simply by $\pm$, while A and B are used instead of $\Pi_\mathrm{v}=1$ and $\Pi_\mathrm{v}=-1$. Also, to get uniform notation, we will in general denote SAB vectors as $|km\Pi\rangle$, where Π stands for the set of all relevant parities, and allow $\Pi=0$ for the vectors being neither even nor odd with respect to the corresponding transformation: e.g., for the representations ${}_{\tilde{k}}E_{\tilde{m}}$, SAB vectors are $|\tilde{k}\tilde{m}0\rangle$ and $|-\tilde{k},-\tilde{m},0\rangle$.

4.3.3 Dimensions and Compatibility Relations

The first family groups are abelian and their representations are one-dimensional. As the induction method gives the representations with either the same or doubled dimension, for the families $2,\dots,8$ the representations may be maximally two-dimensional, while for the remaining families they are at most four-dimesional.

The dimensions are maximal in the interior of the reduced Brillouin zone. In fact, the interior of irreducible domain is covered by several series of the irreducible representations, which, for negative groups, at the boundaries usually become reducible

Table 4.14 Compatibility relations of the irreducible representations of the line groups. The limiting representation $k \to 0, \pi$ for each irreducible band representation ${}_kD_m^{\Pi}$ is found and, if reducible, its decomposition onto the irreducible components is given. Note that for the positive line groups (families 1,6,7, and 8) there are no special points in the Brillouin zone and hence there is no reduction at the Brillouin zone boundaries

F	m or $\tilde{m}$	Dim	$k=0$	$\leftarrow$ IR $\rightarrow$	$k=\pi$
2	$(-\frac{n}{2}, \frac{n}{2}]$	$2 \to 1$	${}_0A_m^+ + {}_0A_m^-$	${}_kE_m$	${}_\pi A_m^+ + {}_\pi A_m^-$
3	$(-\frac{n}{2}, \frac{n}{2}]$	$2 \to 1$	${}_0A_m^+ + {}_0A_m^-$	${}_kE_m$	${}_\pi A_m^+ + {}_\pi A_m^-$
4	$(-n, n]$	$2 \to 1$	${}_0A_m^+ + {}_0A_m^-$	${}_kE_m$	/
5	$0, \frac{q}{2}$	$2 \to 1$	${}_0A_m^+ + {}_0A_m^-$	${}_kE_m$	/
	$-\frac{p}{2}, \frac{q-p}{2}$		/	${}_kE_m$	${}_\pi A_m^+ + {}_\pi A_m^-$
	$0, \frac{n}{2}$	$2 \to 1$	${}_0A_{\tilde{m}}^+ + {}_0A_{\tilde{m}}^-$	${}_{\tilde{k}}E_{\tilde{m}}$	${}_\pi A_{\tilde{m}}^+ + {}_\pi A_{\tilde{m}}^-$
9	0	$2 \to 1$	${}_0A_m^+ + {}_0A_m^-$	${}_kE_m^A$	${}_\pi A_m^+ + {}_\pi A_m^-$
			${}_0B_m^+ + {}_0B_m^-$	${}_kE_m^B$	${}_\pi B_m^+ + {}_\pi B_m^-$
	$(0, \frac{n}{2})$	$4 \to 2$	${}_0E_m^+ + {}_0E_m^-$	${}_kG_m$	${}_\pi E_m^+ + {}_\pi E_m^-$
10	0	$2 \to 1$	${}_0A_m^+ + {}_0A_m^-$	${}_kE_m^A$	/
			${}_0B_m^+ + {}_0B_m^-$	${}_kE_m^B$	/
	$\frac{n}{2}$		/	${}_kE_m^A$	${}_\pi A_m^+ + {}_\pi A_m^-$
			/	${}_kE_m^B$	${}_\pi B_m^+ + {}_\pi B_m^-$
	$(0, \frac{n}{2})$	$4 \to 2$	${}_0E_m^+ + {}_0E_m^-$	${}_kG_m$	${}_\pi E_m^+ + {}_\pi E_m^-$
11	$0, \frac{n}{2}$	$2 \to 1$	${}_0A_m^+ + {}_0A_m^-$	${}_kE_m^A$	${}_\pi A_m^+ + {}_\pi A_m^-$
			${}_0B_m^+ + {}_0B_m^-$	${}_kE_m^B$	${}_\pi B_m^+ + {}_\pi B_m^-$
	$(0, \frac{n}{2})$	$4 \to 2$	${}_0E_m^+ + {}_0E_m^-$	${}_kG_m$	${}_\pi E_m^+ + {}_\pi E_m^-$
12	$0, \frac{n}{2}$	$2 \to 1$	${}_0A_m^+ + {}_0A_m^-$	${}_kE_m^A$	/
			${}_0B_m^+ + {}_0B_m^-$	${}_kE_m^B$	/
	$(0, \frac{n}{2})$	$4 \to 2$	${}_0E_m^+ + {}_0E_m^-$	${}_kG_m$	${}_\pi E_m^+ + {}_\pi E_m^-$
13	$0, n$	$2 \to 1$	${}_0A_m^+ + {}_0A_m^-$	${}_kE_m^A$	/
			${}_0B_m^+ + {}_0B_m^-$	${}_kE_m^B$	/
	$(0, n)$	$4 \to 2$	${}_0E_m^+ + {}_0E_m^-$	${}_kG_m$	/

to some lower dimensional irreducible components. In other words, the boundaries are the only special points in the one-dimensional reduced Brillouin zone, and therefore the corresponding little group is larger than in other points. This results in lower dimension of the corresponding representations. The irreducible components of the band representations at the edges of the irreducible domain give *compatibility relations* (Table 4.14), important in the analysis of topology of the energy bands of nanotubes and polymers 8.1.2.

4.3.4 Reality of Representations

Tables 4.1–4.13 give additional information about the reality of the representations, according to the *Wigner's classification* [1] (Sect. 6.2.1). In fact, the representation $D(\boldsymbol{L})$ is of the *first kind* if all its matrices are real.[2] If this is not the case, but $D(\boldsymbol{L})$ and the conjugated representation $D^*(\boldsymbol{L})$ are still equivalent (as well

[2] More precisely, if there is a nonsingular operator A such that all the matrices $AD(\ell)A^{-1}$ are real.

as for the representations of the first kind there is a nonsingular operator B such that $D(\ell) = BD^*(\ell)B^{-1}$, and the characters of $D(\boldsymbol{L})$ and $D^*(\boldsymbol{L})$ are the same), $D(\boldsymbol{L})$ is of the *second kind*. Otherwise, when $D(\boldsymbol{L})$ and $D^*(\boldsymbol{L})$ are not equivalent (i.e., trace of at least one matrix $D(\ell)$ is not real), the representation is of the *third kind*. In the captions of the tables we give the reality as follows: at first, we list the representations of the first and the second kind (the latter ones are overlined), each of them giving a single real representations. The remaining representations are of the third kind and they are grouped into conjugated pairs, giving a single real representation. This information is sufficient to find real (or physically) irreducible representations, by the method explained in Sect. 6.2.2.

References

1. E.P. Wigner, *Group Theory and its Applications to the Quantum Mechanics of Atomic Spectra* (Academic Press, New York, 1959)
2. L. Jansen, M. Boon, *Theory of Finite Groups. Applications in Physics* (North-Holland, Amsterdam, 1967)
3. I. Milošević, M. Damnjanović, Phys. Rev. B **47**, 7805 (1993) 3
4. I. Božović, M. Vujičić, F. Herbut, J. Phys. Rev. A **11**, 2133 (1978) 3
5. I. Božović, M. Vujičić, J. Phys. A **14**, 777 (1981) 3
6. S.L. Altman, *Band Theory of Solids. An introduction from the Point of View of Symmetry* (Clarendon Press, Oxford, 1991)

Chapter 5
Tensors

Abstract Each physical quantity is a particular type of a line group tensor, i.e., it is changed in a specific way when line group transformations are applied. Quite generally, the transformation is described by a representation of a line group, and its irreducible components single out irreducible tensors. One important class is various functions, in particular quantum mechanical wave functions, among which symmetry-adapted basis consists of harmonics and covariants. Also, irreducible components of polar and axial vectors and second-rank tensors are found.

5.1 Standard Components

Tensor A is a set of physical quantities A_i $(i = 1, \ldots, n)$ which under the action of a spatial transformation ℓ mutually combine according to the linear law

$$\ell A_i = \sum_{j=1}^{n} D_{ji}^{\mathsf{A}}(\ell) A_j. \tag{5.1}$$

The coefficients $D_{ji}^{\mathsf{A}}(\ell)$ form a matrix $D^{\mathsf{A}}(\ell)$. Collecting these matrices for all the transformations ℓ of a line group $\boldsymbol{L}$, we get *tensor representation* (shortly A-representation) D^{A} of the group.

In general, this representation is reducible with irreducible components given in its decomposition:

$$D^{\mathsf{A}} = \sum_{\mu} f^{\mu} D^{(\mu)}, \tag{5.2}$$

where the *frequency number* f^{μ} shows how many times the irreducible component $D^{(\mu)}$ appears in D^{A}. Hence, if in the (5.1) instead of A_i their properly chosen linear combinations are used, a block diagonal form of the matrices D^{A}, with irreducible representations (5.2) on the diagonal, can be obtained. These *standard* or *symmetry-adapted* components A^{μ}_{m,t_μ} $(t_\mu = 1, \ldots, f^{\mu}, m = 1, \ldots, |\mu|)$ transform according to the corresponding irreducible representations:

Damnjanović, M., Milošević, I.: *Tensors*. Lect. Notes Phys. **801**, 65–84 (2010)
DOI 10.1007/978-3-642-11172-3_5

$$\ell A^{\mu}_{m,t_\mu} = \sum_{m'=1}^{|\mu|} D^{(\mu)}_{m'm}(\ell) A^{\mu}_{m',t_\mu}, \tag{5.3}$$

i.e., each transformation ℓ mixes only the standard components associated to the same irreducible representation and the same index t_μ.

Note that since in (5.1) nature of the components A_i is not specified, action of the line group transformations cannot be given explicitly. In the next sections we shall explicitly consider few particularly important cases.

5.2 Functions: Invariants and Covariants

Action of a transformation ℓ on a function over Euclidean space is defined as

$$\ell f(\boldsymbol{r}) \stackrel{\text{def}}{=} f(\ell^{-1}\boldsymbol{r}). \tag{5.4}$$

It is convenient to use cylindrical or helical coordinate system since line group elements ℓ leave radial coordinate ρ invariant: $\ell f(\rho, \varphi, z) = f(\rho, \varphi', z')$.

This way, the line group transformations act nontrivially only on the functions over cylinder and we can temporarily exclude radial coordinate and only afterward give the general form of the invariants and covariants over the entire space.

5.2.1 Harmonics

The simplest are *scalar* or *invariant functions*, i.e., the functions invariant under the line group transformations:

$$f(\ell^{-1}\boldsymbol{r}) = f(\boldsymbol{r}), \quad \forall \ell \in \boldsymbol{L}. \tag{5.5}$$

According to (5.3), invariant functions are single-component tensors, associated to the *identical representation*. They form a subspace in the space of the all functions. Basis functions of this space are called *harmonics* of the line groups [1].

Important examples are fields produced by nanotubes or polymers, i.e., potentials felt by the probes being close to these objects. Indeed, for any pairwise interatomic interaction $v(\boldsymbol{r}, \boldsymbol{r}_i)$ of a probe positioned at $\boldsymbol{r}$ with an atom at $\boldsymbol{r}_i$, the total potential $V(\boldsymbol{r}) = \sum_i v(\boldsymbol{r}, \boldsymbol{r}_i)$ is invariant, since in the transformed potential only the terms are permuted:

$$\ell V(\boldsymbol{r}) = V(\ell^{-1}\boldsymbol{r}) = \sum_i v(\ell^{-1}\boldsymbol{r}, \boldsymbol{r}_i) = \sum_i v(\boldsymbol{r}, \ell\boldsymbol{r}_i). \tag{5.6}$$

In the space spanned by the functions over cylinder, the condition of invariance (5.5) becomes

$$\ell H(\varphi, z) = H(\varphi, z), \quad \ell \tilde{H}(\tilde{\varphi}, \tilde{z}) = \tilde{H}(\tilde{\varphi}, \tilde{z}), \tag{5.7}$$

where the action of any transformation ℓ may be easily derived from (2.23). Although each line group element imposes one condition (5.7), the independent set of the conditions is obtained by taking into account only the generators. Thus, we begin with harmonics of the first family subgroup and then analyze the action of other generators (given in Table 5.1), which will result in two, three, or four equations to be simultaneously solved.

5.2.1.1 Cylindrical Harmonics of the First Family

As the first family line groups are abelian, their generators have common eigenbasis:

$$F^m_\omega(\varphi, z) \stackrel{\text{def}}{=} \mathrm{e}^{im\varphi}\mathrm{e}^{\mathrm{i}2\pi\omega z}, \quad m = 0, \pm 1, \pm 2\ldots, \ \omega \in \mathbb{R}, \tag{5.8}$$

$$\int_0^{2\pi}\int_{-\infty}^{\infty} F^{m^*}_\omega(\varphi, z) F^{m'}_{\omega'}(\varphi, z)\,\mathrm{d}\varphi\,\mathrm{d}z = 2\pi\,\delta_{mm'}\delta(\omega-\omega'), \tag{5.9}$$

where $\delta_{mm'}$ and $\delta(\omega-\omega')$ are Kronecker delta and Dirac delta function, respectively. The eigenvalues of the generators C_n and $(C^r_q|f)$ corresponding to the eigenfunctions $\mathrm{e}^{-\mathrm{i}2\pi m/n}$ and $\mathrm{e}^{-\mathrm{i}2\pi(mr/q+\omega f)}$ define the irreducible subspaces.

The first two equations of (2.23) and the condition (5.5) in the case of the functions (5.8) select harmonics H^M_ω as the eigenfunctions for the eigenvalue one:

$$C_n H^M_\omega(\varphi, z) = H^M_\omega(\varphi, z), \qquad m = nM, \ M = 0, \pm 1, \ldots, \tag{5.10}$$

$$(C_Q|f) H^M_\omega(\varphi, z) = H^M_\omega(\varphi, z), \qquad Mn + \omega f Q = 0, \pm Q, \pm 2Q, \ldots \tag{5.11}$$

Therefore, the solutions are provided only by the special values $\omega = (-Mn/Q + K)/f$ with integer K, i.e., the first family line group harmonics are

$$C^M_K(\varphi, z) \stackrel{\text{def}}{=} F^{nM}_{(K-Mn/Q)/f}(\varphi, z), \quad M, K = 0, \pm 1, \ldots. \tag{5.12}$$

The periodicity of the considered functions allows to define their scalar product as the integral over (compact) domain $\varphi \in [0, 2\pi/n)$ and $z \in [0, f)$, giving the orthogonality relations

$$\int_0^{\frac{2\pi}{n}}\int_0^f C^{M^*}_K(\varphi, z) C^{M'}_{K'}(\varphi, z)\,\mathrm{d}\varphi\,\mathrm{d}z = \frac{2\pi f}{n}\delta_{MM'}\delta_{KK'}. \tag{5.13}$$

Repeating the same procedure for the basis $F^M_\omega(\tilde\varphi, \tilde z)$, or directly substituting cylindrical coordinates in (5.12), we get the invariants expressed in terms of the helical coordinates:

$$C^M_K(\tilde\varphi, \tilde z) = F^{nM}_{Kq/r\sqrt{4\pi^2\rho^2+h^2}}(\tilde\varphi, \tilde z) = \mathrm{e}^{\mathrm{i}nM\tilde\varphi}\mathrm{e}^{\mathrm{i}\frac{2\pi Kq}{r\sqrt{4\pi^2\rho^2+h^2}}\tilde z}, \quad M, K = 0, \pm 1, \ldots, \tag{5.14}$$

where M and K count rotational and helical harmonics over $\tilde\varphi$ and $\tilde z$, respectively.

For the commensurate groups, where q is an integer and $a = \tilde{q} f$, the harmonics become

$$C_K^M(\varphi, z) = F^{nM}_{(-Mr+K\tilde{q})/a}(\varphi, z), \quad M, K = 0, \pm 1, \ldots \tag{5.15a}$$

In the special cases, when screw-axis degenerates to pure translations $\boldsymbol{T}_1^0(a)$ (coordinate helixes $\tilde{z}$ are vertical lines and the condition (5.11) is M-independent) or to $\boldsymbol{T}_{2n}^1(a)$, the cylindrical harmonics are, respectively,

$$T_K^M(\varphi, z) = F^{nM}_{K/a}(\varphi, z), \quad M, K = 0, \pm 1, \ldots, \tag{5.15b}$$

$$Z_K^M(\varphi, z) = F^{nM}_{(-M+2K)/a}(\varphi, z), \quad M, K = 0, \pm 1, \ldots. \tag{5.15c}$$

5.2.1.2 Cylindrical Harmonics of the Other Families

The families 2–13 have elements which do not commute with roto-helical transformations and invariants of these groups form thus a nontrivial subspace of the first family harmonics space.

Moreover, the $\boldsymbol{L}^{(1)}$-invariants (5.12) cannot pertain to the harmonics of the families 2–13, but their suitable linear combinations. In order to get such a basis it is sufficient to find the action of the coset representatives (Table 5.1) onto the roto-helical harmonics. It turns out that in general this action has the form $gC_K^M = \alpha C_{K'}^{M'}$. In particular,

$$C_{2n}\sigma_{\mathrm{h}} C_K^M(\varphi, z) = \mathrm{e}^{-\mathrm{i}\pi M} C^M_{-K+2\frac{nMr}{q}}(\varphi, z), \tag{5.16a}$$

$$(\sigma_{\mathrm{v}}|a/2) C_K^M(\varphi, z) = \mathrm{e}^{\mathrm{i}2\pi\frac{nMr}{q}} C^{-M}_{K-2\frac{nMr}{q}}(\varphi, z), \tag{5.16b}$$

$$U C_K^M(\varphi, z) = C^{-M}_{-K}(\varphi, z), \tag{5.16c}$$

$$\sigma_{\mathrm{h}} C_K^M(\varphi, z) = C^M_{-K+2\frac{nMr}{q}}(\varphi, z), \tag{5.16d}$$

$$\sigma_{\mathrm{v}} C_K^M(\varphi, z) = C^{-M}_{K-2\frac{nMr}{q}}(\varphi, z). \tag{5.16e}$$

While the U-axis can be combined with any helical axis, the rest of the generators are compatible only with the achiral roto-helical subgroups (thus, only harmonics (5.15b) and (5.15c) are to be considered). Further, as the square of an additional generator g is always from the roto-helical subgroup $\boldsymbol{L}^{(1)}$, the harmonic invariant under g (and therefore under the group $\boldsymbol{L}^{(1)} + g\boldsymbol{L}^{(1)}$) is of the form $C_K^M + gC_K^M$. This way we directly find the harmonics of the families 2–8, while for the rest of the line group families the procedure is to be repeated, acting on these new harmonics by the remaining coset representatives. All the line group harmonics are listed in Table 5.1, together with factors making them orthonormal with respect to the scalar product

$$(A, B) = \int_0^{\frac{2\pi}{n}} \int_0^f A^*(\varphi, z) B(\varphi, z)\, \mathrm{d}\varphi\, \mathrm{d}z. \tag{5.17}$$

Table 5.1 Harmonics of the line groups. Below the family number F in the first column are the generators complementing $\boldsymbol{L}^{(1)}$ (Table 2.2). Harmonics H_K^M are given in terms of chiral, translational, or zigzag (C, T, Z) invariants, and explicitly below; the range of K and M are in the second column (for $F = 5$ instead of M, K may be taken non-negative, as well). Normalized harmonics are CH_K^M, with C in the last column: general value (the first line) above exceptions

F Generators	Range M Range K	H_K^M	C
1	$M = 0, \pm 1, \pm 2, \ldots$	$C_K^M = F^{nM}_{(K-\frac{Mn}{Q})/f}$	
	$K = 0, \pm 1, \pm 2, \ldots$	$e^{inM\varphi} e^{i2\pi(K-\frac{Mn}{Q})z/f}$	$\sqrt{\frac{n}{2\pi f}}$
2	$M = 0, \pm 1, \pm 2, \ldots$	$\frac{1}{2}T_K^M + \frac{1}{2}(-)^M T_{-K}^M$	$\sqrt{\frac{n}{\pi f}}$
$C_{2n}\sigma_h$	$K = 0, 1, 2, \ldots$	M even: $e^{inM\varphi}\cos 2\pi Kz/a$	
		M odd: $e^{inM\varphi}\sin 2\pi Kz/a$	$H_0^M : \sqrt{\frac{n}{2\pi f}}$
3	$M = 0, \pm 1, \pm 2, \ldots$	$\frac{1}{2}T_K^M + \frac{1}{2}T_{-K}^M$	$\sqrt{\frac{n}{\pi f}}$
σ_h	$K = 0, 1, 2, \ldots$	$e^{inM\varphi}\cos 2\pi Kz/a$	$H_0^M : \sqrt{\frac{n}{2\pi f}}$
4	$M = 0, \pm 1, \pm 2, \ldots$	$\frac{1}{2}Z_K^M + \frac{1}{2}Z_{M-K}^M$	$\sqrt{\frac{n}{\pi f}}$
σ_h	$K = 0, 1, 2, \ldots$	$e^{inM\varphi}\cos 2\pi(2K - M)z/a$	$H_{M/2}^M : \sqrt{\frac{n}{2\pi f}}$
5	$M = 0, 1, 2, \ldots$	$\frac{1}{2}C_K^M + \frac{1}{2}C_{-K}^{-M}$	$\sqrt{\frac{n}{\pi f}}$
U	$K = 0, \pm 1, \pm 2, \ldots$	$\cos(nM\varphi + 2\pi(K - \frac{Mn}{Q})z/f)$	$H_{Mn/Q}^M : \sqrt{\frac{n}{2\pi f}}$
6	$M = 0, 1, 2, \ldots$	$\frac{1}{2}T_K^M + \frac{1}{2}T_K^{-M}$	$\sqrt{\frac{n}{\pi f}}$
σ_v	$K = 0, \pm 1, \pm 2, \ldots$	$\cos nM\varphi e^{i2\pi Kz/a}$	$H_K^0 : \sqrt{\frac{n}{2\pi f}}$
7	$M = 0, 1, 2, \ldots$	$\frac{1}{2}T_K^M + \frac{1}{2}(-)^K T_K^{-M}$	$\sqrt{\frac{n}{\pi f}}$
$(\sigma_v\vert\frac{1}{2})$	$M = 0, \pm 1, \pm 2, \ldots$	K even: $\cos nM\varphi e^{i2\pi Kz/a}$	
		K odd: $\sin nM\varphi e^{i2\pi Kz/a}$	$H_K^0 : \sqrt{\frac{n}{2\pi f}}$
8	$M = 0, 1, 2, \ldots$	$\frac{1}{2}Z_K^M + \frac{1}{2}Z_{K-M}^{-M}$	$\sqrt{\frac{n}{\pi f}}$
σ_v	$K = 0, \pm 1, \pm 2, \ldots$	$\cos nM\varphi e^{i2\pi(2K-M)z/a}$	$H_K^0 : \sqrt{\frac{n}{2\pi f}}$
9	$M = 0, 1, 2, \ldots$	$\frac{1}{4}T_K^M + \frac{1}{4}(-)^M T_{-K}^M + \frac{1}{4}T_K^{-M} + \frac{1}{4}(-)^M T_{-K}^{-M}$	$\sqrt{\frac{2n}{\pi f}}$
U_d, σ_v	$K = 0, 1, 2, \ldots$	M even: $\cos nM\varphi \cos 2\pi Kz/a$	$H_0^M, H_K^0 : \sqrt{\frac{n}{\pi f}}$
		M odd: $\cos nM\varphi \sin 2\pi Kz/a$	$H_0^0 : \sqrt{\frac{n}{2\pi f}}$
10	$M = 0, 1, 2, \ldots$	$\frac{1}{4}T_K^M + \frac{1}{4}(-)^K T_K^{-M} + \frac{1}{4}(-)^M T_{-K}^{-M} + \frac{1}{4}(-)^{M+K} T_{-K}^M$	$\sqrt{\frac{2n}{\pi f}}$
$(\sigma_v\vert\frac{1}{2}), U_d$	$K = 0, 1, 2, \ldots$	K, M even: $\cos nM\varphi \cos 2\pi Kz/a$	$H_0^M, H_K^0 : \sqrt{\frac{n}{\pi f}}$
		K even, M odd: $\cos nM\varphi \sin 2\pi Kz/a$	$H_0^0 : \sqrt{\frac{n}{2\pi f}}$
		K odd, M even: $\sin nM\varphi \sin 2\pi Kz/a$	
		K, M odd: $\sin nM\varphi \cos 2\pi Kz/a$	
11	$M = 0, 1, 2, \ldots$	$\frac{1}{4}T_K^M + \frac{1}{4}T_{-K}^M + \frac{1}{4}T_K^{-M} + \frac{1}{4}T_{-K}^{-M}$	$\sqrt{\frac{2n}{\pi f}}$
U, σ_v	$K = 0, 1, 2, \ldots$	$\cos nM\varphi \cos 2\pi Kz/a$	$H_0^M, H_K^0 : \sqrt{\frac{n}{\pi f}}$
			$H_0^0 : \sqrt{\frac{n}{2\pi f}}$

Table 5.1 (continued)

12	$M=0,1,2,\ldots$	$\frac{1}{4}T_K^M+\frac{1}{4}(-)^K T_K^{-M}+\frac{1}{4}T_{-K}^{-M}+\frac{1}{4}(-)^K T_{-K}^M$	$\sqrt{\frac{2n}{\pi f}}$
$(\sigma_{\mathrm{v}}\vert\frac{1}{2}),\sigma_{\mathrm{h}}$	$K=0,1,2,\ldots$	K even: $\cos 2\pi Kz/a\cos nM\varphi$	$H_0^M, H_K^0:\sqrt{\frac{n}{\pi f}}$
		K odd: $\sin 2\pi Kz/a\sin nM\varphi$	$H_0^0:\sqrt{\frac{n}{2\pi f}}$
13	$M=0,1,2,\ldots$	$\frac{1}{4}Z_K^M+\frac{1}{4}Z_{K-M}^{-M}+\frac{1}{4}Z_{M-K}^M+\frac{1}{4}Z_{-K}^{-M}$	$\sqrt{\frac{2n}{\pi f}}$
U,σ_{v}	$K=0,1,2,\ldots$	$\cos nM\varphi\cos 2\pi(2K-M)z/a$	$H_{M/2}^M:\sqrt{\frac{n}{\pi f}}$ $H_0^0:\sqrt{\frac{n}{2\pi f}}$

The harmonics with $M=0$ are φ-independent (and also $\tilde{\varphi}$-independent) functions, i.e., they are constants on the circles around z-axis. However, harmonics with $K=0$ are constant along the coordinate helixes $\tilde{z}$, and only when $\boldsymbol{L}^{(1)}=\boldsymbol{T}(a)$ (families 2,3,6,7,9,10,11, and 12) they are constant along the z-axis. For $M\neq 0$, harmonic H_K^M is invariant under $\boldsymbol{C}_{nM}$, as its rotational period is $2\pi/n|M|$: $H_K^M(\varphi+2\pi/n|M|,z)=H_K^M(\varphi,z)$; only for $M=1$ rotational period $2\pi/n$ of the harmonics coincides with the rotational symmetry of the system. Thus, rotational symmetry of harmonic is larger than that of the system, except for $M=\pm 1$ when they coincide. The analogue is true for K and the periodicity in $\tilde{z}$ (or z). In other words, line group $\boldsymbol{L}$ is a subgroup of the symmetry group of its harmonics H_K^M, comprising the full symmetry only of the harmonics with $|M|=|K|=1$.

When an invariant is a real function, their expansion amplitudes satisfy $\alpha_K^M(\rho)=\alpha_{-K}^{-M}(\rho)$. Consequently, in this case for the families 5 and 9–13 the amplitudes are real, while $\alpha_K^M(\rho)=\alpha_K^{-M}(\rho)$ for the families 2, 3, and 4 and $\alpha_K^M(\rho)=\alpha_{-K}^M(\rho)$ for the families 6, 7, and 8.

5.2.1.3 Entire Space Invariants

Having at disposal harmonics in the space of the functions over cylinder, it is easy to write down the general form of the harmonics over the total space:

$$U_{KI}^M(\boldsymbol{r})=R_{IK}^M(\rho)H_K^M(\varphi,z). \tag{5.18}$$

Here, for any fixed M and K the functions $R_{IK}^M(\rho)$ form a basis in the space of functions over ρ. Singularity at $\rho=0$ of the cylindrical and helical coordinates implies that any function $F(\rho,\varphi,z)$ at $\rho=0$ must be φ-independent; therefore, for $M\neq 0$ the functions $R_{IK}^M(\rho)$ vanish at $\rho=0$.

For various applications it is advantageous to use the basis

$$U_{Kb}^M(\boldsymbol{r})=\sqrt{b}J_{|nM|}(b\rho)H_K^M(\varphi,z),\quad b\in\mathbb{R} \tag{5.19}$$

(Bessel functions J_n normalized by $\int_0^\infty J_n(b\rho)J_n(b'\rho)\rho d\rho=\frac{1}{b}\delta(b-b')$), as this is the eigenbasis of kinetic energy:

$$\Delta U^M_{Kb}(\boldsymbol{r}) = -\left(b^2 + \left(2\pi \frac{K\tilde{q} - Mr}{a}\right)^2\right) U^M_{Kb}(\boldsymbol{r}). \tag{5.20}$$

5.2.2 *Covariants*

After analysis of the invariant functions, we can easily find symmetry-adapted basis in the space of functions over $\boldsymbol{r}$, i.e., basis of functions $\Psi^{(\lambda)l}_{t_\lambda}(\boldsymbol{r})$ satisfying (5.3):

$$\ell \Psi^{(\lambda)l}_{t_\lambda}(\boldsymbol{r}) \stackrel{\text{def}}{=} \Psi^{(\lambda)l}_{t_\lambda}(\ell^{-1}\boldsymbol{r}) = \sum_{l'=1}^{|\lambda|} D^{(\lambda)}_{l'l}(\ell)\Psi^{(\lambda)l'}_{t_\lambda}(\boldsymbol{r}), \quad t_\lambda = 1, 2, \ldots, f^\lambda. \tag{5.21}$$

Notably, as the line groups are not abelian (except for the first family groups) the functions are grouped into the *multiplets* of covariants $\Psi^{(\lambda)l}(\boldsymbol{r})$ ($l = 1, \ldots, |\lambda|$) corresponding to the irreducible representations $D^{(\lambda)}$ of the dimension $|\lambda|$ ($|\lambda| = 1, 2, 4$). Finally, t_λ counts independent covariant multiplets transforming by the same rule; the number of which is the frequency number f^λ of the irreducible representation λ.

In order to describe such a basis, we again first consider symmetry-adapted basis $\Phi^{(\lambda)l}_{t_\lambda}(\varphi, z)$ in the space of the functions over cylinder, satisfying the same transformation rules (5.21). The symmetry-adapted basis of the total space is then easily obtained multiplying $\Phi^{(\lambda)l}_{t_\lambda}(\varphi, z)$ by an arbitrary basis of radial functions, just as in the case of invariants.

5.2.2.1 Cylindrical Covariants

Obviously, if the multiplet $\Phi^{(\lambda)l}_{t_\lambda}(\varphi, z)$ is multiplied simultaneously by the same invariant function $H(\varphi, z)$, we obtain functions $\Phi^{(\lambda)l}_{t_\lambda}(\varphi, z)H(\varphi, z)$ satisfying again the transformation rule (5.21). Therefore, fixing one representative multiple $\Phi^{(\lambda)l}_{00}$ for each irreducible representation, we generate symmetry-adapted basis as its product with the basis of harmonics:

$$\Phi^{(\lambda)l}_{KM}(\varphi, z) = \Phi^{(\lambda)l}_{00}(\varphi, z)H^M_K(\varphi, z). \tag{5.22}$$

Notice that t_λ becomes a pair of the indices K and M counting the harmonics. In particular, the invariant functions are covariants of the identical representation $D^{(\mathrm{id})}$; therefore, harmonics themselves are a part of the symmetry-adapted basis if we take $\Phi^{(\mathrm{id})1}_{00}(\varphi, z) = 1$.

Having harmonics tabulated (Table 5.1) it remains to find [2] the representative functions $\Phi^{(\lambda)l}_{00}(\varphi, z)$ for each irreducible representation λ of the line groups. For the first family, the irreducible representations (Table 4.1) are one-dimensional. This transforms the condition (5.21) to the eigensystem with eigenvalues (being matrix elements of the irreducible representations) a priory classified in terms of quasi-momenta:

$$(C_q^r|f)\Phi_{00}^{(\tilde{k}\tilde{m})}(\varphi,z)=\mathrm{e}^{\mathrm{i}\tilde{k}f}\Phi_{00}^{(\tilde{k}\tilde{m})}(\varphi,z),\quad C_n\Phi_{00}^{(\tilde{k}\tilde{m})}(\varphi,z)=\mathrm{e}^{\mathrm{i}2\pi\tilde{m}/n}\Phi_{00}^{(\tilde{k}\tilde{m})}(\varphi,z). \tag{5.23}$$

This way the subspace of the covariants with fixed quasi-momenta is completely determined, and representative functions may be taken in the form

$$\Phi_{KM}^{(km)}(\varphi,z)=\mathrm{e}^{-\mathrm{i}(m\varphi+kz)},\quad \Phi_{00}^{(\tilde{k}\tilde{m})}(\varphi,z)=\mathrm{e}^{-\mathrm{i}\tilde{m}\varphi+\mathrm{i}(\frac{2\pi\tilde{m}}{Qf}-\tilde{k})z}. \tag{5.24}$$

For the other line group families, the representative functions are found (Table 5.2) from the first family ones by the induction procedure. In fact, besides (5.23), the covariants satisfy additional condition (5.3) for each additional generator. These conditions involve the matrices of the irreducible representations, which are obtained by induction from the first family line groups (Sect. 4.2). Recall that when $\boldsymbol{L}'$ is a halving subgroup of $\boldsymbol{L}$, i.e., $\boldsymbol{L}=\boldsymbol{L}'+\ell\boldsymbol{L}'$, the irreducible representations of $\boldsymbol{L}$ are related to those of $\boldsymbol{L}'$ in one of two ways (2).

The halving subgroup in the families 2–8 is the first family group. The whole space $\mathcal{L}$ is decomposed as $\mathcal{L}=\oplus_{\tilde{k}\tilde{m}}\mathcal{L}^{(\tilde{k}\tilde{m})}$ onto covariant subspaces of $\boldsymbol{L}'=\boldsymbol{L}^{(1)}$.

When $D^{(\tilde{k}\tilde{m})}(\boldsymbol{L}')$ is of the first type (i.e., corresponds to λ in (D.2a)), then $\mathcal{L}^{(\tilde{k}\tilde{m})}$ is invariant under the additional generator ℓ, meaning that $\mathcal{L}^{(\tilde{k}\tilde{m})}=\mathcal{L}^{(\tilde{k}\tilde{m}+)}\oplus\mathcal{L}^{(\tilde{k}\tilde{m}-)}$. Therefore, the additional generator decomposes $\mathcal{L}^{(\tilde{k}\tilde{m})}$ onto subspaces with defined parity $\Pi_\ell=\pm1$. This is performed by the group projector (F.2): in the space $\mathcal{L}^{(\tilde{k}\tilde{m})}$ it is reduced to $\boldsymbol{L}^{\tilde{k}\tilde{m}\pm}=\frac{1}{2}(1\pm A)$ (here $A=\pm D^{(\tilde{k}\tilde{m}\pm)}(\ell)$), according to (D.2a), giving (normalization factors are omitted)

$$\Phi^{(\tilde{k}\tilde{m}\pm)}\sim\Phi_t^{(\tilde{k}\tilde{m})}\pm A\Phi_t^{(\tilde{k}\tilde{m})}. \tag{5.25}$$

As for $A=1$ the simplest choice $\Phi_t^{(\tilde{k}\tilde{m})}=\Phi_{00}^{(\tilde{k}\tilde{m})}$ has vanishing projection by $\boldsymbol{L}^{\tilde{k}\tilde{m}\pm}$, for uniqueness of notation and compact presentation, the supergroup invariants are frequently built by $\Phi_{01}^{(\tilde{k}\tilde{m})}$ or $\Phi_{10}^{(\tilde{k}\tilde{m})}$ (and not $\Phi_{00}^{(\tilde{k}\tilde{m})}$).

When $D^{(\tilde{k}\tilde{m})}(\boldsymbol{L}')$ is of the second type, then $D(g)$ maps $\mathcal{L}^{(\tilde{k}\tilde{m})}$ into the space $\mathcal{L}^{(\tilde{k}_\ell\tilde{m}_\ell)}$, causing the doubling of the dimension of $D^{(\tilde{k}\tilde{m})}(\boldsymbol{L})$. When the projector (F.2) is calculated (taking into account the matrices of form (D.2b)), it immediately follows that the simplest representative doublet is $\Phi_{00}^{(\tilde{k}\tilde{m})}$ and $D(g)\Phi_{00}^{(\tilde{k}\tilde{m})}$.

Finally, for the families 9–13, we repeat the procedure taking as $\boldsymbol{L}'$ one of the groups from the families 2–8.

5.2.2.2 Entire Space Covariants

Combining (5.21) and (5.18), we get SAB in the whole space:

Table 5.2 Representative covariant functions of the line group. For each family $\boldsymbol{L}^{(F)}$, corresponding irreducible representations are listed in the column IR

F	IR	Representative function
1	${}_kA_m$	$e^{-i(m\varphi+kz)}$
$\boldsymbol{T}_Q(f)\boldsymbol{C}_n$	${}_{\tilde{k}}A_{\tilde{m}}$	$e^{-i\tilde{m}\varphi+i(\frac{2\pi\tilde{m}}{Qf}-\tilde{k})z}$
2	${}_kA_m^{\Pi_h}$	$e^{-im\varphi}(e^{i(\frac{2\pi}{a}-k)z}+\Pi_U e^{-i(\frac{2\pi}{a}-k)z})$
$\boldsymbol{T}(a)\boldsymbol{S}_{2n}$	${}_kE_m$	$e^{-im\varphi-ikz}$ $e^{i\frac{m\pi}{n}}e^{-im\varphi+ikz}$
3	${}_kA_m^{\Pi_h}$	$e^{-im\varphi}(e^{i(\frac{2\pi}{a}-k)z}+\Pi_h e^{-i(\frac{2\pi}{a}-k)z})$
$\boldsymbol{T}(a)\boldsymbol{C}_{nh}$	${}_kE_m$	$e^{-im\varphi-ikz}$ $e^{-im\varphi+ikz}$
4	${}_0A_m^{\Pi_h}$	$e^{-im\varphi}(e^{i(\frac{4\pi}{a}-k)z}+\Pi_h e^{-i(\frac{4\pi}{a}-k)z})$
$\boldsymbol{T}^1_{2n}(\frac{a}{2})\boldsymbol{C}_{nh}$	${}_kE_m$	$e^{-im\varphi-ikz}$ $e^{-im\varphi+ikz}$
5	${}_kA_m^{\Pi_U}$	$e^{i(n-m)\varphi-i(\frac{2\pi n}{Qf}+k)z}+\Pi_U e^{-i(n-m)\varphi+i(\frac{2\pi n}{Qf}+k)z}$
$\boldsymbol{T}_Q(f)\boldsymbol{D}_n$	${}_kE_m$	$e^{-im\varphi-ikz}$ $e^{im\varphi+ikz}$
	${}_{\tilde{k}}A_{\tilde{m}}^{\Pi_U}$	$e^{i(n-\tilde{m})\varphi-i(\frac{2\pi(n-\tilde{m})}{Qf}+\tilde{k})z}+\Pi_U e^{-i(n-\tilde{m})\varphi+i(\frac{2\pi(n-\tilde{m})}{Qf}+\tilde{k})z}$
	${}_{\tilde{k}}E_{\tilde{m}}$	$e^{-i\tilde{m}\varphi-i(-\frac{2\pi\tilde{m}}{Qf}+\tilde{k})z}$ $e^{i\tilde{m}\varphi+i(-\frac{2\pi\tilde{m}}{Qf}+\tilde{k})z}$
6	${}_kA/B_m$	$e^{-ikz}(e^{i(n-m)\varphi}+\Pi_v e^{-i(n-m)\varphi})$
$\boldsymbol{T}(a)\boldsymbol{C}_{nv}$	${}_kE_m$	$e^{-im\varphi-ikz}$ $e^{im\varphi-ikz}$
7	${}_kA/B_m$	$e^{-ikz}(e^{i(n-m)\varphi}+\Pi_v e^{-i(n-m)\varphi})$
$\boldsymbol{T}'(\frac{a}{2})\boldsymbol{C}_n$	${}_kE_m$	$e^{-im\varphi-ikz}$ $e^{im\varphi-ikz}$
8	${}_kA/B_m$	$e^{-ikz}(e^{i(2n-m)\varphi}+\Pi_v e^{-i(2n-m)\varphi})$
$\boldsymbol{T}^1_{2n}(\frac{a}{2})\boldsymbol{C}_{nv}$	${}_kE_m$	$e^{-im\varphi-ikz}$ $e^{im\varphi-ikz}$
	${}_{\tilde{k}}A/B_0$	$e^{i(\tilde{k}-\frac{2\pi}{a})z}(e^{in\varphi}+\Pi_v e^{-in\varphi})$
	${}_{\tilde{k}}E_{\tilde{m}}$	$e^{-i\tilde{m}\varphi+i(\frac{2\pi\tilde{m}}{na}-\tilde{k})z}$ $e^{i\tilde{m}\varphi-i(\frac{2\pi\tilde{m}}{na}-\tilde{k})z}$
9	${}_kA/B_0^{\Pi_U}$	$e^{in\varphi+i(\frac{2\pi}{a}-k)z}+\Pi_U e^{-in\varphi-i(\frac{2\pi}{a}-k)z}+$ $+\Pi_v e^{-in\varphi+i(\frac{2\pi}{a}-k)z}+\Pi_U\Pi_v e^{in\varphi-i(\frac{2\pi}{a}-k)z}$
$\boldsymbol{T}(a)\boldsymbol{D}_{nd}$	${}_kE_m^{\Pi_h}$	$e^{-im\varphi}(e^{i(\frac{2\pi}{a}-k)z}+\Pi_h e^{-i(\frac{2\pi}{a}-k)z})$ $e^{im\varphi}(e^{i(\frac{2\pi}{a}-k)z}+\Pi_h e^{-i(\frac{2\pi}{a}-k)z})$
	${}_kE_{\frac{n}{2}}$	$\cos m\varphi e^{-ikz}$ $\sin m\varphi e^{-ikz}$
	${}_kE_m^{\Pi_v}$	$e^{-ikz}(e^{i(n-m)\varphi}+(-1)^{\Pi_v}e^{-i(n-m)\varphi})$ $e^{ikz}(e^{i(n-m)\varphi}+\Pi_v e^{-i(n-m)\varphi})$

Table 5.2 (continued)

	${}_kG_m$	$e^{-im\varphi-ikz}$ $e^{im\varphi-ikz}$ $e^{im\varphi+ikz}$ $e^{-im\varphi+ikz}$
10 $\boldsymbol{T}'(\frac{a}{2})\boldsymbol{S}_{2n}$	${}_kA/B_m^{\Pi_U}$	$e^{i(n-m)\varphi+i(\frac{2\pi}{a}-k)z} - \Pi_U\Pi_v e^{i\frac{\pi m}{n}+i\frac{ka}{2}} e^{-i(n-m)\varphi+i(\frac{2\pi}{a}-k)z} +$ $+ \Pi_U e^{i\frac{\pi m}{n}} e^{i(n-m)\varphi-i(\frac{2\pi}{a}-k)z} - \Pi_v e^{i\frac{ka}{2}} e^{-i(n-m)\varphi-i(\frac{2\pi}{a}-k)z}$
	${}_0E_m$	$\cos m\varphi$ $\sin m\varphi$
	${}_\pi E_m$	$e^{-i\frac{\pi}{a}z}$ $e^{i\frac{\pi}{a}z}$
	${}_kE_m^{\Pi_h}$	$e^{-im\varphi}(e^{i(\frac{4\pi}{a}-k)z} + \Pi_h e^{i\frac{ka}{2}} e^{-i(\frac{4\pi}{a}-k)z})$ $e^{im\varphi}(e^{i(\frac{4\pi}{a}-k)z} + \Pi_h e^{i\frac{ka}{2}} e^{-i(\frac{4\pi}{a}-k)z})$
	${}_kE_m^{\Pi_v}$	$e^{-ikz}(e^{i(n-m)\varphi} + \Pi_v e^{i\frac{2\pi m}{n}} e^{-i(n-m)\varphi})$ $e^{ikz}(e^{i(n-m)\varphi} + \Pi_v e^{i\frac{2\pi m}{n}} e^{-i(n-m)\varphi})$
	${}_kG_m$	$e^{-im\varphi-ikz}$ $e^{im\varphi-ikz}$ $e^{-im\varphi+ikz}$ $e^{im\varphi+ikz}$
11 $\boldsymbol{T}(a)\boldsymbol{D}_{nh}$	${}_kA/B_m^{\Pi_h}$	$e^{i(n-m)\varphi+i(\frac{2\pi}{a}-k)z} + \Pi_v e^{-i(n-m)\varphi+i(\frac{2\pi}{a}-k)z} +$ $+ \Pi_h e^{i(n-m)\varphi-i(\frac{2\pi}{a}-k)z} + \Pi_h\Pi_v e^{-i(n-m)\varphi-i(\frac{2\pi}{a}-k)z}$
	${}_kE_m^{\Pi_h}$	$e^{-im\varphi}(e^{i(\frac{2\pi}{a}-k)z} + \Pi_h e^{-i(\frac{2\pi}{a}-k)z})$ $e^{im\varphi}(e^{i(\frac{2\pi}{a}-k)z} + \Pi_h e^{-i(\frac{2\pi}{a}-k)z})$
	${}_kE_m^{\Pi_v}$	$e^{-ikz}(e^{i(n-m)\varphi} + \Pi_v e^{-i(n-m)\varphi})$ $e^{ikz}(e^{i(n-m)\varphi} + \Pi_v e^{-i(n-m)\varphi})$
	${}_kG_m$	$e^{-im\varphi-ikz}$ $e^{im\varphi-ikz}$ $e^{-im\varphi+ikz}$ $e^{im\varphi+ikz}$
12 $\boldsymbol{T}'(\frac{a}{2})\boldsymbol{C}_{nh}$	${}_0A/B_m^{\Pi_h}$	$e^{i(n-m)\varphi+i\frac{4\pi}{a}z} + \Pi_v e^{-i(n-m)\varphi+i\frac{4\pi}{a}z} +$ $+ \Pi_h e^{i(n-m)\varphi-i\frac{4\pi}{a}z} + \Pi_h\Pi_v e^{-i(n-m)\varphi-i\frac{4\pi}{a}z}$
	${}_kE_m^{\Pi_h}$	$e^{-im\varphi}(e^{i(\frac{4\pi}{a}-k)z} + \Pi_h e^{-i(\frac{4\pi}{a}-k)z})$ $e^{im\varphi}(e^{i(\frac{4\pi}{a}-k)z} + \Pi_h e^{-i(\frac{4\pi}{a}-k)z})$
	${}_kE_m^{\Pi_v}$	$e^{-ikz}(e^{i(n-m)\varphi} + \Pi_v e^{-i(n-m)\varphi})$ $e^{ikz}(e^{i(n-m)\varphi} + \Pi_v e^{-i(n-m)\varphi})$
	${}_\pi E_m$	$e^{-ikz}\cos(n-m)\varphi$ $e^{ikz}\cos(n-m)\varphi$
	${}_kG_m$	$e^{-im\varphi-ikz}$ $e^{im\varphi-ikz}$ $e^{-im\varphi+ikz}$ $e^{im\varphi+ikz}$
13 $\boldsymbol{T}^1_{2n}(\frac{a}{2})\boldsymbol{D}_{nh}$	${}_0A/B_m^{\Pi_h}$	$e^{i(2n-m)\varphi+i\frac{4\pi}{a}z} + \Pi_U e^{-i(2n-m)\varphi-i\frac{4\pi}{a}z} +$ $+ \Pi_v e^{-i(2n-m)\varphi+i\frac{4\pi}{a}z} + \Pi_U\Pi_v e^{i(2n-m)\varphi-i\frac{4\pi}{a}z}$
	${}_0E_m^{\Pi_h}$	$e^{-im\varphi}(e^{i\frac{4\pi}{a}z} + \Pi_h e^{-i\frac{4\pi}{a}z})$ $e^{im\varphi}(e^{i\frac{4\pi}{a}z} + \Pi_h e^{-i\frac{4\pi}{a}z})$

Table 5.2 (continued)

${}_kE_m^{A/B}$	$e^{-ikz}(e^{i(2n-m)\varphi}+\Pi_v e^{-i(2n-m)\varphi})$ $e^{ikz}(e^{i(2n-m)\varphi}+\Pi_v e^{-i(2n-m)\varphi})$
${}_\pi E_{n/2}^{\Pi_U}$	$e^{im\varphi+ikz}+\Pi_U e^{-im\varphi-ikz}$ $e^{-im\varphi+ikz}+\Pi_U e^{im\varphi-ikz}$
${}_kG_m$	$e^{-im\varphi-ikz}$ $e^{im\varphi-ikz}$ $e^{-im\varphi+ikz}$ $e^{im\varphi+ikz}$

$$\Psi_{IKM}^{(\lambda)l}(\boldsymbol{r})=\Phi_{00}^{(\lambda)l}(\varphi,z)R_{IK}^{M}(\rho)H_K^M(\varphi,z). \tag{5.26}$$

The result obtained enables expansion of any function $\Psi^{(\lambda)l}(\boldsymbol{r})$ over SAB:

$$\Psi^{(\lambda)l}(\boldsymbol{r})=\sum_{IKM}\alpha_{IKM}\Phi_{00}^{(\lambda)l}(\varphi,z)R_{IK}^{M}(\rho)H_K^M(\varphi,z), \tag{5.27}$$

where the sum is over all allowed values of I, M, and K, while the amplitudes are scalar products: $\alpha_{IKM}=(\Psi_{IKM}^{(\lambda)l},\Psi^{(\lambda)l})=\int\Psi_{IKM}^{(\lambda)l*}(\rho,\varphi,z)\Psi^{(\lambda)l}(\rho,\varphi,z)\rho\,\mathrm{d}\rho\,\mathrm{d}\varphi\,\mathrm{d}z$. We also introduce expansions over harmonics with the radial functions being independent on the choice of $R_{IK}^M(\rho)$:

$$\Psi^{(\lambda)l}(\boldsymbol{r})=\sum_{KM}\alpha_K^M(\rho)H_K^M(\varphi,z), \tag{5.28}$$

$$\alpha_K^M(\rho)=\sum_{I}\alpha_{IKM}R_{IK}^M(\rho)=\rho\int_0^{2\pi}\int_{-\infty}^{\infty}\Phi_{KM}^{(\lambda)l*}(\varphi,z)\Psi^{(\lambda)l}(\rho,\varphi,z)\,\mathrm{d}\varphi\,\mathrm{d}z. \tag{5.29}$$

This generalization (to the line group symmetry) of the famous Bloch theorem (derived for the translational periodicity only) will be considered in Sect. 8.1. Applications to the tight-binding model (Sect. 8.5.1) and several examples of the expansions (5.28) are discussed in the context of carbon nanotubes (Sect. 9.2).

5.3 Vectors

For the sake of the numerous physical applications here we elaborate [3] symmetry-adapted basis of the three dimensional Euclidean space. Recall that a vector quantity $\boldsymbol{v}$, like momentum or electric and magnetic field, has three components, and the action of the group elements (5.1) is realized by three dimensional matrices. There are two types of such quantities, *polar* and *axial vectors*. Both are transformed in the same way under rotations, both are invariant under translations, but the *enantiomorphic transformations* distinguish between them. For instance, spatial inversion changes the sign of the polar vectors and leaves the axial vectors invariant.

Two corresponding types of *vector representations* are called *polar* D^{p} and *axial representation* D^{a}. Note that for the first and fifth family line groups polar and axial representations are identical, as these groups do not contain enantiomorphic elements.

Both vector representations of line groups are reducible, and when the isogonal group principle axis q is greater than two the standard components (making symmetry-adapted basis) are

$$v_0 = v_z, \quad v_\pm = v_x \mp \mathrm{i} v_y. \tag{5.30}$$

Actually, under a line group transformation z-component of a vector either remains invariant or changes the sign. Thus, it is itself a standard component associated to one-dimensional representation. The remaining x- and y-components are standard only in some of the groups with $q < 3$, and in these cases they also correspond to the one-dimensional representations. Otherwise, they are combined into the standard components $v_\pm$, which transform independently, according to the different one-dimensional representations in the families 1–4, while in the other families form a multiplet of a two-dimensional representation. The linear (for $q \geq 3$) and helical (for $n \geq 3$) momenta of the components (5.30) are

$$v_0: \quad k = m = 0; \quad \tilde{k} = \tilde{m} = 0, \tag{5.31a}$$

$$v_\pm: k = 0,\ m \mathrel{\mathring{=}} \pm 1;\ \tilde{k} \mathrel{\mathring{=}} \pm\kappa,\ \tilde{m} \mathrel{\mathring{=}} \pm 1 \quad (\kappa = \frac{2\pi}{fQ}). \tag{5.31b}$$

Here, $\mathring{=}$ denotes equality modulo range of the involved quantities. In particular, for commensurate groups, when $Q = q/r$ and $f = an/q$, this results in $\kappa = \frac{2r\pi}{na}$ if $r \leq q/2$, while otherwise $\kappa = \frac{2(r-q)\pi}{na}$. Also, when $q = 2$ $(n = 2)$ then $m = 1$ $(\tilde{m} = 1)$ for both $v_\pm$, and for $q = 1$ $(n = 1)$ for all the components $m = 0$ $(\tilde{m} = 0)$.

Equations (5.31) single out relevant representations of the first family groups, and for other families the parities should be additionally checked. As for the U-parity, v_z is odd, while $v_\pm$ are (for $q \geq 3$) neither even nor odd. For polar (axial) vector v_0 is even (odd) in vertical and odd (even) in horizontal mirror and glide planes, while $v_\pm$ is even (odd) in horizontal and for $q \geq 3$ neither even nor odd in vertical planes. All this is summarized in Table 5.3, where the standard components of vectors are corresponded to the linear irreducible sub-representations of D^{p} and D^{a}. As $k = 0$ for the vector quantities, the reduction of the vector representations of the commensurate groups is essentially related to the isogonal point groups.

There are many vectorial quantities describing various properties of physical systems. Among them, only those components corresponding to the identity representation may be nonzero in a structure having a line group symmetry. For example, nonzero electrical field, characterizing ferroelectric structures, may appear (spontaneously) only if some of its components are invariant under the symmetry transformations. In this context the invariant components of the vectors are important. The inspection of Table 5.3 shows that z-component of the polar vector is invariant for the families 1, 6, 7, and 8, while the same component of the axial vector is invariant

Table 5.3 Symmetry-adapted components of the polar and axial vectors. In the first column the family numbers F of the line groups $\boldsymbol{L}^{(F)}$ are above the corresponding isogonal group $\boldsymbol{P}_{\text{I}}$. Then the standard components of antisymmetric polar/axial vectors for each order q of the principle axis of the isogonal group are listed below the corresponding irreducible representations. While for the families 1 and 5 all the types of vectors have the same irreducible representations, for other families the irreducible representations of polar vectors differ by parities Π_{h} and Π_{v} from those of the axial ones; respective parities are then denoted by $\pm$ or $\mp$, as well as $X = A/B$ and $Y = B/A$. Linear quantum numbers are used; helical ones are obtained according to the equivalence (5.31) retaining all the parities. For $q = 1, 2$ the listed components correspond to the choice of xz-plane as σ_{v} in $\boldsymbol{C}_{n\text{v}}$, $\boldsymbol{D}_{n\text{d}}$, and $\boldsymbol{D}_{q\text{h}}$ and x-axis as U in $\boldsymbol{D}_q$

F / $\boldsymbol{P}_{\text{I}}$	$q = 1$			$q = 2$			$q = 3, 4, \ldots$		
1	3_0A_0			${}_0A_0$		2_0A_1	${}_0A_0$	${}_0A_1$	${}_0A_{-1}$
$\boldsymbol{C}_q$	v_x, v_y, v_z			v_z		v_x, v_y	v_z	v_+	v_-
2	$3_0A_0^{\mp}$			${}_0A_0^{\mp}$	${}_0A_1^{\pm}$	${}_0A_{-1}^{\pm}$	${}_0A_0^{\mp}$	${}_0A_1^{\pm}$	${}_0A_{-1}^{\pm}$
$\boldsymbol{S}_{2q}$	v_x, v_y, v_z			v_z	v_+	v_-	v_z	v_+	v_-
3, 4	$2_0A_0^{\pm}$		${}_0A_0^{\mp}$	${}_0A_0^{\mp}$		$2_0A_1^{\pm}$	${}_0A_0^{\mp}$	${}_0A_1^{\pm}$	${}_0A_{-1}^{\pm}$
$\boldsymbol{C}_{q\text{h}}$	v_x, v_y		v_z	v_z		v_x, v_y	v_z	v_+	v_-
5	$2_0A_0^-$		${}_0A_0^+$	${}_0A_0^-$	${}_0A_1^+$	${}_0A_1^-$	${}_0A_0^-$	${}_0E_1$	
$\boldsymbol{D}_q$	v_z, v_y		v_x	v_z	v_x	v_y	v_z	(v_+, v_-)	
6 – 8	2_0X_0		${}_0Y_0$	${}_0X_0$	${}_0X_1$	${}_0Y_1$	${}_0X_0$	${}_0E_1$	
$\boldsymbol{C}_{q\text{v}}$	v_z, v_x		v_y	v_z	v_x	v_y	v_z	(v_+, v_-)	
9, 10	$2_0X_0^-$		${}_0Y_0^+$	${}_0X_0^-$		${}_0E_1$	${}_0X_0^-$	${}_0E_1^+$	
$\boldsymbol{D}_{q\text{d}}$	v_z, v_x		v_y	v_z		(v_+, v_-)	v_z	(v_+, v_-)	
11 – 13	${}_0X_0^{\mp}$	${}_0X_0^{\pm}$	${}_0Y_0^{\pm}$	${}_0X_0^{\mp}$	${}_0X_1^{\pm}$	${}_0Y_1^{\pm}$	${}_0X_0^{\mp}$	${}_0E_1^{\pm}$	
$\boldsymbol{D}_{q\text{h}}$	v_z	v_x	v_y	v_z	v_x	v_y	v_z	(v_+, v_-)	

for the families 1, 2, 3, and 4. Except for few trivial cases, where $q = 1, 2$, neither of the other components can be invariant for any line group symmetry.

5.4 Second-Rank Tensors

The second-rank tensor $\mathsf{T} = \mathsf{T}(\boldsymbol{v}_1, \boldsymbol{v}_2)$ relates two vector quantities $\boldsymbol{v}_1$ and $\boldsymbol{v}_2$ in a linear way: $\boldsymbol{v}_1 = \mathsf{T}(\boldsymbol{v}_1, \boldsymbol{v}_2)\boldsymbol{v}_2$. In Cartesian components this reads $v_{1i} = \sum_j T_{ij}(\boldsymbol{v}_1, \boldsymbol{v}_2)v_{2j}$. Therefore, transformation of T under the action of the line group elements follows the transformation of the involved vectors. Denoting by $D^{\boldsymbol{v}_i}$ $(i = 1, 2)$ the corresponding (polar or axial) vector representation, i.e., $\ell \boldsymbol{v}_i = D^{\boldsymbol{v}_i}(\ell)\boldsymbol{v}_i$, it follows that the transformed tensor $\mathsf{T}' = \ell\mathsf{T}$ satisfies

$$\mathsf{T}' = D^{\boldsymbol{v}_1}(\ell^{-1})\mathsf{T}(\boldsymbol{v}_1, \boldsymbol{v}_2)D^{\boldsymbol{v}_2}(\ell), \text{ i.e., } T'_{ij}(\boldsymbol{v}_1, \boldsymbol{v}_2) = \sum_{i',j'=1}^{3} D^{\boldsymbol{v}_1}_{i'i}(\ell)D^{\boldsymbol{v}_2}_{j'j}(\ell)T_{i'j'}(\boldsymbol{v}_1, \boldsymbol{v}_2), \tag{5.32}$$

since the matrices of the vector representations are orthogonal, i.e., $D^{\boldsymbol{v}T} = D^{\boldsymbol{v}-1}$. Therefore, one concludes that the second-rank tensor is transformed according to the direct product representation:

$$D^{\mathsf{T}}(\ell) = D^{v_1}(\ell) \otimes D^{v_2}(\ell). \tag{5.33}$$

The obtained representation D^{T} is reducible, and the standard components are called *irreducible tensor components*.

Substituting axial $\boldsymbol{a}$ and polar $\boldsymbol{p}$ vectors for $\boldsymbol{v}_1$ and $\boldsymbol{v}_2$, we find four types of the second-rank tensors: $\mathsf{T}(\boldsymbol{p}, \boldsymbol{p})$, $\mathsf{T}(\boldsymbol{a}, \boldsymbol{a})$, $\mathsf{T}(\boldsymbol{p}, \boldsymbol{a})$, and $\mathsf{T}(\boldsymbol{a}, \boldsymbol{p})$. Still, the representations $D^{\mathrm{p}} \otimes D^{\mathrm{p}}$ and $D^{\mathrm{a}} \otimes D^{\mathrm{a}}$ are equivalent, as well as $D^{\mathrm{p}} \otimes D^{\mathrm{a}}$ and $D^{\mathrm{a}} \otimes D^{\mathrm{p}}$, i.e., their standard components are the same combinations of the Cartesian ones A_{ij}. As the first pair is invariant under the spatial inversion and the second one changes the sign, such tensors are called *axial* and *polar* tensors, denoted as A and P, respectively. Still, for the groups of the first and fifth families there is no such distinction, as there is no difference between axial and polar vectors.

For each tensor T we define its *symmetric* and *antisymmetric* part $\mathsf{T}^{\pm} = \frac{1}{2}(\mathsf{T} + \mathsf{T}^T)$, which are also second-rank tensors. Their components $T^{\pm}_{ij} = \frac{1}{2}(T_{ji} \pm T A_{ji})$ transform independently (without mixing symmetrical and antisymmetrical ones) under the line group transformations. This enables us to list separately the standard components of the symmetric and antisymmetric parts [4], which is particularly convenient for tensors being themselves symmetric or antisymmetric. In particular, the representation $D^{\mathsf{A}} = D^{a} \otimes D^{a}$ of the axial tensors automatically reduces to the symmetrized and antisymmetrized squares $D^{\mathsf{A}+} = [D^{a}]^2$ and $D^{\mathsf{A}-} = \{D^{a}\}^2$, corresponding to A^+ and A^-.

As for the first family line groups, the standard components are

$$\mathsf{T}^- : a_0 = \begin{bmatrix} 0 & 1 & 0 \\ -1 & 0 & 0 \\ 0 & 0 & 0 \end{bmatrix}, \quad a_{\pm 1} = \begin{bmatrix} 0 & 0 & 1 \\ 0 & 0 & \mp\mathrm{i} \\ -1 & \pm\mathrm{i} & 0 \end{bmatrix}, \tag{5.34a}$$

$$\begin{aligned} \mathsf{T}^+ : s_0 &= \begin{bmatrix} 1 & 0 & 0 \\ 0 & 1 & 0 \\ 0 & 0 & 0 \end{bmatrix}, s_0' = \begin{bmatrix} 0 & 0 & 0 \\ 0 & 0 & 0 \\ 0 & 0 & 1 \end{bmatrix}, s_{\pm 1} = \begin{bmatrix} 0 & 0 & 1 \\ 0 & 0 & \mp\mathrm{i} \\ 1 & \mp\mathrm{i} & 0 \end{bmatrix}, \\ s_{\pm 2} &= \begin{bmatrix} 1 & \mp\mathrm{i} & 0 \\ \mp\mathrm{i} & -1 & 0 \\ 0 & 0 & 0 \end{bmatrix}. \end{aligned} \tag{5.34b}$$

They transform according to the one-dimensional representations, with linear and helical quantum numbers:

$$a_0, s_0 : \quad k = m = 0; \quad \tilde{k} = \tilde{m} = 0, \tag{5.35a}$$

$$a_{\pm 1}, s_{\pm 1} : k = 0,\ m \stackrel{\circ}{=} \pm 1;\ \tilde{k} \stackrel{\circ}{=} \pm\kappa,\ \tilde{m} \stackrel{\circ}{=} \pm 1 \quad \left(\kappa = \frac{2\pi}{fQ}\right), \tag{5.35b}$$

$$s_{\pm 2} : k = 0,\ m \stackrel{\circ}{=} \pm 2;\ \tilde{k} \stackrel{\circ}{=} \pm 2\kappa,\ \tilde{m} \stackrel{\circ}{=} \pm 2. \tag{5.35c}$$

Note that the modular equalities (see comment below (5.31)) provide that for small q (or n) the angular quantum numbers 2 and 1 are effectively lowered.

For the other families, depending on the symmetry group, standard components are either mutually independent (i.e., transforming according to the one dimensional representations) or combine into two dimensional multiplets. Note that for $q < 5$ (Table 5.4) the standard components in some cases are also

Table 5.4 Standard components of the second-rank tensors. In the first column the families F of the line groups $\boldsymbol{L}^{(F)}$ above the corresponding isogonal group $\boldsymbol{P}_{\mathrm{I}}$ are given. Then the standard components of antisymmetric T^- and symmetric T^+ tensors are listed below the corresponding irreducible representations (first, the orders $q = 1, 2$ of the principle axis of the isogonal group are included, and then the cases $q > 2$). For the families 1 and 5 all types of tensors have the same standard components. For the other families the irreducible representations of the axial tensors differ by parities Π_{h} and Π_{v} from those of the polar tensors, as well as some of the standard components; respective parities and tensor components are then denoted by $\pm$ or $\mp$, as well as $X = A/B$ and $Y = B/A$. Linear quantum numbers are presented, and the helical ones are obtained according to the equivalence (5.34), retaining all the parities

F	$\mathsf{A}^-/\mathsf{P}^-$						$\mathsf{A}^+/\mathsf{P}^+$								
$\boldsymbol{P}_{\mathrm{I}}$	$q=1$			$q=2$			$q=1$				$q=2$				
1	3_0A_0			${}_0A_0$	2_0A_1		6_0A_0				4_0A_0			2_0A_1	
$\boldsymbol{C}_q$	a_0, a_1, a_{-1}			a_0	a_1, a_{-1}		$s_0, s_0', s_1, s_{-1}, s_2, s_2$				s_0, s_0', s_2, s_{-2}			s_1, s_{-1}	
2	$3_0A_0^{\pm}$			${}_0A_0^{\pm}$	${}_0A_1^{\mp}$	${}_0A_1^{\pm}$	$6_0A_0^{\pm}$				$2_0A_0^{\pm}$	${}_0A_1^{\mp}$	${}_0A_1^{\pm}$	$2_0A_0^{\mp}$	
$\boldsymbol{S}_{2q}$	a_0, a_1, a_{-1}			a_0	a_1	a_{-1}	$s_0, s_0', s_1, s_{-1}, s_2, s_2$				s_0, s_0'	s_1	s_{-1}	s_2, s_{-2}	
3, 4	${}_0A_0^{\pm}$	$2_0A_0^{\mp}$		${}_0A_0^{\pm}$	$2_0A_1^{\mp}$		$4_0A_0^{\pm}$		$2_0A_0^{\mp}$		$4_0A_0^{\pm}$			$2_0A_1^{\mp}$	
$\boldsymbol{C}_{q\mathrm{h}}$	a_0	a_1, a_{-1}		a_0	a_1, a_{-1}		s_0, s_0', s_2, s_{-2}		s_1, s_{-1}		s_0, s_0', s_2, s_{-2}			s_1, s_{-1}	
5	${}_0A_0^{+}$	$2_0A_0^{-}$		${}_0A_0^{-}$	${}_0A_1^{+}$	${}_0A_1^{-}$	$4_0A_0^{+}$		$2_0A_0^{-}$		$3_0A_0^{+}$		${}_0A_0^{-}$	${}_0A_1^{+}$	${}_0A_1^{-}$
$\boldsymbol{D}_q$	a_{yz}	a_0, a_{xz}		a_0	a_{yz}	a_{xz}	s_0, s_0', s_{yz}, d		s_{xy}, s_{xz}		s_0, s_0', d		s_{xy}	s_{yz}	s_{xz}
6–**8**	${}_0X_0$	2_0Y_0		${}_0Y_0$	${}_0X_1$	${}_0Y_1$	4_0X_0		2_0Y_0		3_0X_0		${}_0Y_0$	${}_0X_1$	${}_0Y_1$
$\boldsymbol{C}_{q\mathrm{v}}$	a_{xz}	a_0, a_{yz}		a_0	a_{xz}	a_{yz}	s_0, s_0', s_{xz}, d		s_{xy}, s_{yz}		s_0, s_0', d		s_{xy}	s_{xz}	s_{yz}
9, 10	${}_0X_0^{+}$	$2_0Y_0^{-}$		${}_0Y_0^{-}$	${}_0E_1$		$4_0X_0^{+}$		$2_0Y_0^{-}$		$2_0X_0^{+}$	${}_0X_0^{-}$	${}_0Y_0^{+}$	${}_0E_1$	
$\boldsymbol{D}_{q\mathrm{d}}$	a_{xz}	a_0, a_{yz}		a_0	$(a_1, \pm a_{-1})$		s_0, s_0', s_{xz}, d		s_{xy}, s_{yz}		s_0, s_0'	d	s_{xy}	$(s_1, \pm s_{-1})$	
11–**13**	${}_0X_0^{\mp}$	${}_0Y_0^{\pm}$	${}_0Y_0^{\mp}$	${}_0Y_0^{\pm}$	${}_0X_1^{\mp}$	${}_0Y_1^{\mp}$	$3_0X_0^{\pm}$	${}_0X_0^{\mp}$	${}_0Y_0^{\pm}$	${}_0Y_0^{\mp}$	$3_0X_0^{\pm}$		${}_0Y_0^{\pm}$	${}_0X_1^{\mp}$	${}_0Y_1^{\mp}$
$\boldsymbol{D}_{q\mathrm{h}}$	a_{xz}	a_0	a_{yz}	a_0	a_{xz}	a_{yz}	s_0, s_0', d	s_{xz}	s_{xy}	s_{yz}	s_0, s_0', d		s_{xy}	s_{xz}	s_{yz}

Table 5.4 (continued)

F	A$^-$/P$^-$			A$^+$/P$^+$														
P_1 q:	$q=3,4,\ldots$			$q=3$					$q=4$					$q=5,6,\ldots$				
1	${}_0A_0$	${}_0A_1$	${}_0A_{-1}$	$2{}_0A_0$	$2{}_0A_1$		$2{}_0A_{-1}$		$2{}_0A_0$	${}_0A_1$	${}_0A_{-1}$	$2{}_0A_2$		$2{}_0A_0$	${}_0A_1$	${}_0A_{-1}$	${}_0A_2$	${}_0A_{-2}$
$\boldsymbol{C}_q$	a_0	a_1	a_{-1}	s_0, s_0'	s_1, s_{-2}		s_{-1}, s_2		s_0, s_0'	s_1	s_{-1}	s_2, s_{-2}		s_0, s_0'	s_1	s_{-1}	s_2	s_{-2}
2	${}_0A_0^\pm$	${}_0A_1^\mp$	${}_0A_{-1}^\mp$	$2{}_0A_0^\pm$	$2{}_0A_1^\mp$		$2{}_0A_{-1}^\mp$		$2{}_0A_0^\pm$	${}_0A_1^\mp$	${}_0A_{-1}^\mp$	${}_0A_2^\pm$	${}_0A_2^\mp$	$2{}_0A_0^\pm$	${}_0A_1^\mp$	${}_0A_{-1}^\mp$	${}_0A_2^\pm$	${}_0A_{-2}^\pm$
$\boldsymbol{S}_{2q}$	a_0	a_1	a_{-1}	s_0, s_0'	s_1, s_{-2}		s_{-1}, s_2		s_0, s_0'	s_1	s_{-1}	s_2	s_{-2}	s_0, s_0'	s_1	s_{-1}	s_2	s_{-2}
3, 4	${}_0A_0^\pm$	${}_0A_1^\mp$	${}_0A_{-1}^\mp$	$2{}_0A_0^\pm$	${}_0A_1^\mp$	${}_0A_{-1}^\mp$	${}_0A_{-1}^\pm$	${}_0A_1^\pm$	$2{}_0A_0^\pm$	${}_0A_1^\mp$	${}_0A_{-1}^\mp$	$2{}_0A_2^\pm$		$2{}_0A_0^\pm$	${}_0A_1^\mp$	${}_0A_{-1}^\mp$	${}_0A_2^\pm$	${}_0A_{-2}^\pm$
$\boldsymbol{C}_{q\mathrm{h}}$	a_0	a_1	a_{-1}	s_0, s_0'	s_1	s_{-1}	s_2	s_{-2}	s_0, s_0'	s_1	s_{-1}	s_2, s_{-2}		s_0, s_0'	s_1	s_{-1}	s_2	s_{-2}
5	${}_0A_0^-$	${}_0E_1$		$2{}_0A_0^+$	$2{}_0E_1$				$2{}_0A_0^+$	${}_0E_1$		${}_0A_2^+$	${}_0A_2^-$	$2{}_0A_0^+$	${}_0E_1$		${}_0E_2$	
$\boldsymbol{D}_q$	a_0	$(a_1, -a_{-1})$		s_0, s_0'	$(s_1, -s_{-1}), (s_{-2}, s_2)$				s_0, s_0'	$(s_1, -s_{-1})$		d	s_{xy}	s_0, s_0'	$(s_1, -s_{-1})$		(s_2, s_{-2})	
6–8	${}_0Y_0$	${}_0E_1$		$2{}_0X_0$	$2{}_0E_1$				$2{}_0X_0$	${}_0E_1$		${}_0X_2$	${}_0Y_2$	$2{}_0X_0$	${}_0E_1$		${}_0E_2$	
$\boldsymbol{C}_{q\mathrm{v}}$	a_0	$(a_1, \pm a_{-1})$		s_0, s_0'	$(s_1, \pm s_{-1}), (s_{-2}, \pm s_2)$				s_0, s_0'	$(s_1, \pm s_{-1})$		d	s_{xy}	s_0, s_0'	$(s_1, \pm s_{-1})$		$(s_2, \pm s_{-2})$	
9, 10	${}_0Y_0^-$	${}_0E_1^-$		$2{}_0X_0^+$	$2{}_0E_1^-$				$2{}_0X_0^+$	${}_0E_1^-$		${}_0E_2$		$2{}_0X_0^+$	${}_0E_1^-$		${}_0E_2^+$	
$\boldsymbol{D}_{q\mathrm{d}}$	a_0	$(a_1, \pm a_{-1})$		s_0, s_0'	$(s_1, \pm s_{-1}), (s_{-2}, \pm s_2)$				s_0, s_0'	$(s_1, \pm s_{-1})$		$(s_2, \pm s_{-2})$		s_0, s_0'	$(s_1, \pm s_{-1})$		$(s_2, \pm s_{-2})$	
11–13	${}_0Y_0^\pm$	${}_0E_1^\mp$		$2{}_0X_0^\pm$	${}_0E_1^\mp$		${}_0E_1^\pm$		$2{}_0X_0^\pm$	${}_0E_1^\mp$		${}_0X_2^\pm$	${}_0Y_2^\pm$	$2{}_0X_0^\pm$	${}_0E_1^\mp$		${}_0E_2^\pm$	
$\boldsymbol{D}_{q\mathrm{h}}$	a_0	(a_1, a_{-1})		s_0, s_0'	$(s_1, \pm s_{-1})$		$(s_{-2}, \pm s_2)$		s_0, s_0'	$(s_1, \pm s_{-1})$		d	s_{xy}	s_0, s_0'	$(s_1, \pm s_{-1})$		$(s_2, \pm s_{-2})$	

$$\mathsf{T}^- : a_{xz} = \begin{bmatrix} 0 & 0 & 1 \\ 0 & 0 & 0 \\ -1 & 0 & 0 \end{bmatrix}, \quad a_{yz} = \begin{bmatrix} 0 & 0 & 0 \\ 0 & 0 & 1 \\ 0 & -1 & 0 \end{bmatrix}, \tag{5.36a}$$

$$\mathsf{T}^+ : d = \begin{bmatrix} 1 & 0 & 0 \\ 0 & -1 & 0 \\ 0 & 0 & 0 \end{bmatrix}, s_{xy} = \begin{bmatrix} 0 & 1 & 0 \\ 1 & 0 & 0 \\ 0 & 0 & 0 \end{bmatrix}, s_{xz} = \begin{bmatrix} 0 & 0 & 1 \\ 0 & 0 & 0 \\ 1 & 0 & 0 \end{bmatrix}, s_{yz} = \begin{bmatrix} 0 & 0 & 0 \\ 0 & 0 & 1 \\ 0 & 1 & 0 \end{bmatrix}. \tag{5.36b}$$

The standard components of the symmetric and antisymmetric tensors, classified according to the irreducible representations of the line groups, are given in Table 5.4.

Applications in physics frequently require to know the form of the invariant tensors: obviously, tensorial physical quantity T produced by system with symmetry group $\boldsymbol{L}$ can be nonzero only if it is invariant under the transformation of $\boldsymbol{L}$. Therefore, the general form of such a tensor is obtained as the sum $a\mathsf{T}_1 + b\mathsf{T}_2 + \ldots$ over the all standard components T_i corresponding to the identity representations,

Table 5.5 Invariant second-rank tensors. In the first column are the family numbers F of the line groups above the corresponding isogonal group $\boldsymbol{P}_\mathrm{I}$. The forms of antisymmetric and symmetric invariant tensors are given separately for the order q of principle axis of the isogonal group being 1, 2, and larger. The first line refers to the invariant axial tensors and the second one to the polar ones, except for the families 1 and 5, for which there is no difference

		$q=1$		$q=2$		$q=3,4,\ldots$	
F	$\boldsymbol{P}_\mathrm{I}$	$\mathsf{A}^-/\mathsf{P}^-$	$\mathsf{A}^+/\mathsf{P}^+$	$\mathsf{A}^-/\mathsf{P}^-$	$\mathsf{A}^+/\mathsf{P}^+$	$\mathsf{A}^-/\mathsf{P}^-$	$\mathsf{A}^+/\mathsf{P}^+$
1	$\boldsymbol{C}_q$	$\begin{bmatrix} 0 & a & b \\ -a & 0 & l \\ -b & -c & 0 \end{bmatrix}$	$\begin{bmatrix} d & e & f \\ e & g & h \\ f & h & i \end{bmatrix}$	$\begin{bmatrix} 0 & a & 0 \\ -a & 0 & 0 \\ 0 & 0 & 0 \end{bmatrix}$	$\begin{bmatrix} d & e & 0 \\ e & g & 0 \\ 0 & 0 & i \end{bmatrix}$	$\begin{bmatrix} 0 & a & 0 \\ -a & 0 & 0 \\ 0 & 0 & 0 \end{bmatrix}$	$\begin{bmatrix} d & 0 & 0 \\ 0 & d & 0 \\ 0 & 0 & i \end{bmatrix}$
2	$\boldsymbol{S}_{2q}$	$\begin{bmatrix} 0 & a & b \\ -a & 0 & l \\ -b & -c & 0 \end{bmatrix}$	$\begin{bmatrix} d & e & f \\ e & g & h \\ f & h & i \end{bmatrix}$	$\begin{bmatrix} 0 & a & 0 \\ -a & 0 & 0 \\ 0 & 0 & 0 \end{bmatrix}$	$\begin{bmatrix} d & 0 & 0 \\ 0 & d & 0 \\ 0 & 0 & i \end{bmatrix}$	$\begin{bmatrix} 0 & a & 0 \\ -a & 0 & 0 \\ 0 & 0 & 0 \end{bmatrix}$	$\begin{bmatrix} d & 0 & 0 \\ 0 & d & 0 \\ 0 & 0 & i \end{bmatrix}$
		None	None	None	$\begin{bmatrix} d & e & 0 \\ e & -d & 0 \\ 0 & 0 & 0 \end{bmatrix}$	None	None
3, 4	$\boldsymbol{C}_{q\mathrm{h}}$	$\begin{bmatrix} 0 & a & 0 \\ -a & 0 & 0 \\ 0 & 0 & 0 \end{bmatrix}$	$\begin{bmatrix} d & e & 0 \\ e & g & 0 \\ 0 & 0 & i \end{bmatrix}$	$\begin{bmatrix} 0 & a & 0 \\ -a & 0 & 0 \\ 0 & 0 & 0 \end{bmatrix}$	$\begin{bmatrix} d & e & 0 \\ e & g & 0 \\ 0 & 0 & i \end{bmatrix}$	$\begin{bmatrix} 0 & a & 0 \\ -a & 0 & 0 \\ 0 & 0 & 0 \end{bmatrix}$	$\begin{bmatrix} d & 0 & 0 \\ 0 & d & 0 \\ 0 & 0 & i \end{bmatrix}$
		$\begin{bmatrix} 0 & 0 & b \\ 0 & 0 & l \\ -b & -c & 0 \end{bmatrix}$	$\begin{bmatrix} 0 & 0 & f \\ 0 & 0 & h \\ f & h & 0 \end{bmatrix}$	None	None	None	None
5	$\boldsymbol{D}_q$	$\begin{bmatrix} 0 & 0 & 0 \\ 0 & 0 & l \\ 0 & -c & 0 \end{bmatrix}$	$\begin{bmatrix} d & 0 & 0 \\ 0 & g & h \\ 0 & h & i \end{bmatrix}$	None	$\begin{bmatrix} d & 0 & 0 \\ 0 & g & 0 \\ 0 & 0 & i \end{bmatrix}$	None	$\begin{bmatrix} d & 0 & 0 \\ 0 & d & 0 \\ 0 & 0 & i \end{bmatrix}$
6–8	$\boldsymbol{C}_{q\mathrm{v}}$	$\begin{bmatrix} 0 & 0 & b \\ 0 & 0 & 0 \\ -b & 0 & 0 \end{bmatrix}$	$\begin{bmatrix} d & 0 & f \\ 0 & g & 0 \\ f & 0 & i \end{bmatrix}$	None	$\begin{bmatrix} d & e & 0 \\ e & d & 0 \\ 0 & 0 & i \end{bmatrix}$	None	$\begin{bmatrix} d & 0 & 0 \\ 0 & d & 0 \\ 0 & 0 & i \end{bmatrix}$
		$\begin{bmatrix} 0 & a & 0 \\ -a & 0 & l \\ 0 & -c & 0 \end{bmatrix}$	$\begin{bmatrix} 0 & e & 0 \\ e & 0 & h \\ 0 & h & 0 \end{bmatrix}$	$\begin{bmatrix} 0 & a & 0 \\ -a & 0 & 0 \\ 0 & 0 & 0 \end{bmatrix}$	$\begin{bmatrix} 0 & e & 0 \\ e & 0 & 0 \\ 0 & 0 & 0 \end{bmatrix}$	$\begin{bmatrix} 0 & a & 0 \\ -a & 0 & 0 \\ 0 & 0 & 0 \end{bmatrix}$	None
9, 10	$\boldsymbol{D}_{q\mathrm{d}}$	$\begin{bmatrix} 0 & 0 & 0 \\ 0 & 0 & l \\ 0 & -c & 0 \end{bmatrix}$	$\begin{bmatrix} d & e & f \\ e & d & 0 \\ f & 0 & i \end{bmatrix}$	None	$\begin{bmatrix} d & 0 & 0 \\ 0 & d & 0 \\ 0 & 0 & i \end{bmatrix}$	None	$\begin{bmatrix} d & 0 & 0 \\ 0 & d & 0 \\ 0 & 0 & i \end{bmatrix}$
		None	None	None	$\begin{bmatrix} 0 & e & 0 \\ e & 0 & 0 \\ 0 & 0 & 0 \end{bmatrix}$	None	None
11–13	$\boldsymbol{D}_{q\mathrm{h}}$	None	$\begin{bmatrix} d & 0 & 0 \\ 0 & g & 0 \\ 0 & 0 & i \end{bmatrix}$	None	$\begin{bmatrix} d & 0 & 0 \\ 0 & g & 0 \\ 0 & 0 & i \end{bmatrix}$	None	$\begin{bmatrix} d & 0 & 0 \\ 0 & d & 0 \\ 0 & 0 & i \end{bmatrix}$
		$\begin{bmatrix} 0 & 0 & 0 \\ 0 & 0 & l \\ 0 & -c & 0 \end{bmatrix}$	$\begin{bmatrix} 0 & 0 & 0 \\ 0 & 0 & h \\ 0 & h & 0 \end{bmatrix}$	None	None	None	None

with the coefficients $a, b, \ldots$ characterizing the quantity. When this summation is performed, the general form of invariant tensors is found (results are presented in Table 5.5).

5.5 Application: Clebsch–Gordan Coefficients and Selection Rules

Within quantum mechanical formalism, states of the considered system are vectors in the Hilbert space $\mathcal{S}$, while physical quantities are hermitian operators. Geometrical transformations ℓ are unitary operators $D(\ell)$, giving (infinite dimensional, in general reducible) representation D of the symmetry group. Accordingly, there is a *symmetry-adapted basis* standard vectors $|\mu m t_\mu\rangle$ of which satisfy (5.3):

$$D(\ell)\,|\mu t_\mu m\rangle = \sum_{m'=1}^{|\mu|} D^{(\mu)}_{m'm}(\ell)\,|\mu m' t_\mu\rangle. \tag{5.37}$$

Such an action yields transformation of the operators of the second-rank tensors: $\ell Q = D^{-1}(\ell) Q D(\ell)$, (5.32). The standard components are again defined by (5.3), and in this context they are called *irreducible tensor operators*. This way, both the states and the operators are assigned by irreducible representations of the symmetry group, i.e., by conserved quantum numbers.

As a consequence, expanding vectors over SAB and the operators over irreducible tensors, arbitrary matrix element becomes a sum of the matrix elements $\langle \mu m t_\mu \mid Q^\nu_{nt_\nu} \mid \lambda l t_\lambda\rangle$ with the symmetry-adapted ingredients only. Clearly, these entities are to a large extent determined by symmetry, as it is enlighten by the Wigner–Eckart theorem:

$$\langle \mu m t_\mu |\, Q^\nu_{nt_\nu}\, |\lambda l t_\lambda\rangle = (\mu t_\mu \| Q^\nu_{t_\nu} \| \lambda t_\lambda)\langle \mu m | \nu n; \lambda l\rangle. \tag{5.38}$$

The first factor, called *reduced matrix element*, depends only on the involved irreducible representations, not on the indices m, n, and l. Therefore, it can be calculated or even measured just once, irrespective of the dimensions of the involved representations. On the other hand, the *Clebsch–Gordan coefficients* $\langle \mu m \mid \nu n; \lambda l\rangle$ do not depend neither on the representation counters (t_μ, t_ν and t_λ) nor on the physical content of the quantity Q. They are completely determined by the symmetry of the system and can be tabulated a priori.

For many applications it is important to know only whether a particular Clebsch–Gordan coefficient vanishes. Quite generally, Clebsch–Gordan coefficient is zero unless the quantum numbers of the involved representations satisfy *selection rules* (for helical and linear quantum numbers):

$$\Delta\tilde{k} = \tilde{k}_\mathrm{f} - \tilde{k}_\mathrm{i} \stackrel{\circ}{=} \tilde{k}, \quad \Delta\tilde{m} = \tilde{m}_\mathrm{f} - \tilde{m}_\mathrm{i} \stackrel{\circ}{=} \tilde{m}, \quad \Pi_\mathrm{f}\Pi\Pi_\mathrm{i} \neq -1, \tag{5.39a}$$

$$\Delta k = k_\mathrm{f} - k_\mathrm{i} \stackrel{\circ}{=} k, \quad \Delta m = m_\mathrm{f} - m_\mathrm{i} \stackrel{\circ}{=} m + Kp, \quad \Pi_\mathrm{f}\Pi\Pi_\mathrm{i} \neq -1. \tag{5.39b}$$

The dotted equalities are taken modulo the ranges $(-\pi/f, \pi/f]$ and $(-n/2, n/2]$ of $\tilde{k}$ and $\tilde{m}$, i.e., $(-\pi/a, \pi/a]$ and $(-q/2, q/2]$ of k and m, respectively; as for K, it is the unique integer making $k_\mathrm{f} - k_\mathrm{i} - 2\pi K/a$ being from the Brillouin zone. The selection rules correspond to the physical conservation laws of momenta and parities (if defined). In particular, the second rule in (5.39b) transparently reveals that m is not a conserved quantum number as $\Delta m \neq m$ when $K \neq 0$ (*Umklapp processes*). According to the Wigner–Eckart theorem, matrix elements vanish whenever the quantum numbers do not satisfy the selection rules (5.39). However, note that they can also vanish for some other, not symmetry-based reasons. In other words, a matrix element may happen to be zero even if the relevant Clebsch–Gordan coefficient is positive. In such a case it is the reduced matrix element which vanishes. This is due to the particular operator $Q^{\nu}_{nt_\nu}$ and vectors $|\mu m t_\mu\rangle$ and $|\lambda l t_\lambda\rangle$ and cannot be predicted by symmetry.

Particular values of Clebsch–Gordan coefficients are easy to find for the first family line groups $\boldsymbol{T}_Q(f)\boldsymbol{C}_n$. In the helical and linear (for commensurate groups $\boldsymbol{L}q_p = \boldsymbol{T}^r_q(na/q)\boldsymbol{C}_n$, only) quantum numbers they read

$$\langle \tilde{k}_f \tilde{m}_f | \tilde{k}\tilde{m}; \tilde{k}_i \tilde{m}_i\rangle = \begin{cases} 1 & \text{if } \Delta\tilde{k} = \tilde{k}_\mathrm{f} - \tilde{k}_\mathrm{i} \stackrel{\circ}{=} \tilde{k} \text{ and } \Delta\tilde{m} = \tilde{m}_\mathrm{f} - \tilde{m}_\mathrm{i} \stackrel{\circ}{=} \tilde{m}, \\ 0 & \text{otherwise.} \end{cases} \quad (5.40\text{a})$$

$$\langle k_f, m_f | km; k_i m_i\rangle = \begin{cases} 1 & \text{if } \Delta k = k_\mathrm{f} - k_\mathrm{i} \stackrel{\circ}{=} k \text{ and } \Delta m = m_\mathrm{f} - m_\mathrm{i} \stackrel{\circ}{=} m + Kp, \\ 0 & \text{otherwise.} \end{cases} \quad (5.40\text{b})$$

Clebsch–Gordan coefficients for other line groups can be calculated straightforwardly from the irreducible representations (e.g., using modified group projector technique, Appendix F). The results are too lengthy for tabulation, but the numerical code calculating each particular Clebsch–Gordan coefficient for any line group is available.

As an illustration relevant for the *chiral nanotubes*, we list Clebsch–Gordan coefficients for fifth family line groups. First, Clebsch–Gordan coefficients vanish whenever any of the selection rules (5.39b) with $\Pi = \Pi_U$ is not satisfied. Otherwise, all the coefficients are equal to 1, except the following ones:

$$\langle \tilde{k}_f, \tilde{m}_f | \tilde{k}\tilde{m}; \tilde{k}_i \tilde{m}_i -\rangle = -1, \text{ if } \tilde{k} < 0, \text{ or } \tilde{k} = 0, \frac{\pi}{f} \text{ and } \tilde{m} < 0,$$

$$\langle \tilde{k}_f, \tilde{m}_f | \tilde{k}\tilde{m}-; \tilde{k}_i \tilde{m}_i\rangle = -1, \text{ if } \tilde{k}_i < 0, \text{ or } \tilde{k}_i = 0, \frac{\pi}{f} \text{ and } \tilde{m}_i < 0,$$

$$\langle \tilde{k}_f, \tilde{m}_f, \pm | \tilde{k}, \tilde{m}; \tilde{k}_i, \tilde{m}_i\rangle = \begin{cases} \pm\frac{1}{\sqrt{2}}, & \tilde{k} < 0, \text{ or } \tilde{k} = 0, \frac{\pi}{f} \text{ and } \tilde{m} < 0, \\ \frac{1}{\sqrt{2}}, & \text{otherwise,} \end{cases} \quad (5.41\text{a})$$

$$\langle k_f m_f 0 | km0; k_i m_i, -1\rangle = -1, \text{ if } \begin{matrix} k < 0, \text{ or } k = 0 \text{ and } m < 0, \text{ or} \\ k = \frac{\pi}{a} \text{ and } m \notin [-\frac{p}{2}, \frac{q-p}{2}], \end{matrix}$$

$$\langle k_f m_f 0 | km, -1; k_i m_i 0\rangle = -1, \text{ if } \begin{matrix} k_i < 0, \text{ or } k_i = 0 \text{ and } m_i < 0, \text{ or} \\ k_i = \frac{\pi}{a} \text{ and } m_i \notin [-\frac{p}{2}, \frac{q-p}{2}], \end{matrix}$$

$$\langle k_f m_f, \pm 1 | km0; k_i m_i 0\rangle = \begin{cases} \pm\frac{1}{\sqrt{2}}, & \text{if } \begin{matrix} k<0, \text{ or } k=0 \text{ and } m<0, \text{ or} \\ k=\frac{\pi}{a} \text{ and } m_i \notin [-\frac{p}{2}, \frac{q-p}{2}], \end{matrix} \\ \frac{1}{\sqrt{2}}, & \text{otherwise} \end{cases} \tag{5.41b}$$

for the helical (5.41a) and linear (5.41b) quantum numbers.

References

1. I. Milŏsević, B. Dakić, M. Damnjanović, J. Phys. A: Math. Gen. **39**, 11833 (2006)
2. B. Dakić, M. Damnjanović, I.M. Sević, J. Phys. A: Math. Gen. **42**, 125202 (2009)
3. M. Damnjanović, Phys. Lett. A **94**, 337 (1983)
4. I. Miloševiĺ, Phys. Lett. A **204**, 63 (1995)

Chapter 6
Magnetic Line Groups

Abstract In this chapter we consider magnetic line groups, which besides geometrical symmetries of quasi-one-dimensional regular systems include also *time reversal* operation Θ. These groups and their *co-representations* besides being relevant in the context of spin ordering and magnetic phenomena, give insight into the properties of energy spectrum in some other dynamical problems.

6.1 Classification

In this section we will derive all the magnetic line groups. It turns out that to this end one has to construct all the halving subgroups of all the line groups.

6.1.1 Structure

As time reversal changes the sign of the time coordinate, and leaves the spatial coordinates unchanged, it is an involution, $\Theta^2 = (I|0)$ which commutes with all spatial transformations ℓ. Consequently, product of two space–time transformations is a spatial transformation, $\Theta\ell_1\Theta\ell_2 = \ell_1\ell_2$, implying that spatial transformations form a halving subgroup of a magnetic line group, i.e., general structure of a magnetic line group is $\boldsymbol{L}' + \Theta\ell^*\boldsymbol{L}'$, where $\boldsymbol{L}'$ is a line group. By setting Θ to identity the line group $\boldsymbol{L} = \boldsymbol{L}' + \ell^*\boldsymbol{L}'$ is obtained. If ℓ^* belongs to $\boldsymbol{L}'$ then the coset $\Theta\ell^*\boldsymbol{L}'$ is simply $\Theta\boldsymbol{L}'$, and *gray group* is constructed:

$$\boldsymbol{L}^*(\boldsymbol{L}) = \boldsymbol{L} \otimes \{e, \Theta\} = \boldsymbol{L} + \Theta\boldsymbol{L}. \tag{6.1a}$$

Otherwise, when $\boldsymbol{L}'$ does not contain ℓ^*, $\boldsymbol{L}'$ is a halving subgroup of $\boldsymbol{L}$ and *black-and-white magnetic line group* is obtained:

$$\boldsymbol{L}^*(\boldsymbol{L}') = \boldsymbol{L}' + \Theta\ell^*\boldsymbol{L}'. \tag{6.1b}$$

Thus, to each line group $\boldsymbol{L}$ a class of magnetic line groups [1] $\boldsymbol{L}^*$ is associated: line group itself, gray group, and a set of black-and-white groups.

Damnjanović, M., Milošević, I.: *Magnetic Line Groups*. Lect. Notes Phys. **801**, 85–93 (2010)
DOI 10.1007/978-3-642-11172-3_6

6.1.2 Construction

As shown in the last section, in order to construct all magnetic line groups, one should find all halving subgroups of line groups. A quite general method of construction of the magnetic groups, developed within the space group theory, exists. It is based on the alternating representations. However, it will not be applied here because the line group factorization (Sect. 2.1) offers more efficient approach, namely, it has been shown [2] that an index-two subgroup of the line group $\boldsymbol{ZP}$ takes one of the following three forms: (a) $\boldsymbol{Z}'_j\boldsymbol{P}$, (b) $\boldsymbol{ZP}'_i$, and (c) $\boldsymbol{Z}'_j\boldsymbol{P}'_i + \boldsymbol{Z}^{\prime c}_j\boldsymbol{P}^{\prime c}_i$, where $\boldsymbol{Z}'_j$ and $\boldsymbol{P}'_i$ are halving subgroups of $\boldsymbol{Z}$ and $\boldsymbol{P}$, respectively, while $\boldsymbol{Z}^{c}_j = \boldsymbol{Z}\backslash\boldsymbol{Z}'_j$ and $\boldsymbol{P}^{\prime c}_i = \boldsymbol{P}\backslash\boldsymbol{P}'_i$ are their complementary cosets. So, in order to construct halving subgroups of a line group, we derive halving subgroups of the factors $\boldsymbol{Z}$ and $\boldsymbol{P}$ and then construct sets (a)–(c). Finally, among them we chose only those which satisfy the condition that $\boldsymbol{Z}$ or $\boldsymbol{Z}'_j$ commute with $\boldsymbol{P}'_i$ or $\boldsymbol{P}$, i.e., the prerequisite of a subgroup structure.

6.1.2.1 Point Factor

The halving subgroups of the point and of the magnetic point groups are well known [1]; they can be found [2] using the described algorithm and the factorization (Table 2.1).

6.1.2.2 Generalized Translations

Being cyclic, generalized translational group $\boldsymbol{Z}$ contains only one subgroup of index two, infinite cyclic group $\boldsymbol{Z}'$, generated by $Z' = Z^2$ of the generator of $\boldsymbol{Z}$ (Sect. 2.1).

For the helical group $\boldsymbol{T}_Q(f)$ squared generator is $(C_Q|f)^2 = (C_{Q/2}|2f)$. Therefore, applying the convention $Q' \geq 1$, we obtain helical group:

$$\boldsymbol{Z} = \boldsymbol{T}_Q(f): \quad \boldsymbol{Z}_1 = \boldsymbol{T}_{Q'}(2f), \quad Q' = \begin{cases} \frac{Q}{2}, & \text{if } Q \geq 2; \\ \frac{Q}{2-Q}, & \text{otherwise.} \end{cases} \tag{6.2a}$$

In particular, halving subgroup of a commensurate helical group $\boldsymbol{T}^r_q(f)$ is also commensurate and generated by $(C^{r_1}_{q_1}|2f = a_1/q_1)$:

$$\boldsymbol{Z} = \boldsymbol{T}^r_q(f): \; \boldsymbol{Z}' = \boldsymbol{T}^{r'}_{q'}\left(\frac{a'}{q'}\right), \; q' = \begin{cases} \frac{q}{2}, & q \text{ even}, \\ q, & q \text{ odd}, \end{cases} \; r' = q'\left\{\frac{2r}{q}\right\}, \; a' = \begin{cases} a, & q \text{ even}, \\ 2a & q \text{ odd}. \end{cases} \tag{6.2b}$$

Note that for q even commensurate helical group and its halving subgroup have the same period.

In the special cases of achiral helical groups and glide plane we get

$$\boldsymbol{Z} = \boldsymbol{T}(a) : \quad \boldsymbol{Z}' = \boldsymbol{T}(2a), \tag{6.3}$$

$$\boldsymbol{Z} = \boldsymbol{T}_2^1(a/2) : \quad \boldsymbol{Z}' = \boldsymbol{T}(a), \tag{6.4}$$

$$\boldsymbol{Z} = \boldsymbol{T}'(a/2) : \quad \boldsymbol{Z}' = \boldsymbol{T}(a). \tag{6.5}$$

6.1.3 Results and Notation

It turns out that it is convenient to elaborate results for the magnetic groups related to the first family line groups and to consider parities afterward. This is one more manifestation of the specific structure of the line groups.

6.1.3.1 First Family Line Groups

Halving subgroups of the first family group $\boldsymbol{T}_Q(f)\boldsymbol{C}_n$ are easily found applying the above-described procedure and convention (2.11):

$$\boldsymbol{T}_{Q'}(2f)\boldsymbol{C}_n : \quad Q' = \begin{cases} \frac{Q}{2}, & \text{if } Q \geq 2n; \\ \frac{nQ}{2n+Q+Q\left[-\frac{2n}{Q}\right]}, & \text{otherwise.} \end{cases} \tag{6.6a}$$

$$\boldsymbol{T}_Q(f)\boldsymbol{C}_{n/2}; \tag{6.6b}$$

$$\boldsymbol{T}_{Q'}(f)\boldsymbol{C}_{n/2} : \quad Q' = \frac{Qn}{Q+n}. \tag{6.6c}$$

Commensurate line groups have commensurate halving subgroups. After applying convention C1 to (6.6), various values for q', r', and p' are obtained depending on the combinations of the parities of $\tilde{q} = q/n$, $\tilde{p} = p/n$, and $(r\tilde{p} - 1)/\tilde{q}$ (see Table C.1, taking into account that within C1, q and r cannot be simultaneously even). These results are exhaustively listed in Table 6.1. Summarizing table, we note that for n even three black-and-white magnetic groups $\boldsymbol{L}q_p^*(a)$, $\boldsymbol{L}_c q_p(2a)$, and $\boldsymbol{L}_s q_p(2a)$ are obtained, while for n odd only one of them, depending on the parities of the parameters $\tilde{q}$, $\tilde{p}$, and A.

6.1.3.2 Other Families

The above given algorithm of construction of the magnetic line groups can be now performed easily as all the relevant subgroups are determined. All 13 classes of these groups are presented in Table 6.2. In the total of 81 families [3] of the magnetic line groups, there are 13 families of the ordinary as well as the gray groups, while the remaining 55 families are the black-and-white magnetic line groups.

Table 6.1 Black-and-white groups of the commensurate first family line groups. For each of the subgroups (6.6) conventional coset representative ℓ^* (two alternatives in the last case), the resulting black-and-white group $\boldsymbol{L}^*(\boldsymbol{L}')$ (international notation) and the halving subgroup $\boldsymbol{L}'$ (international notation and parameters r', n', and f'; r' and $\tilde{p}'$ should be taken modulo q'/n') are given. Corresponding conditions on the parities of $\tilde{q}$, $\tilde{p}$, and $A = (r\tilde{p} - 1)/\tilde{q}$ are denoted by "e" and "o" for even and odd; data in bold face are main conditions, i.e., those which determine parities of other parameters (entry is empty when both parities are allowed, while $-x$ denotes the parity opposite to that of x)

Type	Conditions					Halving subgroup					
ℓ^*	n	$\tilde{q}$	$\tilde{p}$	A	q	$\boldsymbol{L}' = \boldsymbol{L}q'_{p'}$	r'	n'	f'	a'	$\boldsymbol{L}^*(\boldsymbol{L}')$
(6.6a)		**e**	o		e	$\boldsymbol{L}(\frac{q}{2})_p$	r	n	$2f$	a	$\boldsymbol{L}q^*_p(a)$
$(C_Q\vert f)$		**o**	**e**	o		$\boldsymbol{L}q_{p/2}$	$2r$	n	$2f$	$2a$	$\boldsymbol{L}_c q_p(2a)$
		o	**o**	e		$\boldsymbol{L}q_{(p+q)/2}$	$2r$	n	$2f$	$2a$	$\boldsymbol{L}_s q_p(2a)$
(6.6b)	**e**		o	**e**	e	$\boldsymbol{L}q_{p/2}$	r	$n/2$	f	$2a$	$\boldsymbol{L}_c q_p(2a)$
C_n	**e**	-$\tilde{p}$	-$\tilde{q}$	**o**	e	$\boldsymbol{L}q_{(p+q)/2}$	r	$n/2$	f	$2a$	$\boldsymbol{L}_s q_p(2a)$
(6.6c)	**e**	**e**	o	**e**	e	$\boldsymbol{L}q_{(p+q)/2}$	$r+\tilde{q}$	$n/2$	f	$2a$	$\boldsymbol{L}_s q_p(2a)$
C_n (or	**e**	**e**	o	**o**	e	$\boldsymbol{L}q_{p/2}$	$r+\tilde{q}$	$n/2$	f	$2a$	$\boldsymbol{L}_c q_p(2a)$
$(C_Q\vert f)$)	**e**	**o**			e	$\boldsymbol{L}(\frac{q}{2})_p$	$(r+\tilde{q})/2$	$n/2$	f	a	$\boldsymbol{L}q^*_p(a)$

6.1.3.3 Notation

The international symbol of a commensurate magnetic line group $\boldsymbol{L}^*$ can be uniquely derived from the symbol $\boldsymbol{L}P$ of the corresponding ordinary line group $\boldsymbol{L}$ (Sect. 2.3.4). Gray group in the $\boldsymbol{L}$-class is represented as $\boldsymbol{L}P\,1^*$. Notation of black-and-white groups $\boldsymbol{L}^*(\boldsymbol{L}')$ manifests one of the two possible situations:

1. The translational periods of $\boldsymbol{L}$ and $\boldsymbol{L}'$ are equal, while the isogonal group $\boldsymbol{P}'_I$ of $\boldsymbol{L}'$ halves the isogonal group $\boldsymbol{P}_\mathrm{I}$ of $\boldsymbol{L}$. Corresponding magnetic group is denoted by $\boldsymbol{L}P^*(P')$, where $P^*(P')$ is the international symbol of the magnetic point group $\boldsymbol{P}^*_I(\boldsymbol{P}'_I)$.
2. Translational period of $\boldsymbol{L}'$ is twice the period of $\boldsymbol{L}$, while $\boldsymbol{P}_\mathrm{I}$ is equal to $\boldsymbol{P}'_I$. In this case we add a subscript to the symbol of the ordinary line group in order to indicate the halving subgroup:

   a. If ℓ^* is from $\boldsymbol{L}^{(1)}$, then the first family subgroup $\boldsymbol{L}'^{(1)}$ of $\boldsymbol{L}'$ halves $\boldsymbol{L}^{(1)}$; as it follows from Table 6.1, there are two choices $\boldsymbol{L}'^{(1)} = \boldsymbol{L}q_{p/2}$ or $\boldsymbol{L}'^{(1)} = \boldsymbol{L}q_{(n+p)/2}$ of the halving subgroups with the isogonal group of $\boldsymbol{L}^{(1)}$ (i.e., with $q' = q$). The corresponding magnetic groups are denoted as $\boldsymbol{L}_c P$ and $\boldsymbol{L}_s P$, respectively.
   b. If ℓ^* is not from $\boldsymbol{L}^{(1)}$, then $\boldsymbol{L}^{(1)} = \boldsymbol{L}'^{(1)}$. The only remaining way to double the period of $\boldsymbol{L}'$ (and still to preserve the isogonal group) is $\ell^* = (\sigma_\mathrm{v}|0)$, which transforms mirror plane σ_v of $\boldsymbol{L}$ into the glide plane $\ell^* = (\sigma_\mathrm{v}|a = a'/2)$ of $\boldsymbol{L}'$ and the resulting magnetic group $\boldsymbol{L}^*(\boldsymbol{L}')$ is denoted as $\boldsymbol{L}_m P$.

Table 6.2 Commensurate magnetic line groups. Each of the 81 families belongs to a class of an ordinary group (first listed; gray group is the last). Its ordinary halving subgroup is at the right (denoted by $\boldsymbol{L}''$ if its period is doubled initial group period). For families 2 and 24, $p' = \frac{q}{2}\{\frac{2p}{q}\}$

	Magnetic group n even		n odd	Halving subgroup n even		n odd
1		$\boldsymbol{L}q_p$			Ordinary group $\boldsymbol{L}^{(1)} = \boldsymbol{T}_Q(f)\boldsymbol{C}_n$	
2	$\boldsymbol{L}q_p^*$		a	$\boldsymbol{L}(\frac{q}{2})_{p'}$		a
3	$\boldsymbol{L}_c q_p$		a	$\boldsymbol{L}''q_{p/2}$		a
4	$\boldsymbol{L}_s q_p$		a	$\boldsymbol{L}''q_{(p+q)/2}$		a
5		$\boldsymbol{L}q_p 1^*$			$\boldsymbol{L}q_p$	
6	$\boldsymbol{L}\overline{2n}$		$\boldsymbol{L}\overline{n}$		Ordinary group $\boldsymbol{L}^{(2)} = \boldsymbol{T}\boldsymbol{S}_{2n}$	
7	$\boldsymbol{L}\overline{2n}^*$		$\boldsymbol{L}\overline{n}^*$	$\boldsymbol{L}n$		$\boldsymbol{L}n$
8	$\boldsymbol{L}_c(\overline{2n})$		$\boldsymbol{L}_c\overline{n}$	$\boldsymbol{L}''(\overline{2n})$		$\boldsymbol{L}''\overline{n}$
9	$\boldsymbol{L}\overline{2n}\,1^*$		$\boldsymbol{L}\overline{n}\,1^*$	$\boldsymbol{L}\overline{2n}$		$\boldsymbol{L}\overline{n}$
10	$\boldsymbol{L}n/m$		$\boldsymbol{L}\overline{2n}$		Ordinary group $\boldsymbol{L}^{(3)} = \boldsymbol{T}\boldsymbol{C}_{n\mathrm{h}}$	
11	$\boldsymbol{L}n/m^*$		$\boldsymbol{L}\overline{2n}^*$	$\boldsymbol{L}n$		$\boldsymbol{L}n$
12	$\boldsymbol{L}n^*/m^*$			$\boldsymbol{L}2\overline{\tfrac{n}{2}}$, $\boldsymbol{L}\overline{\tfrac{n}{2}}$		
13	$\boldsymbol{L}n^*/m$			$\boldsymbol{L}\frac{n}{2}/m$, $\boldsymbol{L}2\overline{\tfrac{n}{2}}$		
14	$\boldsymbol{L}_c n/m$		$\boldsymbol{L}_c(\overline{2n})$	$\boldsymbol{L}''n/m$		$\boldsymbol{L}''(\overline{2n})$
15	$\boldsymbol{L}_s n/m$			$\boldsymbol{L}''n_{\frac{n}{2}}/m$		
16	$\boldsymbol{L}n/m\,1^*$		$\boldsymbol{L}\overline{2n}\,1^*$	$\boldsymbol{L}n/m$		$\boldsymbol{L}\overline{2n}$
17		$\boldsymbol{L}2n_n/m$			Ordinary group $\boldsymbol{L}^{(4)} = \boldsymbol{T}^1_{2n}(\frac{a}{2})\boldsymbol{C}_{n\mathrm{h}}$	
18		$\boldsymbol{L}2n_n/m^*$			$\boldsymbol{L}2n_n$	
19		$\boldsymbol{L}2n_n^*/m^*$		$\boldsymbol{L}\overline{2n}$		$\boldsymbol{L}\overline{n}$
20		$\boldsymbol{L}2n_n^*/m$		$\boldsymbol{L}n/m$		$\boldsymbol{L}\overline{2n}$
21		$\boldsymbol{L}2n_n/m\,1^*$			$\boldsymbol{L}2n_n/m$	
22	$\boldsymbol{L}q_p 22$		$\boldsymbol{L}q_p 2$		Ordinary group $\boldsymbol{L}^{(5)} = \boldsymbol{T}_Q(f)\boldsymbol{D}_n$	
23	$\boldsymbol{L}q_p 2^*2^*$		$\boldsymbol{L}q_p 2^*$	$\boldsymbol{L}q_p$		$\boldsymbol{L}q_p$
24	$\boldsymbol{L}q_p^* 22^*$		b	$\boldsymbol{L}(\frac{q}{2})_{p'}22$, $\boldsymbol{L}(\frac{q}{2})_{p'}2$		b
25	$\boldsymbol{L}_c q_p 22$		b	$\boldsymbol{L}''q_{p/2}22$		b
26	$\boldsymbol{L}_s q_p 22$		b	$\boldsymbol{L}''q_{(q+p)/2}22$		b
27	$\boldsymbol{L}q_p 22\,1^*$		$\boldsymbol{L}q_p 2\,1^*$	$\boldsymbol{L}q_p 22$		$\boldsymbol{L}q_p 2$
28	$\boldsymbol{L}nmm$		$\boldsymbol{L}nm$		Ordinary group $\boldsymbol{L}^{(6)} = \boldsymbol{T}(a)\boldsymbol{C}_{n\mathrm{v}}$	
29	$\boldsymbol{L}nm^*m^*$		$\boldsymbol{L}nm^*$	$\boldsymbol{L}n$		$\boldsymbol{L}n$
30	$\boldsymbol{L}n^*mm$			$\boldsymbol{L}\frac{n}{2}mm$, $\boldsymbol{L}\frac{n}{2}m$		
31	$\boldsymbol{L}_c nmm$		$\boldsymbol{L}_c nm$	$\boldsymbol{L}''nmm$		$\boldsymbol{L}''nm$
32	$\boldsymbol{L}_m nmm$		$\boldsymbol{L}_m nm$	$\boldsymbol{L}''ncc$		$\boldsymbol{L}''nc$
33	$\boldsymbol{L}_s nmm$			$\boldsymbol{L}''n_{n/2}mc$		
34	$\boldsymbol{L}nmm\,1^*$		$\boldsymbol{L}nm\,1^*$	$\boldsymbol{L}nmm$		$\boldsymbol{L}nm$
35	$\boldsymbol{L}ncc$		$\boldsymbol{L}nc$		Ordinary group $\boldsymbol{L}^{(7)} = \boldsymbol{T}'(\frac{a}{2})\boldsymbol{C}_n$	
36	$\boldsymbol{L}nc^*c^*$		$\boldsymbol{L}nc^*$	$\boldsymbol{L}n$		$\boldsymbol{L}n$
37	$\boldsymbol{L}n^*cc$			$\boldsymbol{L}\frac{n}{2}cc$, $\boldsymbol{L}\frac{n}{2}c$		
38	$\boldsymbol{L}ncc\,1^*$		$\boldsymbol{L}nc\,1^*$	$\boldsymbol{L}ncc$		$\boldsymbol{L}nc$
39		$\boldsymbol{L}2n_n mc$			Ordinary group $\boldsymbol{L}^{(8)} = \boldsymbol{T}^1_{2n}(\frac{a}{2})\boldsymbol{C}_{n\mathrm{v}}$	
40		$\boldsymbol{L}2n_n m^*c^*$			$\boldsymbol{L}2n_n$	
41		$\boldsymbol{L}2n_n^* mc^*$		$\boldsymbol{L}nmm$		$\boldsymbol{L}nm$
42		$\boldsymbol{L}2n_n^* m^*c$		$\boldsymbol{L}ncc$		$\boldsymbol{L}nc$
43		$\boldsymbol{L}2n_n mc\,1^*$			$\boldsymbol{L}2n_n mc$	
44	$\boldsymbol{L}\overline{2n}2m$		$\boldsymbol{L}\overline{n}m$		Ordinary group $\boldsymbol{L}^{(9)} = \boldsymbol{T}(a)\boldsymbol{D}_{n\mathrm{d}}$	
45	$\boldsymbol{L}\overline{2n}2^*m^*$		$\boldsymbol{L}\overline{n}m^*$	$\boldsymbol{L}\overline{2n}$		$\boldsymbol{L}\overline{n}$
46	$\boldsymbol{L}\overline{2n}^*2m^*$		$\boldsymbol{L}\overline{n}^*m^*$	$\boldsymbol{L}n22$		$\boldsymbol{L}n2$

Table 6.2 (continued)

	Magnetic group n even	n odd	Halving subgroup n even	n odd
47	$L\overline{2n}^*2^*m$	$L\bar{n}^*m$	$Lnmm$	Lnm
48	$L_c\overline{2n}2m$	$L_c\bar{n}m$	$L''\overline{2n}2m$	$L''\bar{n}m$
49	$L_m\overline{2n}2m$	$L_m\bar{n}m$	$L''\overline{2n}2c$	$L''\bar{n}c$
50	$L\overline{2n}2m\,1^*$	$L\bar{n}m\,1^*$	$L\overline{2n}2m$	$L\bar{n}m$
51	$L\overline{2n}2c$	$L\bar{n}c$	Ordinary group $L^{(10)} = T'(\frac{a}{2})S_{2n}$	
52	$L\overline{2n}2^*c^*$	$L\bar{n}c^*$	$L\overline{2n}$	$L\bar{n}$
53	$L\overline{2n}^*2c^*$	$L\bar{n}^*c^*$	$Ln22$	$Ln2$
54	$L\overline{2n}^*2^*c$	$L\bar{n}^*c$	$Lncc$	Lnc
55	$L\overline{2n}2c\,1^*$	$L\bar{n}c\,1^*$	$L\overline{2n}2c$	$L\bar{n}c$
56	Ln/mmm	$L\overline{2n}2m$	Ordinary group $L^{(11)} = T(a)D_{nh}$	
57	Ln/mm^*m^*	$L\overline{2n}2^*m^*$	Ln/m	$L\overline{2n}$
58	$Ln/m^*m^*m^*$	$L\overline{2n}^*2m^*$	$Ln22$	$Ln2$
59	Ln/m^*mm	$L\overline{2n}^*2^*m$	$Lnmm$	Lnm
60	Ln^*/m^*mm^*		$L\overline{2\frac{n}{2}}2m$, $L\overline{\frac{n}{2}}m$	
61	Ln^*/mmm^*		$L\frac{n}{2}/mmm$, $L\overline{2\frac{n}{2}}2m$	
62	L_cn/mmm	$L_c(\overline{2n})2m$	$L''n/mmm$	$L''(\overline{2n})2m$
63	L_sn/mmm		$L''n_{n/2}/mcm$	
64	L_mn/mmm	$L_m(\overline{2n})2m$	$L''n/mcc$	$L''(\overline{2n})2c$
65	$Ln/mmm\,1^*$	$L\overline{2n}2m\,1^*$	Ln/mmm	$L\overline{2n}2m$
66	Ln/mcc	$L\overline{2n}2c$	Ordinary group $L^{(12)} = T'(\frac{a}{2})C_{nh}$	
67	Ln/mc^*c^*	$L\overline{2n}2^*c^*$	Ln/m	$L\overline{2n}$
68	$Ln/m^*c^*c^*$	$L\overline{2n}^*2c^*$	$Ln22$	$Ln2$
69	Ln/m^*cc	$L\overline{2n}^*2^*c$	$Lncc$	Lnc
70	Ln^*/m^*cc^*		$L\overline{2\frac{n}{2}}2c$, $L\overline{\frac{n}{2}}c$	
71	Ln^*/mcc^*		$L\frac{n}{2}/mcc$, $L\overline{2\frac{n}{2}}2c$	
72	$Ln/mcc\,1^*$	$L\overline{2n}2c\,1^*$	Ln/mcc	$L\overline{2n}2c$
73	$L2n_n/mcm$		Ordinary group $L^{(13)} = T^1_{2n}(\frac{a}{2})D_{nh}$	
74	$L2n_n/mc^*m^*$		$L2n_n/m$	
75	$L2n_n/m^*c^*m^*$		$L2n_n22$	
76	$L2n_n/m^*cm$		$L2n_nmc$	
77	$L2n_n^*/m^*c^*m$		$L\overline{2n}2m$	$L\bar{n}m$
78	$L2n_n^*/m^*cm^*$		$L\overline{2n}2c$	$L\bar{n}c$
79	$L2n_n^*/mc^*m$		Ln/mmm	$L\overline{2n}2m$
80	$L2n_n^*/mcm^*$		Ln/mcc	$L\overline{2n}2c$
81	$L2n_n/mcm\,1^*$		$L2n_n/mcm$	

[a] Allowed only one of $Lq_p^*(a)$, $L_cq_p(2a)$, or $L_sq_p(2a)$, according to Table 6.1 for n odd.
[b] Allowed only one of $Lq_p^*(a)22'$, $L_cq_p(2a)22$, or $L_sq_p(2a)22$, according to Table 6.1 for n odd.

6.2 Co-representations

When in the Schrödinger equation $i\hbar\frac{\partial\psi}{\partial t} = H\psi$ for a conservative system (i.e., the hamiltonian H is time independent) t is substituted by $-t$, which is the effect of the action of the time reversal, the minus sign appears at the left. The same effect has conjugation, meaning that the time reversal is represented by conjugation K. It is an antilinear operator, and therefore the halving subgroup $\boldsymbol{L}$ of the purely geometrical operations is represented by linear operators, while antilinear operators represent the coset $\Theta\ell^*\boldsymbol{L}$ in the form $K D(\ell'\ell)$. Such a representation is called linear–antilinear representation of the magnetic group, while matrices obtained by omitting the conjugation operator K are known as *co-representation*.

6.2.1 Irreducible Co-representations

Theory of co-representations is quite similar to that of ordinary representations. As far as the line groups are considered, the irreducible co-representations are known [4, 5]. The method of construction, the so-called $*$-*induction*, uses irreducible representations of the halving ordinary line group $\boldsymbol{L}$ (for a brief review see Appendix D.2).

Its main part is Wigner's classification of the irreducible representations of the geometric subgroup $\boldsymbol{L}'$ into three types. Precisely, $D(\boldsymbol{L})$ is an irreducible representation of the *first* or *second kind* if it is equivalent to its $*$-ℓ'-conjugated representation $D^*_{\ell'}(\ell) = D^*(\ell'^{-1}\ell\ell')$, and the matrix A realizing this equivalence satisfies $AA^* = D(\ell'^2)$ and $AA^* = -D(\ell'^2)$, respectively. If $D(\boldsymbol{L})$ and $D^*_{\ell'}(\boldsymbol{L})$ are not equivalent, the representation is of the *third kind*. Each representation of the first kind generates one co-representation of $\boldsymbol{L}^*(\boldsymbol{L})$ of the same dimension; as for the second kind, this is also true; only the generated co-representation is of the doubled dimension. For the third kind, the pair of $*$-ℓ-conjugated representations of $\boldsymbol{L}$ generates single co-representation of $\boldsymbol{L}^*(\boldsymbol{L})$ of the doubled dimension.

6.2.2 Gray Groups and Reality of Representations

While black-and-white groups are related to particular systems with specific ordering of ionic spins, gray groups refer to a considerable wide class of systems being invariant under the time reversal. In particular, this includes subsystems, like phonons or electrons, with dynamics described by real hamiltonian, i.e., hamiltonian which commutes with the time reversal. Therefore, here we restrict our attention to gray groups only.

It turns out that co-representations of gray groups are neatly related to the reality of the halving subgroup representations namely, when ℓ^* is identity, $*$-ℓ^*-conjugation is actually $*$-conjugation, and Wigner's criterion essentially examines how the representation and the conjugated one are related. Precisely, when there is A such that $D^*(\ell) = A^{-1}D^*(\ell)A$ for each ℓ, then either $AA^* = I$ (identity matrix) or $AA^* = -I$, i.e., the representation is either of the first (real) or of the second (quaternionic) kind. In all other cases, the representation is of the third (complex) kind. Only the representations of the first kind have real matrices. The second-kind representations have only real characters, while both characters and matrices of the third class representations are complex.

6.2.3 Real or Physical Representations

In many physical problems there is a need to deal with real irreducible representations only. For example, analysis of vibronic instability (Sect. 8.3.4) is based on such representations, dimension of which gives insight into the systematic degeneracy of the energies (Sect. 8.1) if hamiltonian is a real operator (this is the most frequent situation when spin can be neglected and magnetic field absent). Their importance in physics is the reason for which these representations are also called *physically irreducible*.

Construction of the co-representations of the gray line groups automatically gives physically irreducible representations of the ordinary line groups. Precisely, the matrices $\bar{D}(\boldsymbol{L})$ representing halving subgroup are real representations, and for the complete set of irreducible co-representations of $\boldsymbol{L} + \Theta\boldsymbol{L}$, these matrices give the complete set of irreducible real representations of $\boldsymbol{L}$. Thus, Theorem 2, when applied to gray groups gives direct and simple method for construction of physically irreducible representations. Furthermore, it is only necessary to classify the representations according to their kind. Then, the explicit form of the irreducible representation is easy to found. This classification is given in caption of Tables 4.1–4.13 listing irreducible representations of the line groups (representations of the second and third kind are singled out, assuming that all others are of the first kind).

6.3 Application: Spin Ordering

As a kind of angular momentum, spin is reversed by time reversal. Therefore, if the ions are not spin neutral, it is convenient to use magnetic groups to describe such systems, i.e., magnetic groups and their co-representations take role of the groups and representations. Still, in many cases it is more convenient to work with halving subgroup of spatial transformations and incorporate time reversal a posteriori. Therefore, applications of magnetic line groups will be illustrated here only by an example of ordering of spins, but it should be stressed out that all examples of usage of line groups commented in Chap. 8 are easily adapted to magnetic groups.

To illustrate magnetic ordering in quasi-one-dimensional systems, we consider orbit a_1 of the group $\boldsymbol{L}^5_{24}(f)\boldsymbol{C}_4$ (the first conformation class in Fig. 3.2). The parameters of this group are

$$\boldsymbol{L}24_{20}: \quad q = 24, \quad r = 5, \quad n = 4, \quad p = 20, \quad \tilde{q} = 6, \quad \tilde{p} = 5, \quad A = 6. \tag{6.7}$$

As n is even there are three halving subgroups, with parameters given in Table 6.1 (convention C0):

$$\boldsymbol{L}24^*_{20}: \quad q = 12, \quad r = 5, \quad n = 4, \quad p = 8, \quad \tilde{q} = 3, \quad \tilde{p} = 2; \tag{6.8a}$$

$$\boldsymbol{L}_c24_{20}: \quad q = 24, \quad r = 5, \quad n = 2, \quad p = 10, \quad \tilde{q} = 12, \quad \tilde{p} = 5; \tag{6.8b}$$

$$\boldsymbol{L}_s24_{20}: q = 24, \quad r = 11, \quad n = 2, \quad p = 22, \quad \tilde{q} = 12, \quad \tilde{p} = 11. \tag{6.8c}$$

While fractional translations are unchanged in the last two cases and doubled in the first one, period is unchanged in the first case and doubled in the remaining ones. In Fig. 6.1 we illustrate possible spin orderings corresponding to these groups.

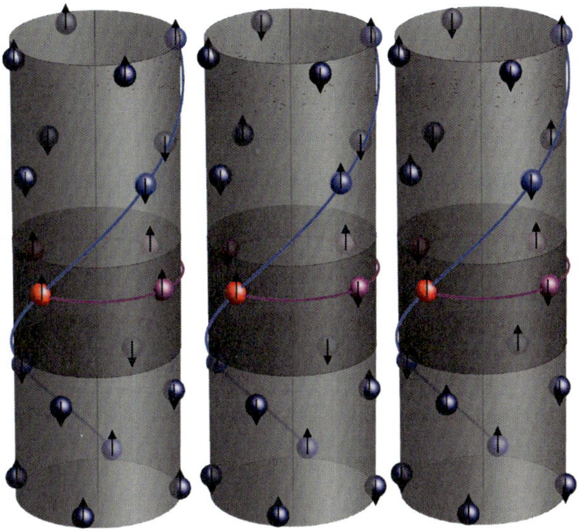

Fig. 6.1 Spin arrangements (*arrows*) with symmetry described by the magnetic groups (6.8). Depicted helical and rotational symmetries refer to the atoms without spin

References

1. W. Opechowski, *Crystallographic and Metacrystallographic Groups* (North-Holland, Amsterdam, 1986)
2. M. Damnjanović, M. Vujičić, J. Phys. A **14**, 1055 (1981)
3. M. Damnjanović, M. Vujičić, Phys. Rev. B **25**, 6987 (1982) 3
4. M. Damnjanović, I. Milošević, M. Vujičić, Phys. Rev. B **39**, 4610 (1989)
5. M. Damnjanović, I. Milošević, Phys. Rev. B **43**, 13482 (1991)

Chapter 7
Vibrational Analysis

Abstract In 1930 Wigner showed that symmetry offers efficient method for systematic classification of normal vibrational modes, which are valuable tool in many physical problems (e.g., Raman or infrared activity, vibronic (in)stability). This task has been completed for small molecules (with the help of the point groups), quasi-one-dimensional systems (using line groups), and layers (utilizing diperiodic groups). For three-dimensional crystals (space groups) only partial results exist in literature. Here we discuss normal modes of the systems with line group symmetry. The results are also applicable to three-dimensional crystals, since some of the line groups are subgroups of the relevant space groups. Classification of normal modes will be performed for all the orbits of the line groups. In fact, there are only several orbit types for each line group, and any system with a line group symmetry consists of such simple subsystems. Symmetry assignation of normal modes is achieved through reduction of the dynamical representation of the system onto its irreducible components. This representation is sum of the dynamical representations of the constituting orbits. Therefore, with decompositions of the dynamical representations of the orbits presented here, the result for any system can be easily obtained by summation over the orbits included in the considered system.

7.1 Dynamical Representation

Normal vibrational modes of the system S, having the symmetry group $\boldsymbol{L}$ and containing $\mid S \mid$ atoms, are found by the well-known Wigner's method [1]. Its main part is the reduction of the *dynamical representation* [2] $D^{\mathrm{dyn}}(\boldsymbol{L})$ to the irreducible components. In order to construct $D^{\mathrm{dyn}}(\boldsymbol{L})$, a basis $\{e_x^s, e_y^s, e_z^s\}$ is associated to each atom s of S. All these vectors span three $\mid S \mid$-dimensional displacements space of the collective modes of the system. Action of the group in this space is given by the matrix representation $D^{\mathrm{dyn}}(\boldsymbol{L}) = S(\boldsymbol{L}) \otimes D^{\mathrm{p}}(\boldsymbol{L})$. Here, $D^{\mathrm{p}}(\boldsymbol{L})$ is (three-dimensional) *polar-vector representation* (Sect. 5.3) of the symmetry group, while $S(\boldsymbol{L})$ is $\mid S \mid$-dimensional *permutational representation* of $\boldsymbol{L}$, manifesting the action [3] of $\boldsymbol{L}$ on S: to atom s corresponds sth row and sth column of the matrix $S(\ell)$ with the matrix elements $S_{s's}(\ell) = \delta_{s',\ell s}$ (equal to 1 if $\ell s = s'$ and 0 otherwise). Thus, this is *permutational matrix*, having in each row and each column

Damnjanović, M., Milošević, I.: *Vibrational Analysis*. Lect. Notes Phys. **801**, 95–111 (2010)
DOI 10.1007/978-3-642-11172-3_7

single nonzero element equal to 1. An important property of the representation $S(\boldsymbol{L})$ is that it is automatically given in the partially reduced form $S(\boldsymbol{L}) = \oplus_{i=1}^{K} N_i S_i(\boldsymbol{L})$ for arbitrary system $\boldsymbol{L}[N_1 S_1, \ldots, N_K S_K]$ containing several orbits (symmetry notation, Sect. 3.5). Then the dynamical representation is additive and reduces over the orbits:

$$D_S^{\mathrm{dyn}}(\boldsymbol{L}) = \oplus_{i=1}^{K} N_i D_{S_i}^{\mathrm{dyn}}(\boldsymbol{L}). \tag{7.1}$$

This manifests that group acts on the orbits independently and in particular that $D_S^{\mathrm{dyn}}(\boldsymbol{L})$ does not depend on the relative position of the orbits of S. Hence, with the orbits of the line groups derived in Sect. 3.1, we construct and reduce $D_S^{\mathrm{dyn}}(\boldsymbol{L})$ to the irreducible components for each orbit S_i. Precisely, according to (7.1), the frequency number f^λ of the irreducible representation $D^{(\lambda)}(\boldsymbol{L})$ is sum of the orbital ones, f_i^λ:

$$D_S^{\mathrm{dyn}}(\boldsymbol{L}) = \oplus_\lambda f^\lambda D^{(\lambda)}(\boldsymbol{L}), \quad f^\lambda = \sum_{i=1}^{K} N_i f_i^\lambda. \tag{7.2}$$

In order to find decomposition onto the irreducible components of an orbital dynamical representation, i.e., frequency numbers f_i^λ, we use modified group projector technique for induced representations (Appendix F). Indeed, the dynamical representation of each orbit S_i is equivalent to the induced [4] vector representation of the orbit stabilizer [3] $\boldsymbol{L}_i$ onto the whole group: $D_{S_i}^{\mathrm{dyn}}(\boldsymbol{L}) = D^{\mathrm{p}}(\boldsymbol{L}_i \uparrow \boldsymbol{L})$. Then the frequency number f_i^λ is equal to the intertwining number of the representations $D^{(\lambda)}(\boldsymbol{L} \downarrow \boldsymbol{L}_i)$ (subduced irreducible representation on the stabilizer) and $D^{\mathrm{p}}(\boldsymbol{L}_i)$:

$$f_i^\lambda = \frac{1}{\mid \boldsymbol{L}_i \mid} \sum_{\ell \in \boldsymbol{L}_i} d^{\mathrm{p}}(\ell) d^{(\lambda)}(\ell). \tag{7.3}$$

Here, $d^{\mathrm{p}}(\ell)$ and $d^{(\lambda)}(\ell)$ are characters (sum of the diagonal elements) of the matrices $D^{\mathrm{p}}(\ell)$ and $D^{(\lambda)}(\ell)$.

With the data from Table 3.1, this expression is systematically calculated for all the orbits and irreducible representations of each line group and the results are presented in the Tables 7.1–7.13. To include incommensurate systems, helical quantum numbers are used for the families 1 and 5.

Table 7.1 Dynamical representations of the orbits of the line groups $\boldsymbol{L}^{(1)} = \boldsymbol{T}_Q(f)\boldsymbol{C}_n$. Sum is over $\tilde{k} \in (-\frac{\pi}{f}, \frac{\pi}{f}]$, $\tilde{m} \in (-\frac{n}{2}, \frac{n}{2}]$

Orbit	point	Dynamical representation
a_1	(ρ, φ, z)	$3\sum_{\tilde{k},\tilde{m}} {}_{\tilde{k}}A_{\tilde{m}}$
b_1	$(0, 0, z)$	$\sum_{\tilde{k}} ({}_{\tilde{k}}A_0 + {}_{\tilde{k}}A_1 + {}_{\tilde{k}}A_{-1})$

Table 7.2 Dynamical representations of the orbits of the line groups $\boldsymbol{L}^{(2)} = \boldsymbol{T}(a)\boldsymbol{S}_{2n}$. Sum is over $k \in (0, \frac{\pi}{a})$, $m \in (-\frac{n}{2}, \frac{n}{2}]$

Orbit	point	Dynamical representation
a_1	(ρ, φ, z)	$3\sum_m ({}_0A_m^- + {}_0A_m^+ + {}_\pi A_m^- + {}_\pi A_m^+) + 6\sum_{k,m} {}_kE_m$
b_1	$(0, 0, z)$	$n = 1:\ 3({}_0A_0^- + {}_0A_0^+ + {}_\pi A_0^- + {}_\pi A_0^+) + 6\sum_k {}_kE_0$
		$n = 2:\ {}_0A_0^- + {}_0A_0^+ + {}_\pi A_0^- + {}_\pi A_0^+ + 2({}_0A_1^- + {}_0A_1^+ + {}_\pi A_1^- + {}_\pi A_1^+) + \sum_k (2{}_kE_0 + 4{}_kE_1)$
		$n > 2:\ {}_0A_0^- + {}_0A_0^+ + {}_\pi A_0^- + {}_\pi A_0^+ + {}_0A_1^- + {}_0A_1^+ + {}_0A_{-1}^- + {}_0A_{-1}^+ + {}_\pi A_1^- + {}_\pi A_1^+ + {}_\pi A_{-1}^- + {}_\pi A_{-1}^+ + 2\sum_k ({}_kE_0 + {}_kE_1 + {}_kE_{-1})$
b_2	$(0, 0, \frac{a}{2})$	$n = 1:\ 3({}_0A_0^- + {}_\pi A_0^+) + 3\sum_k {}_kE_0$
		$n = 2:\ {}_0A_0^- + {}_\pi A_0^+ + {}_0A_1^+ + {}_\pi A_1^- + {}_0A_1^- + {}_\pi A_1^+ + \sum_k ({}_kE_0 + 2{}_kE_1)$
		$n > 2:\ {}_0A_0^- + {}_\pi A_0^+ + {}_0A_1^+ + {}_0A_{-1}^+ + {}_\pi A_1^- + {}_\pi A_{-1}^- + \sum_k ({}_kE_0 + {}_kE_1 + {}_kE_{-1})$
c_1	$(0, 0, 0)$	$n = 1:\ 3({}_0A_0^- + {}_\pi A_0^-) + 3\sum_k {}_kE_0$
		$n = 2:\ {}_0A_0^- + {}_\pi A_0^- + {}_0A_1^+ + {}_\pi A_1^+ + {}_0A_1^- + {}_\pi A_1^- + \sum_k ({}_kE_0 + 2{}_kE_1)$
		$n > 2:\ {}_0A_0^- + {}_\pi A_0^- + {}_0A_1^+ + {}_0A_{-1}^+ + {}_\pi A_1^+ + {}_\pi A_{-1}^+ + \sum_k ({}_kE_0 + {}_kE_1 + {}_kE_{-1})$

Table 7.3 Dynamical representations of the orbits of the line groups $\boldsymbol{L}^{(3)} = \boldsymbol{T}(a)\boldsymbol{C}_{n\mathrm{h}}$. Sum is over $k \in (0, \frac{\pi}{a})$, $m \in (-\frac{n}{2}, \frac{n}{2}]$

Orbit	point	Dynamical representation
a_1	(ρ, φ, z)	$3\sum_m ({}_0A_m^- + {}_0A_m^+ + {}_\pi A_m^- + {}_\pi A_m^+) + 6\sum_{k,m} {}_kE_m$
a_2	$(\rho, \varphi, \frac{a}{2})$	$\sum_m \left[{}_0A_m^- + {}_\pi A_m^+ + 2({}_0A_m^+ + {}_\pi A_m^-)\right] + 3\sum_{k,m} {}_kE_m$
b_1	$(\rho, \varphi, 0)$	$\sum_m \left[{}_0A_m^- + {}_\pi A_m^- + 2({}_0A_m^+ + {}_\pi A_m^+)\right] + 3\sum_{k,m} {}_kE_m$
c_1	$(0, 0, z)$	$n = 1:\ 3({}_0A_0^- + {}_0A_0^+ + {}_\pi A_0^- + {}_\pi A_0^+) + 6\sum_k {}_kE_0$
		$n = 2:\ {}_0A_0^- + {}_0A_0^+ + {}_\pi A_0^- + {}_\pi A_0^+ + 2({}_0A_1^- + {}_0A_1^+ + {}_\pi A_1^- + {}_\pi A_1^+) + 2\sum_k ({}_kE_0 + 2_kE_1)$
		$n > 2:\ {}_0A_0^- + {}_0A_0^+ + {}_\pi A_0^- + {}_\pi A_0^+ + {}_0A_1^- + {}_0A_1^+ + {}_0A_{-1}^- + {}_0A_{-1}^+ + {}_\pi A_1^- + {}_\pi A_1^+ + {}_\pi A_{-1}^- + {}_\pi A_{-1}^+ + 2\sum_k ({}_kE_0 + {}_kE_1 + {}_kE_{-1})$
c_2	$(0, 0, \frac{a}{2})$	$n = 1:\ 2({}_0A_0^+ + {}_\pi A_0^-) + {}_0A_0^- + {}_\pi A_0^+ + 3\sum_k {}_kE_0$
		$n = 2:\ {}_0A_0^- + {}_\pi A_0^+ + 2({}_0A_1^+ + {}_\pi A_1^-) + \sum_k ({}_kE_0 + 2_kE_1)$
		$n > 2:\ {}_0A_0^- + {}_\pi A_0^+ + {}_0A_1^+ + {}_0A_{-1}^+ + {}_\pi A_1^- + {}_\pi A_{-1}^- + \sum_k ({}_kE_0 + {}_kE_1 + {}_kE_{-1})$
d_1	$(0, 0, 0)$	$n = 1:\ 2({}_0A_0^+ + {}_\pi A_0^+) + {}_0A_0^- + {}_\pi A_0^- + 3\sum_k {}_kE_0$
		$n = 2:\ {}_0A_0^- + {}_\pi A_0^- + 2({}_0A_1^+ + {}_\pi A_1^+) + \sum_k ({}_kE_0 + 2_kE_1)$
		$n > 2:\ {}_0A_0^- + {}_\pi A_0^- + {}_0A_1^+ + {}_0A_{-1}^+ + {}_\pi A_1^+ + {}_\pi A_{-1}^+ + \sum_k ({}_kE_0 + {}_kE_1 + {}_kE_{-1})$

Table 7.4 Dynamical representations of the orbits of the line groups $\boldsymbol{L}^{(4)} = \boldsymbol{T}^1_{2n}(\frac{a}{2})\boldsymbol{C}_{n\mathrm{h}}$. Sum is over $k \in (0, \frac{\pi}{a})$; in the primed sums $m \in (0, n]$ and otherwise $m \in (-n, n]$

Orbit	point	Dynamical representation
a_1	(ρ, φ, z)	$3\sum_m ({}_0A^+_m + {}_0A^-_m) + 6\sum_{k,m} {}_kE_m + 6\sum_m' {}_\pi E_m$
b_1	$(\rho, \varphi, 0)$	$\sum_m ({}_0A^-_m + 2{}_0A^+_m) + 3\sum_{k,m} {}_kE_m + 3\sum_m' {}_\pi E_m$
c_1	$(0, 0, z)$	$n = 1:\ 3({}_0A^-_0 + {}_0A^+_0 + {}_0A^-_1 + {}_0A^+_1) + 6\sum_k ({}_kE_0 + {}_kE_1) + 6{}_\pi E^0_1$
		$n > 1:\ {}_0A^-_0 + {}_0A^+_0 + {}_0A^-_1 + {}_0A^+_1 + {}_0A^-_{-1} + {}_0A^+_{-1} + {}_0A^-_{n-1} + {}_0A^+_{n-1} + {}_0A^-_{-n+1} + {}_0A^+_{-n+1} + {}_0A^+_n + {}_0A^-_n + 2({}_\pi E_{n-1} + {}_\pi E_1 + {}_\pi E_n) +$ $+2\sum_k ({}_kE_0 + {}_kE_1 + {}_kE_{-1} + {}_kE_n + {}_kE_{n-1} + {}_kE_{-n+1})$
c_2	$(0, 0, \frac{a}{4})$	$n = 1:\ 3({}_0A^+_1 + {}_0A^-_0) + 3\sum_k ({}_kE_0 + {}_kE_1) + 3{}_\pi E^0_1$
		$n > 1:\ {}_0A^-_0 + {}_0A^+_1 + {}_0A^+_{-1} + {}_0A^+_n + {}_0A^-_{n-1} + {}_0A^-_{-n+1} + {}_\pi E_{n-1} + {}_\pi E_1 + {}_\pi E_n + \sum_k ({}_kE_0 + {}_kE_1 + {}_kE_{-1} + {}_kE_n + {}_kE_{n-1} + {}_kE_{-n+1})$
d_1	$(0, 0, 0)$	$n = 1:\ 2({}_0A^+_0 + {}_0A^+_1) + {}_0A^-_0 + {}_0A^-_1 + 3\sum_k ({}_kE_0 + {}_kE_1) + {}_\pi E^0_1$
		$n > 1:\ {}_0A^-_0 + {}_0A^+_1 + {}_0A^+_{-1} + {}_0A^-_n + {}_0A^+_{n-1} + {}_0A^+_{-n+1} + {}_\pi E_{n-1} + {}_\pi E_1 + {}_\pi E_n + \sum_k ({}_kE_0 + {}_kE_1 + {}_kE_{-1} + {}_kE_n + {}_kE_{n-1} + {}_kE_{-n+1})$

Table 7.5 Dynamical representations of the orbits of the line groups $\boldsymbol{L}^{(5)} = \boldsymbol{T}_Q(f)\boldsymbol{D}_n$. Sum is over $\tilde{k} \in (0, \frac{\pi}{f})$; in the primed sums $\tilde{m} \in (0, \frac{n}{2})$ and otherwise $\tilde{m} \in (-\frac{n}{2}, \frac{n}{2})$

Orbit	point	Dynamical representation
a_1	(ρ, φ, z)	$3({}_0A_0^+ + {}_0A_0^- + {}_\pi A_0^+ + {}_\pi A_0^-) + 6{\sum_{\tilde{m}}}'({}_\pi E_{\tilde{m}} + {}_0E_{\tilde{m}}) + 6\sum_{\tilde{k},\tilde{m}} {}_\pi A^-_{\frac{n}{2}} +$
	n even	$+3({}_0A^+_{\frac{n}{2}} + {}_0A^-_{\frac{n}{2}} + {}_\pi A^+_{\frac{n}{2}} + {}_\pi A^-_{\frac{n}{2}}) + 6\sum_{\tilde{k}} {}_{\tilde{k}}E_{\frac{n}{2}}$
a_2	$(\rho, \frac{\pi}{Q}, \frac{f}{2})$	${}_0A_0^+ + {}_\pi A_0^- + 2({}_0A_0^- + {}_\pi A_0^+) + 3{\sum_{\tilde{m}}}'({}_\pi E_{\tilde{m}} + {}_0E_{\tilde{m}}) + 3\sum_{\tilde{k},\tilde{m}} {}_\pi A^-_{\frac{n}{2}} +$
	n even	$+{}_0A^+_{\frac{n}{2}} + {}_\pi A^-_{\frac{n}{2}} + 2({}_0A^-_{\frac{n}{2}} + {}_\pi A^+_{\frac{n}{2}}) + 3\sum_{\tilde{k}} {}_{\tilde{k}}E_{\frac{n}{2}}$
a_3^*	$(\rho, \frac{\pi}{Q} - \frac{\pi}{n}, \frac{f}{2})$	${}_0A_0^+ + {}_0A^-_{\frac{n}{2}} + {}_\pi A_0^- + {}_\pi A^+_{\frac{n}{2}} + 2({}_0A_0^- + {}_0A^+_{\frac{n}{2}} + {}_\pi A_0^+ + {}_\pi A^-_{\frac{n}{2}}) + 3\sum_{\tilde{k}} {}_{\tilde{k}}E_{\frac{n}{2}} + 3{\sum_{\tilde{m}}}'({}_\pi E_{\tilde{m}} + {}_0E_{\tilde{m}}) + 3\sum_{\tilde{k},\tilde{m}} {}_\pi A^-_{\frac{n}{2}}$
b_1	$(\rho, 0, 0)$	${}_0A_0^+ + {}_\pi A_0^+ + 2({}_0A_0^- + {}_\pi A_0^-) + 3{\sum_{\tilde{m}}}'({}_\pi E_{\tilde{m}} + {}_0E_{\tilde{m}}) + 3\sum_{\tilde{k},\tilde{m}} {}_\pi A^-_{\frac{n}{2}} +$
	n even	$+{}_0A^+_{\frac{n}{2}} + {}_\pi A^+_{\frac{n}{2}} + 2({}_0A^-_{\frac{n}{2}} + {}_\pi A^-_{\frac{n}{2}}) + 3\sum_{\tilde{k}} {}_{\tilde{k}}E_{\frac{n}{2}}$
c_1^*	$(\rho, \frac{\pi}{n}, 0)$	${}_0A_0^+ + {}_0A^-_{\frac{n}{2}} + {}_\pi A_0^+ + {}_\pi A^-_{\frac{n}{2}} + 2({}_0A_0^- + {}_0A^+_{\frac{n}{2}} + {}_\pi A_0^- + {}_\pi A^+_{\frac{n}{2}}) + 3\sum_{\tilde{k}} {}_{\tilde{k}}E_{\frac{n}{2}} + 3{\sum_{\tilde{m}}}'({}_\pi E_{\tilde{m}} + {}_0E_{\tilde{m}}) + 3\sum_{\tilde{k},\tilde{m}} {}_\pi A^-_{\frac{n}{2}}$
d_1	$(0, 0, z)$	$n = 1:\ 6\sum_{\tilde{k}} {}_{\tilde{k}}E_0 + 3({}_0A_0^+ + {}_0A_0^- + {}_\pi A_0^+ + {}_\pi A_0^-)$
		$n = 2:\ \sum_{\tilde{k}}(2_{\tilde{k}}E_0 + 4_{\tilde{k}}E_1) + {}_0A_0^+ + {}_0A_0^- + {}_\pi A_0^+ + {}_\pi A_0^- + 2({}_0A_1^+ + {}_0A_1^- + {}_\pi A_1^+ + {}_\pi A_1^-)$
		$n > 2:\ 2\sum_{\tilde{k}}({}_{\tilde{k}}E_0 + {}_{\tilde{k}}E_1 + {}_{\tilde{k}}E_{-1}) + {}_0A_0^+ + {}_0A_0^- + {}_\pi A_0^+ + {}_\pi A_0^- + 2({}_\pi E_1 + {}_0E_1)$
d_2	$(0, 0, \frac{f}{2})$	$n = 1:\ 3\sum_{\tilde{k}} {}_{\tilde{k}}E_0 + {}_0A_0^+ + {}_\pi A_0^- + 2({}_0A_0^- + {}_\pi A_0^+)$
		$n = 2:\ \sum_{\tilde{k}}({}_{\tilde{k}}E_0 + 2_{\tilde{k}}E_1) + {}_0A_0^- + {}_\pi A_0^+ + {}_0A_1^+ + {}_0A_1^- + {}_\pi A_1^+ + {}_\pi A_1^-$
		$n > 2:\ \sum_{\tilde{k}}({}_{\tilde{k}}E_0 + {}_{\tilde{k}}E_1 + {}_{\tilde{k}}E_{-1}) + {}_0A_0^- + {}_\pi A_0^+ + {}_\pi E_1 + {}_0E_1$
e_1	$(0, 0, 0)$	$n = 1:\ 3\sum_{\tilde{k}} {}_{\tilde{k}}E_0 + {}_0A_0^+ + {}_\pi A_0^+ + 2({}_0A_0^- + {}_\pi A_0^-)$
		$n = 2:\ \sum_{\tilde{k}}({}_{\tilde{k}}E_0 + 2_{\tilde{k}}E_1) + {}_0A_0^- + {}_\pi A_0^- + {}_0A_1^+ + {}_0A_1^- + {}_\pi A_1^+ + {}_\pi A_1^-$
		$n > 2:\ \sum_{\tilde{k}}({}_{\tilde{k}}E_0 + {}_{\tilde{k}}E_1 + {}_{\tilde{k}}E_{-1}) + {}_0A_0^- + {}_\pi A_0^- + {}_\pi E_1 + {}_0E_1$

* Orbit exists for n even only.

Table 7.6 Dynamical representations of the orbits of the line groups $\boldsymbol{L}^{(6)} = \boldsymbol{T}(a)\boldsymbol{C}_{n\mathrm{v}}$. Sum is over $k \in (-\frac{\pi}{a}, \frac{\pi}{a}]$, $m \in (0, \frac{n}{2})$

Orbit	point	Dynamical representation
a_1	(ρ, φ, z)	$3\sum_k({}_kA_0 + {}_kB_0) + 6\sum_{k,m} {}_kE_m +$
	n even	$+3\sum_k({}_kA_{\frac{n}{2}} + {}_kB_{\frac{n}{2}})$
b_1	$(\rho, 0, z)$	$\sum_k(2{}_kA_0 + {}_kB_0) + 3\sum_{k,m} {}_kE_m +$
	n even	$+\sum_k(2{}_kA_{\frac{n}{2}} + {}_kB_{\frac{n}{2}})$
c_1^*	$(\rho, \frac{\pi}{n}, z)$	$\sum_k\left[2({}_kA_0 + {}_kB_{\frac{n}{2}}) + {}_kB_0 + {}_kA_{\frac{n}{2}})\right] + 3\sum_{k,m} {}_kE_m$
d_1	$(0, 0, z)$	$n = 1: \ \sum_k(2{}_kA_0 + {}_kB_0)$
		$n = 2: \ \sum_k({}_kA_0 + {}_kA_1 + {}_kB_1)$
		$n > 2: \ \sum_k({}_kA_0 + {}_kE_1)$

* Orbit exists for n even only.

Table 7.7 Dynamical representations of the orbits of the line groups $\boldsymbol{L}^{(7)} = \boldsymbol{T}'(\frac{a}{2})\boldsymbol{C}_n$. Sum is over $k \in (-\frac{\pi}{a}, \frac{\pi}{a}]$, $m \in (0, \frac{n}{2})$

Orbit	point	Dynamical representation
a_1	(ρ, φ, z)	$3\sum_k({}_kA_0 + {}_kB_0) + 6\sum_{k,m} {}_kE_m +$
	n even	$+3\sum_k({}_kA_{\frac{n}{2}} + {}_kB_{\frac{n}{2}})$
b_1	$(0, 0, z)$	$n = 1: \ 3\sum_k({}_kA_0 + {}_kB_0)$
		$n = 2: \ \sum_k[{}_kA_0 + {}_kB_0 + 2({}_kA_1 + {}_kB_1)]$
		$n > 2: \ \sum_k[{}_kA_0 + {}_kB_0 + 2{}_kE_1]$

Table 7.8 Dynamical representations of the orbits of the line groups $\boldsymbol{L}^{(8)} = \boldsymbol{T}^1_{2n}(\frac{a}{2})\boldsymbol{C}_{n\mathrm{v}}$. Sum is over $k \in (-\frac{\pi}{a}, \frac{\pi}{a}]$, $m \in (0, n)$

Orbit	point	Dynamical representation
a_1	(ρ, φ, z)	$3\sum_k({}_kA_0 + {}_kB_0 + {}_kA_n + {}_kB_n) + 6\sum_{k,m} {}_kE_m$
b_1	$(\rho, 0, z)$	$\sum_k\left[2({}_kA_0 + {}_kA_{\frac{n}{2}}) + {}_kB_0 + {}_kB_{\frac{n}{2}}\right] + 3\sum_{k,m} {}_kE_m$
d_1	$(0, 0, z)$	$n = 1: \ \sum_k[2({}_kA_0 + {}_kA_1) + {}_kB_0 + {}_kB_1]$
		$n > 1: \ \sum_k({}_kA_0 + {}_kA_{\frac{n}{2}} + {}_kE_{n-1} + {}_kE_1)$

Table 7.9 Dynamical representations of the orbits of the line groups $\boldsymbol{L}^{(9)} = \boldsymbol{T}(a)\boldsymbol{D}_{nd}$. Sum is over $k \in (0, \frac{\pi}{a})$ and $m \in (0, \frac{n}{2})$

Orbit	point	Dynamical representation
a_1	(ρ, φ, z)	$3({}_0A_0^+ + {}_0A_0^- + {}_0B_0^+ + {}_0B_0^- + {}_\pi A_0^+ + {}_\pi A_0^- + {}_\pi B_0^+ + {}_\pi B_0^-) + 6\sum_m({}_0E_m^+ + {}_0E_m^- + {}_\pi E_m^+ + {}_\pi E_m^-) + 6\sum_k({}_kE_0^A + {}_kE_0^B) + 12\sum_{k,m} {}_kG_m +$
	n even	$+6({}_0E_{\frac{n}{2}} + {}_\pi E_{\frac{n}{2}}) + 6\sum_k({}_kE_{\frac{n}{2}}^A + {}_kE_{\frac{n}{2}}^B)$
a_2	$(\rho, \frac{\pi}{2n}, \frac{a}{2})$	${}_0A_0^+ + {}_0B_0^+ + {}_\pi A_0^- + {}_\pi B_0^- + 2({}_0A_0^- + {}_0B_0^- + {}_\pi A_0^+ + {}_\pi B_0^+) + 3\sum_m({}_0E_m^+ + {}_0E_m^- + {}_\pi E_m^+ + {}_\pi E_m^-) + 3\sum_k({}_kE_0^A + {}_kE_0^B) + 6\sum_{k,m} {}_kG_m +$
	n even	$+3({}_0E_{\frac{n}{2}} + {}_\pi E_{\frac{n}{2}}) + 3\sum_k({}_kE_{\frac{n}{2}}^A + {}_kE_{\frac{n}{2}}^B)$
b_1	$(\rho, 0, z)$	$2({}_0A_0^+ + {}_0A_0^- + {}_\pi A_0^+ + {}_\pi A_0^-) + {}_\pi B_0^+ + {}_\pi B_0^- + {}_0B_0^+ + {}_0B_0^- + 3\sum_m({}_0E_m^+ + {}_0E_m^- + {}_\pi E_m^+ + {}_\pi E_m^-) + \sum_k(4{}_kE_0^A + 2{}_kE_0^B) + 6\sum_{k,m} {}_kG_m +$
	n even	$+3({}_0E_{\frac{n}{2}} + {}_\pi E_{\frac{n}{2}}) + 3\sum_k 3({}_kE_{\frac{n}{2}}^A + {}_kE_{\frac{n}{2}}^B)$
c_1	$(\rho, \frac{\pi}{2n}, 0)$	${}_0A_0^+ + {}_0B_0^+ + {}_\pi A_0^+ + {}_\pi B_0^+ + 2({}_0A_0^- + {}_0B_0^- + {}_\pi A_0^- + {}_\pi B_0^-) + 3\sum_m({}_0E_m^+ + {}_0E_m^- + {}_\pi E_m^+ + {}_\pi E_m^-) + 3\sum_k({}_kE_0^A + {}_kE_0^B) + 6\sum_{k,m} {}_kG_m +$
	n even	$+3({}_0E_{\frac{n}{2}} + {}_\pi E_{\frac{n}{2}}) + 3\sum_k({}_kE_{\frac{n}{2}}^A + {}_kE_{\frac{n}{2}}^B)$
d_1	$(0, 0, z)$	$n = 1: \ 2({}_0A_0^+ + {}_0A_0^- + {}_\pi A_0^+ + {}_\pi A_0^-) + {}_0B_0^+ + {}_0B_0^- + {}_\pi B_0^+ + {}_\pi B_0^- + \sum_k(4{}_kE_0^A + 2{}_kE_0^B)$
		$n = 2: \ {}_0A_0^+ + {}_0A_0^- + {}_\pi A_0^+ + {}_\pi A_0^- + 2\sum_k({}_kE_0^A + {}_kE_{\frac{n}{2}}^B + {}_kE_{\frac{n}{2}}^A) + 2({}_0E_1 + {}_\pi E_1)$
		$n > 2: \ {}_0A_0^+ + {}_0A_0^- + {}_\pi A_0^+ + {}_\pi A_0^- + {}_0E_1^+ + {}_0E_1^- + {}_0E_1^+ + {}_\pi E_1^- + 2\sum_k({}_kE_0^A + {}_kG_1)$
d_2	$(0, 0, \frac{a}{2})$	$n = 1: \ 2({}_0A_0^- + {}_\pi A_0^+) + {}_0B_0^+ + {}_\pi B_0^- + \sum_k(2{}_kE_0^A + {}_kE_0^B)$
		$n = 2: \ {}_0A_0^- + {}_\pi A_0^+ + \sum_k({}_kE_0^A + {}_kE_{\frac{n}{2}}^B + {}_kE_{\frac{n}{2}}^A) + {}_0E_1 + {}_\pi E_1$
		$n > 2: \ {}_0A_0^- + {}_\pi A_0^+ + {}_0E_1^+ + {}_\pi E_1^- + \sum_k({}_kE_0^A + {}_kG_1)$
e_1	$(0, 0, 0)$	$n = 1: \ 2({}_0A_0^- + {}_\pi A_0^-) + {}_0B_0^+ + {}_\pi B_0^+ + \sum_k(2{}_kE_0^A + {}_kE_0^B)$
		$n = 2: \ {}_0A_0^- + {}_\pi A_0^- + \sum_k({}_kE_0^A + {}_kE_{\frac{n}{2}}^B + {}_kE_{\frac{n}{2}}^A) + {}_0E_1 + {}_\pi E_1$
		$n > 2: \ {}_0A_0^- + {}_\pi A_0^- + {}_0E_1^+ + {}_0E_1^+ + \sum_k({}_kE_0^A + {}_kG_1)$

Table 7.10 Dynamical representations of the orbits of the line groups $\boldsymbol{L}^{(10)} = \boldsymbol{T}'(\frac{a}{2})\boldsymbol{S}_{2n}$. Sum is over $k \in (0,\pi)$, $m \in (0,\frac{n}{2})$

Orbit	point	Dynamical representation
a_1	(ρ,φ,z)	$3({}_0A_0^+ + {}_0A_0^- + {}_0B_0^+ + {}_0B_0^-) + 6_\pi E_0 + 6\sum_m({}_0E_m^+ + {}_0E_m^- + {}_\pi E_m^+ + {}_\pi E_m^-) + 6\sum_k({}_kE_0^A + {}_kE_0^B) + 12\sum_{k,m}{}_kG_m +$
	n even	$+3({}_\pi A_{\frac{n}{2}}^+ + {}_\pi A_{\frac{n}{2}}^- + {}_\pi B_{\frac{n}{2}}^+ + {}_\pi B_{\frac{n}{2}}^-) + 6_0E_{\frac{n}{2}} + 6\sum_k({}_kE_{\frac{n}{2}}^A + {}_kE_{\frac{n}{2}}^B)$
a_2	$(\rho,\frac{\pi}{2n},\frac{a}{4})$	${}_0A_0^+ + {}_0B_0^+ + 2({}_0B_0^- + {}_0A_0^-) + 3_\pi E_0 + 3\sum_m({}_0E_m^+ + {}_0E_m^- + {}_\pi E_m^+ + {}_\pi E_m^-) + 3\sum_k({}_kE_0^A + {}_kE_0^B) + 6\sum_{k,m}{}_kG_m$
	n even	$+{}_0E_{\frac{n}{2}} + {}_\pi A_{\frac{n}{2}}^+ + 2_\pi A_{\frac{n}{2}}^- + {}_\pi B_{\frac{n}{2}}^+ + 2_\pi B_{\frac{n}{2}}^-$
a_3^*	$(\rho,\frac{\pi}{2n},\frac{a}{4})$	${}_0A_0^+ + {}_0B_0^+ + 2({}_0B_0^- + {}_0A_0^-) + 3_\pi E_0 + 3\sum_m({}_0E_m^+ + {}_0E_m^- + {}_\pi E_m^+ + {}_\pi E_m^-) + 3\sum_k({}_kE_0^A + {}_kE_0^B) + 6\sum_{k,m}{}_kG_m$
		$+{}_0E_{\frac{n}{2}} + 2_\pi A_{\frac{n}{2}}^+ + {}_\pi A_{\frac{n}{2}}^- + 2_\pi B_{\frac{n}{2}}^+ + {}_\pi B_{\frac{n}{2}}^-$
b_1	$(0,0,z)$	$n=1:\ 3({}_0A_0^+ + {}_0A_0^- + {}_0B_0^+ + {}_0B_0^-) + 6_\pi E_0 + 6\sum_k({}_kE_0^A + {}_kE_0^B)$
		$n=2:\ {}_0A_0^+ + {}_0A_0^- + {}_0B_0^+ + {}_0B_0^- + 2({}_\pi E_0 + {}_\pi A_1^+ + {}_\pi A_1^- + {}_\pi B_1^+ + {}_\pi B_1^-) + 4_0E_1 + \sum_k\left[4({}_kE_{\frac{n}{2}}^B + {}_kE_{\frac{n}{2}}^A) + 2({}_kE_0^A + {}_kE_0^B)\right]$
		$n>2:\ {}_0A_0^+ + {}_0A_0^- + {}_0B_0^+ + {}_0B_0^- + 2({}_\pi E_0 + {}_0E_1^+ + {}_0E_1^- + {}_0E_1^+ + {}_\pi E_1^-) + \sum_k\left[4_kG_1 + 2({}_kE_0^A + {}_kE_0^B)\right]$
b_2	$(0,0,\frac{a}{4})$	$n=1:\ {}_0A_0^+ + {}_0B_0^+ + 2({}_0A_0^- + {}_0B_0^-) + 3_\pi E_0 + 3\sum_k({}_kE_0^A + {}_kE_0^B)$
		$n=2:\ {}_0A_0^- + {}_0B_0^- + {}_\pi E_0 + {}_\pi A_1^+ + {}_\pi A_1^- + {}_\pi B_1^+ + {}_\pi B_1^- + 2_0E_1 + \sum_k\left[2({}_kE_{\frac{n}{2}}^B + {}_kE_{\frac{n}{2}}^A) + {}_kE_0^A + {}_kE_0^B\right]$
		$n>2:\ {}_0A_0^- + {}_0B_0^- + {}_\pi E_0 + {}_0E_1^+ + {}_0E_1^- + {}_0E_1^+ + {}_\pi E_1^- + \sum_k(2_kG_1 + {}_kE_0^A + {}_kE_0^B)$
c_1	$(0,0,0)$	$n=1:\ 3({}_0A_0^- + {}_0B_0^+) + 3_\pi E_0 + 3\sum_k({}_kE_0^A + {}_kE_0^B)$
		$n=2:\ {}_0A_0^- + {}_0B_0^+ + {}_\pi E_0 + {}_\pi A_1^+ + {}_\pi A_1^- + {}_\pi B_1^+ + {}_\pi B_1^- + 2_0E_1 + \sum_k\left[2({}_kE_{\frac{n}{2}}^B + {}_kE_{\frac{n}{2}}^A) + {}_kE_0^A + {}_kE_0^B\right]$
		$n>2:\ {}_0A_0^- + {}_0B_0^+ + {}_\pi E_0 + 2_0E_1^+ + {}_0E_1^+ + {}_\pi E_1^- + \sum_k(2_kG_1 + {}_kE_0^A + {}_kE_0^B)$

* Orbit exists for n even only.

Table 7.11 Dynamical representations of the orbits of the line groups $\boldsymbol{L}^{(11)} = \boldsymbol{T}(a)\boldsymbol{D}_{nh}$. Sum is over $k \in (0, \pi)$, $m \in (0, \frac{n}{2})$

Orbit	point	Dynamical representation
a_1	(ρ, φ, z)	$3({}_0A_0^+ + {}_0A_0^- + {}_0B_0^+ + {}_0B_0^- + {}_\pi A_0^+ + {}_\pi A_0^- + {}_\pi B_0^+ + {}_\pi B_0^-) + 6\sum_k({}_kE_0^A + {}_kE_0^B) + 6\sum_m({}_\pi E_m^+ + {}_\pi E_m^- + {}_0E_m^+ + {}_0E_m^-) + 12\sum_{k,m}{}_kG_m +$
	n even	$+3({}_0A_{\frac{n}{2}}^+ + {}_0A_{\frac{n}{2}}^- + {}_0B_{\frac{n}{2}}^+ + {}_0B_{\frac{n}{2}}^- + {}_\pi A_{\frac{n}{2}}^+ + {}_\pi A_{\frac{n}{2}}^- + {}_\pi B_{\frac{n}{2}}^+ + {}_\pi B_{\frac{n}{2}}^-) + 6\sum_k({}_kE_{A_{\frac{n}{2}}} + {}_kE_{B_{\frac{n}{2}}})$
a_2	$(\rho, \varphi, \frac{a}{2})$	$2({}_0A_0^+ + {}_0B_0^+ + {}_\pi A_0^- + {}_\pi B_0^-) + {}_0A_0^- + {}_0B_0^- + {}_\pi A_0^+ + {}_\pi B_0^+ + 3\sum_k({}_kE_0^A + {}_kE_0^B) + \sum_m\left[4({}_\pi E_m^- + {}_0E_m^+) + 2({}_\pi E_m^+ + {}_0E_m^-)\right] + 6\sum_{k,m}{}_kG_m +$
	n even	$+2({}_0A_{\frac{n}{2}}^+ + {}_0B_{\frac{n}{2}}^+ + {}_\pi A_{\frac{n}{2}}^- + {}_\pi B_{\frac{n}{2}}^-) + {}_0A_{\frac{n}{2}}^- + {}_0B_{\frac{n}{2}}^- + {}_\pi A_{\frac{n}{2}}^+ + {}_\pi B_{\frac{n}{2}}^+ + 3\sum_k({}_kE_{A_{\frac{n}{2}}} + {}_kE_{B_{\frac{n}{2}}})$
b_1	$(\rho, 0, z)$	$2({}_0A_0^+ + {}_0A_0^- + +{}_\pi A_0^+ + {}_\pi A_0^-) + {}_0B_0^+ + {}_0B_0^- + {}_\pi B_0^+ + {}_\pi B_0^- + 6\sum_{k,m}{}_kG_m + \sum_k(2{}_kE_0^B + 4{}_kE_0^A) + 3\sum_m({}_\pi E_m^+ + {}_\pi E_m^- + {}_0E_m^+ + {}_0E_m^-)$
	n even	$+2({}_0A_{\frac{n}{2}}^+ + {}_0A_{\frac{n}{2}}^- + {}_\pi A_{\frac{n}{2}}^+ + {}_\pi A_{\frac{n}{2}}^-) + {}_0B_{\frac{n}{2}}^+ + {}_0B_{\frac{n}{2}}^- + {}_\pi B_{\frac{n}{2}}^+ + {}_\pi B_{\frac{n}{2}}^- + \sum_k(2{}_kE_{B_{\frac{n}{2}}} + 4{}_kE_{A_{\frac{n}{2}}})$
b_2	$(\rho, 0, \frac{a}{2})$	${}_0A_0^+ + {}_0A_0^- + {}_\pi A_0^- + {}_\pi A_0^+ + {}_0B_0^+ + {}_\pi B_0^- + \sum_k({}_kE_0^B + 2{}_kE_0^A) + \sum_m\left[2({}_\pi E_m^- + {}_0E_m^+) + {}_\pi E_m^+ + {}_0E_m^-\right] + 3\sum_{k,m}{}_kG_m +$
	n even	$+{}_0A_{\frac{n}{2}}^+ + {}_0A_{\frac{n}{2}}^- + {}_\pi A_{\frac{n}{2}}^- + {}_\pi A_{\frac{n}{2}}^+ + {}_0B_{\frac{n}{2}}^+ + {}_\pi B_{\frac{n}{2}}^- + \sum_k({}_kE_{B_{\frac{n}{2}}} + 2{}_kE_{A_{\frac{n}{2}}})$
c_1^*	$(\rho, \frac{\pi}{n}, z)$	$2({}_0A_0^+ + {}_0A_0^- + {}_0B_{\frac{n}{2}}^+ + {}_0B_{\frac{n}{2}}^- + {}_\pi B_{\frac{n}{2}}^+ + {}_\pi B_{\frac{n}{2}}^- + {}_\pi A_0^+ + {}_\pi A_0^-) + {}_0B_0^+ + {}_0B_0^- + {}_0A_{\frac{n}{2}}^+ + {}_0A_{\frac{n}{2}}^- + {}_\pi B_0^+ + {}_\pi B_0^- + {}_\pi A_{\frac{n}{2}}^+ + {}_\pi A_{\frac{n}{2}}^-$
		$+\sum_k\left[2({}_kE_0^B + {}_kE_{A_{\frac{n}{2}}}) + 4({}_kE_0^A + {}_kE_{B_{\frac{n}{2}}})\right] + 3\sum_m({}_\pi E_m^+ + {}_\pi E_m^- + {}_0E_m^+ + {}_0E_m^-) + 6\sum_{k,m}{}_kG_m$
c_2^*	$(\rho, \frac{\pi}{n}, \frac{a}{2})$	${}_0A_0^- + {}_0B_{\frac{n}{2}}^+ + {}_0B_{\frac{n}{2}}^- + {}_\pi B_{\frac{n}{2}}^+ + {}_\pi B_{\frac{n}{2}}^- + {}_\pi A_0^+ + {}_\pi A_0^- + {}_0B_0^+ + {}_0A_{\frac{n}{2}}^+ + {}_\pi B_0^- + {}_\pi A_{\frac{n}{2}}^- + 3\sum_{k,m}{}_kG_m + \sum_k\left[{}_kE_0^B + {}_kE_{A_{\frac{n}{2}}} + 2({}_kE_0^A + {}_kE_{B_{\frac{n}{2}}})\right]$
		$+\sum_m\left[2({}_\pi E_m^- + {}_0E_m^+) + {}_\pi E_m^+ + {}_0E_m^-\right] + {}_0A_0^+$
d_1	$(\rho, \varphi, 0)$	$2({}_0A_0^+ + {}_0B_0^+ + {}_\pi A_0^+ + {}_\pi B_0^+) + {}_0A_0^- + {}_0B_0^- + {}_\pi A_0^- + {}_\pi B_0^- + 3\sum_k({}_kE_0^A + {}_kE_0^B) + \sum_m\left[4({}_\pi E_m^+ + {}_0E_m^+) + 2({}_\pi E_m^- + {}_0E_m^-)\right] + 6\sum_{k,m}{}_kG_m +$
	n even	$+2({}_\pi A_{\frac{n}{2}}^+ + {}_\pi B_{\frac{n}{2}}^+ + {}_0A_{\frac{n}{2}}^+ + {}_0B_{\frac{n}{2}}^+) + {}_\pi A_{\frac{n}{2}}^- + {}_\pi B_{\frac{n}{2}}^- + {}_0A_{\frac{n}{2}}^- + {}_0B_{\frac{n}{2}}^- + 3\sum_k({}_kE_{A_{\frac{n}{2}}} + {}_kE_{B_{\frac{n}{2}}})$
e_1	$(\rho, 0, 0)$	${}_0A_0^+ + {}_0A_0^- + {}_\pi A_0^+ + {}_\pi A_0^- + {}_0B_0^+ + {}_\pi B_0^+ + \sum_k({}_kE_0^B + 2{}_kE_0^A) + 3\sum_{k,m}{}_kG_m + \sum_m\left[2({}_\pi E_m^+ + {}_0E_m^+) + {}_\pi E_m^- + {}_0E_m^-\right] +$
	n even	$+{}_0A_{\frac{n}{2}}^+ + {}_0A_{\frac{n}{2}}^- + {}_\pi A_{\frac{n}{2}}^+ + {}_\pi A_{\frac{n}{2}}^- + {}_0B_{\frac{n}{2}}^+ + {}_\pi B_{\frac{n}{2}}^+ + \sum_k({}_kE_{B_{\frac{n}{2}}} + 2{}_kE_{A_{\frac{n}{2}}})$

Table 7.11 (continued)

f_1^*	$(\rho, \frac{\pi}{n}, 0)$	${}_0A_0^- + {}_0B_{\frac{n}{2}}^+ + {}_0B_{\frac{n}{2}}^- + {}_\pi B_{\frac{n}{2}}^+ + {}_\pi B_{\frac{n}{2}}^- + {}_\pi A_0^+ + {}_\pi A_0^- + {}_0B_0^+ + {}_0A_{\frac{n}{2}}^+ + {}_\pi B_0^+ + {}_\pi A_{\frac{n}{2}}^+ + 3\sum_{k,m} {}_kG_m$
		$+\sum_k \left[{}_kE_0^B + {}_kE_{A_{\frac{n}{2}}} + 2({}_kE_0^A + {}_kE_{B_{\frac{n}{2}}})\right] + \sum_m \left[2({}_\pi E_m^+ + {}_0E_m^+) + {}_\pi E_m^- + {}_0E_m^-\right] + {}_0A_0^+$
g_1	$(0, 0, z)$	$n = 1:\ 2({}_0A_0^+ + {}_0A_0^- + {}_\pi A_0^+ + {}_\pi A_0^-) + {}_0B_0^+ + {}_0B_0^- + {}_\pi B_0^+ + {}_\pi B_0^- + \sum_k (4{}_kE_0^A + 2{}_kE_0^B)$
		$n = 2:\ {}_0A_0^- + {}_0A_0^+ + {}_\pi A_0^- + {}_\pi A_0^+ + {}_0A_1^+ + {}_0A_1^- + {}_\pi A_1^+ + {}_\pi A_1^- + {}_0B_1^+ + {}_0B_1^- + {}_\pi B_1^+ + {}_\pi B_1^- + 2\sum_k ({}_kE_0^A + {}_kE_1^A + {}_kE_1^B)$
		$n > 2:\ {}_0A_0^+ + {}_0A_0^- + {}_\pi A_0^+ + {}_\pi A_0^- + {}_0E_1^+ + {}_0E_1^- + {}_0E_1^+ + {}_\pi E_1^- + 2\sum_k ({}_kE_0^A + {}_kG_1)$
g_2	$(0, 0, \frac{a}{2})$	$n = 1:\ {}_0A_0^+ + {}_0A_0^- + {}_\pi A_0^+ + {}_\pi A_0^- + {}_0B_0^+ + {}_\pi B_0^- + \sum_k (2{}_kE_0^A + {}_kE_0^B)$
		$n = 2:\ {}_0A_0^- + {}_\pi A_0^+ + {}_0A_1^+ + {}_\pi A_1^- + {}_0B_1^+ + {}_\pi B_1^- + \sum_k ({}_kE_0^A + {}_kE_1^A + {}_kE_1^B)$
		$n > 2:\ {}_0A_0^- + {}_\pi A_0^+ + {}_0E_1^+ + {}_\pi E_1^- + \sum_k ({}_kE_0^A + {}_kG_1)$
h_1	$(0, 0, 0)$	$n = 1:\ {}_0A_0^+ + {}_0A_0^- + {}_\pi A_0^+ + {}_\pi A_0^- + {}_0B_0^+ + {}_\pi B_0^+ + \sum_k (2{}_kE_0^A + {}_kE_0^B)$
		$n = 2:\ {}_0A_0^- + {}_\pi A_0^- + {}_0A_1^+ + {}_\pi A_1^+ + {}_0B_1^+ + {}_\pi B_1^+ + \sum_k ({}_kE_0^A + {}_kE_1^A + {}_kE_1^B)$
		$n > 2:\ {}_0A_0^- + {}_\pi A_0^- + {}_0E_1^+ + {}_0E_1^+ + \sum_k ({}_kE_0^A + {}_kG_1)$

* Orbit exists for n even only.

Table 7.12 Dynamical representations of the orbits of the line groups $\boldsymbol{L}^{(12)} = \boldsymbol{T}'(\frac{a}{2})\boldsymbol{C}_{nh}$. Sum is over $k \in (0, \frac{\pi}{a})$, $m \in (0, \frac{n}{2})$

Orbit	point	Dynamical representation
a_1	(ρ, φ, z)	$3({}_0A_0^+ + {}_0A_0^- + {}_0B_0^+ + {}_0B_0^-) + 6{}_\pi E_0 + 6\sum_k({}_kE_0^A + {}_kE_0^B) + 6\sum_m({}_\pi E_m^+ + {}_\pi E_m^- + {}_0E_m^+ + {}_0E_m^-) + 12\sum_{k,m}{}_kG_m+$
	n even	$+3({}_0A_{\frac{n}{2}}^+ + {}_0A_{\frac{n}{2}}^- + {}_0B_{\frac{n}{2}}^+ + {}_0B_{\frac{n}{2}}^-) + 6{}_\pi E_{\frac{n}{2}} + 6\sum_k({}_kE_{A_{\frac{n}{2}}} + {}_kE_{B_{\frac{n}{2}}})$
a_2	$(\rho, 0, \frac{a}{4})$	$2({}_0A_0^- + {}_0B_0^+) + {}_0A_0^+ + {}_0B_0^- + 3{}_\pi E_0 + 3\sum_k({}_kE_0^A + {}_kE_0^B) + 6\sum_{k,m}{}_kG_m + 3\sum_m({}_\pi E_m^+ + {}_\pi E_m^- + {}_0E_m^+ + {}_0E_m^-)+$
	$n$even	$+2({}_0A_{\frac{n}{2}}^- + {}_0B_{\frac{n}{2}}^+) + {}_0A_{\frac{n}{2}}^+ + {}_0B_{\frac{n}{2}}^- + 3{}_\pi E_{\frac{n}{2}} + 3\sum_k({}_kE_{A_{\frac{n}{2}}} + {}_kE_{B_{\frac{n}{2}}})$
a_3^*	$(\rho, \frac{\pi}{n}, \frac{a}{4})$	$2({}_0A_0^- + {}_0B_{\frac{n}{2}}^- + {}_0B_0^+ + {}_0A_{\frac{n}{2}}^+) + {}_0A_0^+ + {}_0B_{\frac{n}{2}}^+ + {}_0B_0^- + {}_0A_{\frac{n}{2}}^- + 3({}_\pi E_0 + {}_\pi E_{\frac{n}{2}})+$ $6\sum_{k,m}{}_kG_m + 3\sum_k({}_kE_0^A + {}_kE_{A_{\frac{n}{2}}} + {}_kE_0^B + {}_kE_{B_{\frac{n}{2}}}) + 3\sum_m({}_\pi E_m^+ + {}_\pi E_m^- + {}_0E_m^+ + {}_0E_m^-)$
b_1	$(\rho, \varphi, 0)$	$2({}_0A_0^+ + {}_0B_0^+) + {}_0A_0^- + {}_0B_0^- + 3{}_\pi E_0 + 3\sum_k({}_kE_0^A + {}_kE_0^B) + \sum_m\left[3({}_\pi E_m^+ + {}_\pi E_m^-) + 4{}_0E_m^+ + 2{}_0E_m^-\right] + 6\sum_{k,m}{}_kG_m+$
	n even	$+2({}_0A_{\frac{n}{2}}^+ + {}_0B_{\frac{n}{2}}^+) + {}_0A_{\frac{n}{2}}^- + {}_0B_{\frac{n}{2}}^- + 3{}_\pi E_{\frac{n}{2}} + 3\sum_k({}_kE_{A_{\frac{n}{2}}} + {}_kE_{B_{\frac{n}{2}}})$
c_1	$(0, 0, z)$	$n = 1:\ 3({}_0A_0^+ + {}_0A_0^- + {}_0B_0^+ + {}_0B_0^-) + 6{}_\pi E_0 + 6\sum_k({}_kE_0^B + {}_kE_0^A)$
		$n = 2:\ {}_0A_0^+ + {}_0A_0^- + {}_0B_0^+ + {}_0B_0^- + 2({}_0A_1^+ + {}_0A_1^- + {}_0B_1^+ + {}_0B_1^-) + 2{}_\pi E_0 + 4{}_\pi E_1 + \sum_k\left[2({}_kE_0^B + {}_kE_0^A) + 4({}_kE_1^A + {}_kE_1^B)\right]$
		$n > 2:\ {}_0A_0^+ + {}_0A_0^- + {}_0B_0^+ + {}_0B_0^- + 2({}_\pi E_0 + {}_0E_1^+ + {}_0E_1^- + {}_0E_1^+ + {}_\pi E_1^-) + \sum_k\left[2({}_kE_0^B + {}_kE_0^A) + 4{}_kG_1\right];\ n > 2$
c_2	$(0, 0, \frac{a}{4})$	$n = 1:\ 2({}_0A_0^- + {}_0B_0^+) + {}_0A_0^+ + {}_0B_0^- + 3{}_\pi E_0 + 3\sum_k({}_kE_0^B + {}_kE_0^A)$
		$n = 2:\ {}_0A_0^- + {}_0B_0^+ + {}_0A_1^+ + {}_0A_1^- + {}_0B_1^+ + {}_0B_1^- + {}_\pi E_0 + 2{}_\pi E_1 + \sum_k\left[{}_kE_0^B + {}_kE_0^A + 2({}_kE_1^A + {}_kE_1^B)\right]$
		$n > 2:\ {}_0A_0^- + {}_0B_0^+ + {}_\pi E_0 + {}_0E_1^+ + {}_0E_1^- + {}_0E_1^+ + {}_\pi E_1^- + \sum_k({}_kE_0^B + {}_kE_0^A + 2{}_kG_1)$
d_1	$(0, 0, 0)$	$n = 1:\ 2({}_0A_0^+ + {}_0B_0^+) + {}_0A_0^- + {}_0B_0^- + 3{}_\pi E_0 + 3\sum_k({}_kE_0^B + {}_kE_0^A)$
		$n = 2:\ {}_0A_0^- + {}_0B_0^- + 2({}_0A_1^+ + {}_0B_1^+) + {}_\pi E_0 + 2{}_\pi E_1 + \sum_k\left[{}_kE_0^B + {}_kE_0^A + 2({}_kE_1^A + {}_kE_1^B)\right]$
		$n > 2:\ {}_0A_0^- + {}_0B_0^- + {}_\pi E_0 + 2{}_0E_1^+ + {}_0E_1^+ + {}_\pi E_1^- + \sum_k({}_kE_0^B + {}_kE_0^A + 2{}_kG_1)$

* Orbit exists for n even only.

Table 7.13 Dynamical representations of the orbits of the line groups $\boldsymbol{L}^{(13)} = \boldsymbol{T}^1_{2n}(\frac{a}{2})\boldsymbol{D}_{nh}$. Sum is over $k \in (0, \frac{\pi}{a})$. In the primed sum m is from $(0, \frac{n}{2})$, otherwise from $(0, n)$

Orbit	point	Dynamical representation
a_1	(ρ, φ, z)	$3({}_0A_0^+ + {}_0A_0^- + {}_0B_0^+ + {}_0B_0^- + {}_0A_n^+ + {}_0A_n^- + {}_0B_n^+ + {}_0B_n^-) + 12\sum_m{}' {}_\pi G_m + 12\sum_{k,m} {}_kG_m + 6({}_\pi E_A + {}_\pi E_B) + 6\sum_m({}_0E_m^+ + {}_0E_m^-)$
		$+6\sum_k({}_kE_0^A + {}_kE_0^B + {}_kE_n^A + {}_kE_n^B)+$
	n even	$+6({}_\pi E_{\frac{n}{2}}^+ + {}_\pi E_{\frac{n}{2}}^-)$
a_2	$(\rho, \frac{\pi}{2n}, \frac{a}{4})$	${}_0A_0^+ + {}_0B_n^+ + {}_0B_0^- + {}_0A_n^- + 2({}_0A_0^- + {}_0B_0^+ + {}_0B_n^- + {}_0A_n^+) + 3({}_\pi E_A + {}_\pi E_B + {}_\pi E_{\frac{n}{2}}^+ + {}_\pi E_{\frac{n}{2}}^-)$
		$+3\sum_k({}_kE_0^A + {}_kE_0^B + {}_kE_n^A + {}_kE_n^B) + 3\sum_m({}_0E_m^+ + {}_0E_m^-) + 6\sum_m{}' {}_\pi G_m + 6\sum_{k,m} {}_kG_m$
b_1	$(\rho, 0, z)$	$2({}_0A_0^+ + {}_0A_0^- + {}_0A_n^+ + {}_0A_n^- + {}_\pi E_B) + {}_0B_0^+ + {}_0B_0^- + {}_0B_n^+ + {}_0B_n^- + 4{}_\pi E_A + 3\sum_m({}_0E_m^+ + {}_0E_m^-)$
		$+\sum_k\left[2({}_kE_0^B + {}_kE_n^B) + 4({}_kE_0^A + {}_kE_n^A)\right] + 6\sum_m{}' {}_\pi G_m + 6\sum_{k,m} {}_kG_m+$
	n even	$+3({}_\pi E_{\frac{n}{2}}^+ + {}_\pi E_{\frac{n}{2}}^-)$
d_1	$(\rho, \varphi, 0)$	$2({}_0A_0^+ + {}_0B_0^+ + {}_0A_n^+ + {}_0B_n^+) + {}_0A_0^- + {}_0B_0^- + {}_0A_n^- + {}_0B_n^- + 3({}_\pi E_A + {}_\pi E_B) + 6\sum_{k,m} {}_kG_m + 3\sum_k({}_kE_0^A + {}_kE_0^B + {}_kE_n^A + {}_kE_n^B)$
		$+\sum_m(4{}_0E_m^+ + 2{}_0E_m^-) + 6\sum_m{}' {}_\pi G_m+$
	n even	$+3({}_\pi E_{\frac{n}{2}}^+ + {}_\pi E_{\frac{n}{2}}^-)$
e_1	$(\rho, 0, 0)$	${}_0A_0^+ + {}_0A_0^- + {}_0A_n^+ + {}_0A_n^- + {}_0B_0^+ + {}_0B_n^+ + 2{}_\pi E_A + {}_\pi E_B + 3\sum_m{}' {}_\pi G_m + \sum_k\left[{}_kE_0^B + {}_kE_n^B + 2({}_kE_0^A + {}_kE_n^A)\right]$
		$+\sum_m(2{}_0E_m^+ + {}_0E_m^-) + 3\sum_{k,m} {}_kG_m+$
	n even	$+2{}_\pi E_{\frac{n}{2}}^- + {}_\pi E_{\frac{n}{2}}^+$
f_1^*	$(\rho, \frac{\pi}{n}, 0)$	${}_0A_0^+ + {}_0A_0^- + {}_0A_n^+ + {}_0B_n^+ + {}_0B_0^+ + {}_0A_n^- + 2({}_\pi E_A + {}_\pi E_{\frac{n}{2}}^+) + {}_\pi E_B + {}_\pi E_{\frac{n}{2}}^- + 3\sum_{k,m} {}_kG_m + \sum_k\left[{}_kE_0^B + {}_kE_n^B + 2({}_kE_0^A + {}_kE_n^A)\right]$
		$+\sum_m(2{}_0E_m^+ + {}_0E_m^-) + 3\sum_m{}' {}_\pi G_m$

Table 7.13 (continued)

Orbit	point	Dynamical representation
g_1	$(0, 0, z)$	$n = 1:\ 2({}_0A_0^+ + {}_0A_0^- + {}_0A_1^+ + {}_0A_1^- + {}_\pi E_B) + {}_0B_0^+ + {}_0B_0^- + {}_0B_1^+ + {}_0B_1^- + 4{}_\pi E_A + \sum_k [4({}_kE_0^A + {}_kE_1^A) + 2({}_kE_0^B + {}_kE_1^B)]$
		$n = 2:\ {}_0A_0^+ + {}_0A_0^- + {}_0A_2^+ + {}_0A_2^+ + 2({}_\pi E_A + {}_0E_1^+ + {}_\pi E_1^- + {}_0E_1^+ + {}_0E_1^-) + \sum_k [2({}_kE_0^A + {}_kE_2^A) + 4{}_kG_1]$
		$n > 2:\ {}_0A_0^+ + {}_0A_0^- + {}_0A_n^+ + {}_0A_n^- + 2({}_\pi E_A + {}_\pi G_1) + {}_0E_1^+ + {}_0E_1^- + {}_0E_{n-1}^+ + {}_0E_{n-1}^- + 2\sum_k ({}_kE_0^A + {}_kE_n^A + {}_kG_{n-1} + {}_kG_1)$
g_2	$(0, 0, \frac{a}{4})$	$n = 1:\ {}_0A_0^+ + {}_0A_0^- + {}_0A_1^+ + {}_0A_1^- + {}_0B_0^+ + {}_0B_1^- + {}_\pi E_B + 2{}_\pi E_A + \sum_k [2({}_kE_0^A + {}_kE_1^A) + {}_kE_0^B + {}_kE_1^B]$
		$n = 2:\ {}_0A_0^- + {}_0A_2^+ + {}_\pi E_A + {}_0E_1^+ + {}_\pi E_1^- + {}_0E_1^+ + {}_0E_1^- + \sum_k ({}_kE_0^A + {}_kE_2^A + 2{}_kG_1)$
		$n > 2:\ {}_0A_0^- + {}_0A_n^+ + {}_\pi E_A + {}_0E_1^+ + {}_0E_{n-1}^- + {}_\pi G_1 + \sum_k ({}_kE_0^A + {}_kE_n^A + {}_kG_{n-1} + {}_kG_1)$
h_1	$(0, 0, 0)$	$n = 1:\ {}_0A_0^+ + {}_0A_0^- + {}_0A_1^+ + {}_0A_1^- + {}_0B_0^+ + {}_0B_1^+ + {}_\pi E_B + 2{}_\pi E_A + \sum_k [2({}_kE_0^A + {}_kE_1^A) + {}_kE_0^B + {}_kE_1^B]$
		$n = 2:\ {}_0A_0^- + {}_0A_2^+ + {}_\pi E_A + {}_0E_1^+ + {}_\pi E_1^- + 2{}_0E_1^+ + \sum_k ({}_kE_0^A + {}_kE_2^A + 2{}_kG_1)$
		$n > 2:\ {}_0A_0^- + {}_0A_n^- + {}_\pi E_A + {}_0E_1^+ + {}_0E_{n-1}^+ + {}_\pi G_1 + \sum_k ({}_kE_0^A + {}_kE_n^A + {}_kG_{n-1} + {}_kG_1)$

* Orbit exists for n even only.

7.2 Normal Modes

Frequency numbers of the irreducible components in the dynamical representations of all the orbits of the line groups are sufficient to classify normal vibrations of the systems with line group symmetry. However, to determine the exact form of normal displacements of the concrete system, the group projector technique [5, 6] is to be used. As for a single orbit S, we can again exploit the inductive structure of the displacements space and reduce the calculations to the stabilizer and to the orbit representative. However, this method cannot be directly applied to a general system consisting of several orbits of a line group, because different orbits may have different stabilizers and transversals. In other words, in such a situation the dynamical representation of the system cannot be induced from the dynamical representation of the symcell. To overcome this difficulty, a common transversal of the symcell atoms should be found. However, instead of symcell a larger set of atoms is to be involved in the induction procedure. Nevertheless, this method is the most efficient, although it strongly depends on the system, i.e., on the orbit types it consists of.

Still, the line group structure enables to find quite general method to solve the proposed task [7]. In fact, the group of the generalized translations $\boldsymbol{Z}$ generates the whole system from the monomer. Therefore, it is possible to subduce the dynamical representation onto the symmetry group $\boldsymbol{M}$ of the monomer M. Then, irrespective of the types of orbit building the polymer, cyclic group $\boldsymbol{Z}$ is the transversal for the induction from $D_M^{\text{dyn}}(\boldsymbol{M})$ to $D_S^{\text{dyn}}(\boldsymbol{L})$. The relevant Wigner operator corresponding to the mth row of the irreducible representation $D^{(\mu)}(\boldsymbol{L})$ in the total displacements space is infinite matrix with three$|M|$-dimensional blocks:

$$[P_{\mu m}^{\text{dyn}}]_{ij} = \frac{|\mu|}{|\boldsymbol{L}|}\sum_{\ell\in\boldsymbol{M}} D_{m1}^{(\mu)*}(z^i\ell z^{-j})D_M^{\text{dyn}}(\ell). \tag{7.4}$$

Block indices i and j enumerate monomers.

There are f^μ normal modes corresponding to the μth representation, and for $f^\mu > 1$, the group projector technique gives only a subspace of their linear combinations. Exact displacements can be found by diagonalization of the vibrational hamiltonian (i.e., dynamical matrix) H in this subspace. To build H parameters of the concrete system (configuration, force constants) are to be known. Hence, the procedure of finding normal modes is the following. The f^μ-dimensional subspace $\mathcal{S}_{\mu 1}$ of the eigenvectors of $P_{\mu 1}^{\text{dyn}}$ for the eigenvalue 1 is to be found.[1] Then, the first row normal modes $|\mu 1; i\rangle$ ($i = 1, \dots, f^\mu$) are the vectors of this subspace being also the eigenvectors of H; such vectors form basis of $\mathcal{S}_{\mu 1}$. The rest of the μth modes

[1] Group projector $P_{\mu 1}^{\text{dyn}}$ (only for $m = 1$ Wigner operator is projector) commutes with H, and these two operators have common eigenbasis.

are obtained by acting on this basis by the operators $P^{\rm dyn}_{\mu m}$ for $m = 2, \ldots, |\ \mu\ |$: $|\mu m; i\rangle = P^{\rm dyn}_{\mu m}\ |\mu 1; i\rangle$.

In order to get classification of purely vibrational modes, the *acoustic modes*, describing rigid body motions of the system should not be counted. Three *translational acoustic* modes correspond to purely translational degrees of freedom of arbitrary system. As for the rotational degrees, they are allowed only in the systems finite in the directions perpendicular to the rotational axis (otherwise, the atoms infinitely far away from the axis get infinite velocity). Typical systems described by line group symmetry (like nanotubes and polymers) are infinite along z-axis, but finite in other directions. Hence, rigid body rotation around z-axis becomes a *rotational* (also called *twisting*) mode [8]. However, when a line group is used as a subgroup of the space group to describe three-dimensional crystal there is no such mode; also, for *chains* (one-dimensional structure), such a mode is not degree of freedom, as rotations around z-axis do not change configuration. Translational acoustic mode corresponds to a *polar vector* and twisting mode to the z-component of *axial vector*. The irreducible components of these representations (Table 5.3) are to be subtracted from $D^{\rm dyn}_S(\boldsymbol{L})$.

Among infinitely many modes, the *symmetric* ones are of particular interest. These correspond to the ionic displacements preserving the symmetry of the considered system, i.e., they transform according to the identity representation. There are always at least two such modes: *breathing mode* (simultaneous in-phase radial displacements) and *stretching mode* (longitudinal vibrations of the ions). However, they may not be the normal modes, i.e., specific interaction of ions may couple them (mutually or with other symmetric modes) into another symmetrical eigenvector of the dynamical matrix.

7.3 Example: Polyacetylene

Trans-polyacetylene $(\mathrm{CH})_x$ (Fig. 3.4a) has symmetry described by the 13th family line group $\boldsymbol{L}2_1/mcm = \boldsymbol{T}^1_2(f)\boldsymbol{C}_1$. Carbon and hydrogen atoms form orbits of the e_1-type. We consider normal mode corresponding to the irreducible representation $D^{(\mu)} = {}_0A^-_1$ (also denoted by B_{2g} [9]). Group projector (7.4) is the only Wigner operator for one-dimensional representations: $P^{\rm dyn}_{\mu} = P^{\rm dyn}_{\mu 1C} \oplus P^{\rm dyn}_{\mu 1H}$. From Table 3.1 we see that the stabilizer of the orbit e_1 is $\boldsymbol{D}_{1\rm h} = \{e, \sigma_{xz}, \sigma_{\rm h}, U_x\}$. Matrices of the monomer dynamical representation of $\boldsymbol{D}_{1\rm h}$ are those of the polar-vector representation. Noticing that the generator of $\boldsymbol{Z}$ is $(C_2|f)$, the projector becomes $[P^{\rm dyn}_{\mu}]_{ij} = \frac{(-1)^{i+j}}{|\boldsymbol{Z}|}\begin{bmatrix}0&0&0\\0&0&0\\0&0&1\end{bmatrix}$ and the resulting normal displacements are $x_i = y_i = 0$ and $z_i = (-1)^i z_o$, i.e., the mode is alternating, distortive, and longitudinal. As each of the orbits (C, H) has such a mode, polyacetylene normal modes are the combinations of them (Fig. 7.1).

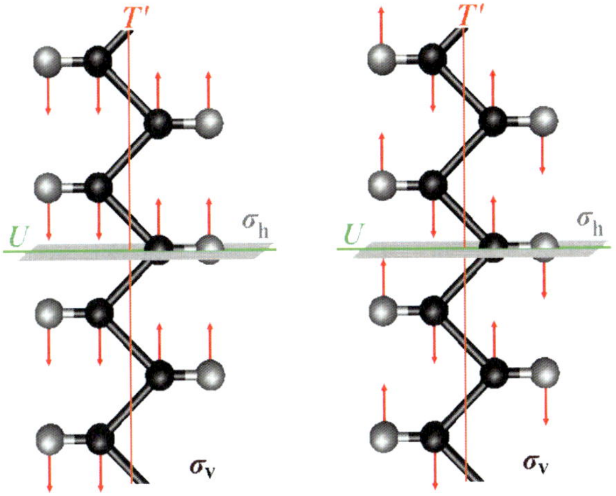

Fig. 7.1 Two *trans*-polyacetylene ${}_0A_1^-$ modes: carbon and hydrogen orbits (see Fig. 3.4) are displaced (along *arrows*) in-phase (*left*) and out-of-phase (*right*)

References

1. E. Wigner, Mat.-Fys. Kl.: Nachrichten, Akad. der Wiss. **13**, 133 (1930)
2. J.L. Birman, *Theory of Crystal Space Groups and Infra-Red and Raman Lattice Processes of Insulting Crystals* (Springer-Verlag, Berlin, 1974)
3. L. Michel, Rev. Mod. Phys **52**, 617 (1980)
4. S.L. Altmann, *Induced Representations in Crystals and Molecules* (Academic Press, London, 1977)
5. E.P. Wigner, *Group Theory and its Applications to the Quantum Mechanics of Atomic Spectra* (Academic Press, New York, 1959)
6. M. Damnjanović, I. Milošević, J. Phys. A **27**, 4859 (1994)
7. I. Milošević, M. Damnjanović, Phys. Rev. B **47**, 7805 (1993)
8. M. Born, K. Huang, *Dynamical Theory of Crystal Lattices* (Clarendon Press, Oxford, 1954)
9. J.C.W. Chien, *Polyacetylene* (Academic Press, New York, 1984)

Chapter 8
Applications

Abstract Standard framework of the symmetry application in physics has three main parts. First, the description of the system is formalized within the *state space* $\mathcal{S}$; it is a vector space, with the vectors representing states of the system. Actually, a physical problem under consideration determines the state space $\mathcal{S}$. For instance, in quantum mechanical studies, $\mathcal{S}$ is a Hilbert space of the wave functions of the system, while in the vibrational analyses it is the space of the all possible atomic displacements. The second part is symmetry group $\boldsymbol{L}$ and its representation $D(\boldsymbol{L})$ in the state space; i.e., the set of the operators $D(\ell)$ in $\mathcal{S}$ corresponding to the group elements. Finally, the main task is usually to solve the eigenproblem of the hamiltonian H, a hermitian operator in the state space which governs the dynamics. Choice of H depends on the specific problem considered, as well as on the approximations involved in the physical model. Hamiltonian and symmetry commute: $D(\ell)H = HD(\ell)$ for each symmetry transformation ℓ.

8.1 Energy Bands and Bloch Functions

The fact that hamiltonian and symmetry commute is the source of any application of symmetry of physics, namely, due to this fact, quantum numbers, i.e., irreducible representations, correspond to the eigenenergies of the hamiltonian. Recall that for commensurate systems there are two sets of quantum numbers, linear and helical. Generally, while the helical ones are favorable choice for studying the processes in the isolated chiral systems, the linear quantum numbers are easier to handle when the interaction with an external field is present. Here, the analysis is illustrated in terms of the helical quantum numbers. Naturally, by repeating the argumentation the energy bands over Brillouin zone assigned by the linear quantum numbers are obtained.

8.1.1 Eigenproblem and Bands

Quantum numbers $\tilde{k}$ and $\tilde{m}$ correspond to physical quantities of helical and z-component angular quasi-momenta. These quantities commute, and, as emerging

Damnjanović, M., Milošević, I.: *Applications*. Lect. Notes Phys. **801**, 113–141 (2010)
DOI 10.1007/978-3-642-11172-3_8

from symmetry, they commute with the hamiltonian H, too. Accordingly, they may be used to reduce eigenproblem of H; namely, all the states with given $\tilde{k}$ and $\tilde{m}$ values of these quantities form a subspace $\mathcal{S}_{\tilde{m}}(\tilde{k})$ of the system state space $\mathcal{S}$, and in each such subspace H acts independently as the reduced operator $H_{\tilde{m}}(\tilde{k})$. Obviously, total spectrum of the hamiltonian consists of the eigenvalues $\varepsilon^{i}_{\tilde{m}}(\tilde{k})$ of the sub-hamiltonians $H_{\tilde{m}}(\tilde{k})$.

8.1.1.1 Energy Bands

Taking into account continuity of physical quantities, for fixed $\tilde{k}$ and $\tilde{m}$ we can count eigenvalues $\varepsilon^{i}_{\tilde{m}}(\tilde{k})$ (i.e., order them by superscript i) in the way that $\varepsilon^{i}_{\tilde{m}}(\tilde{k})$ are continuous functions over helical Brillouin zone. Therefore, the energies are grouped into the *energy bands*: each band is assigned by the angular momentum $\tilde{m}$, and if there are several bands with same $\tilde{m}$, they are distinguished by i. Corresponding eigenvectors are denoted as $|\tilde{k}\tilde{m}, i\rangle$. It is important that this continuity also means that for each $\tilde{k}$ there is the same number of these vectors,[1] i.e., all the spaces $\mathcal{S}(\tilde{k}) = \sum_{\tilde{m}} \mathcal{S}_{\tilde{m}}(\tilde{k})$ are of the same dimension independently of $\tilde{k}$.

The additional symmetries, parities Π, do commute with the hamiltonian, but do not with the momenta. Thus, when applied to an eigenstate of H they produce another eigenstate with the same energy, giving rise to degeneracy. Precisely, $H\Pi \mid \tilde{k}\tilde{m}, i\rangle = \varepsilon^{i}_{\tilde{m}}(\tilde{k})\Pi \mid \tilde{k}\tilde{m}, i\rangle$, but generally $\Pi \mid \tilde{k}\tilde{m}, i\rangle$ and $\mid \tilde{k}\tilde{m}, i\rangle$ are physically different states, except for some special values of the quasi-momenta. These vectors span subspaces of irreducible representations. As a result, to each energy band corresponds a *representation band* (Sect. 4.3.1), dimension of which gives the degeneracy of energy band. The symmetry-adapted eigenbasis becomes $|\tilde{k}\tilde{m}\Pi; i\rangle$, with the quantum numbers specifying the irreducible representation band, and additional counter i distinguishing between the energy bands corresponding to the same representation band.

8.1.1.2 Eigenstates: Bloch Functions

Bloch theorem asserts that (quasi)particle eigenfunctions of the system translationally periodic along z-axis are of the form $\Psi_k(\boldsymbol{r}) = \mathrm{e}^{ikz}u(\boldsymbol{r})$, where $u(\boldsymbol{r})$ is invariant (i.e., periodic) function: $u(\rho, \varphi, z + ta) = u(\rho, \varphi, z)$. Obviously, the first factor defines the rule of the transformation under translations, $\Psi_k(\rho, \varphi, z + ta) = \mathrm{e}^{ikat}\Psi_k(\rho, \varphi, z)$, singling out the irreducible representation $D^{(k)}(I|ta) = \mathrm{e}^{ikat}$ of the translational group, while only the second factor specifies the function obeying this transformation rule.

In order to generalize this concept to full-line group symmetry, we note that the translation group is abelian, having only one-dimensional irreducible representations, and that this property is shared solely by the first family line groups; all the

---

[1] This is easy to prove when the space $\mathcal{S}$ is inductive, as within tight-binding model (Sect. 8.5.1) or in the framework of normal modes (Sect. 7.1).

other families are not abelian, thus having two- and/or four-dimensional irreducible representations as well. Therefore, the eigenstates $|\tilde{k}\tilde{m}\Pi; i\rangle$ are covariants associated to irreducible representations of the line group and obeying general rule (5.3):

$$D(\ell)\,|\lambda l; i\rangle = \sum_{l'} D^{(\lambda l)}(\ell)\,|\lambda l'; i\rangle. \tag{8.1}$$

Thus, we may utilize the derived classification of the covariants (Sect. 5.2.2) to generalize *Bloch theorem*. Indeed, the sum over K and M in (5.28) refers to the harmonics and amplitudes only, resulting in an invariant function $u(\boldsymbol{r})$. Substituting λ and l by the quantum numbers, we get Bloch theorem adapted to the line group symmetry:

$$\psi^{\tilde{k}\tilde{m}\Pi}(\boldsymbol{r}) = \Phi_{00}^{\tilde{k}\tilde{m}\Pi}(\varphi, z)u(\boldsymbol{r}), \tag{8.2}$$

where, the representative functions $\Phi_{00}^{\tilde{k}\tilde{m}\Pi}(\varphi, z)$ are given in Table 5.2. Obviously, this form is similar to the Bloch form for pure translational symmetry, with the representative functions taking role of the plane waves.

8.1.2 Band Topology

Here we discuss topological properties of the energy bands (e.g., band joining and van Hove singularities) as a consequence of the line group symmetry (being thus system independent).

8.1.2.1 Linear Bands and Helical Symmetry

When linear quantum numbers are used to describe band structure of a commensurate system with symmetry given by the line group with nontrivial helical axis, there is very pronounced band sticking at the edges of the irreducible domain ($k = 0, \pi/a$). This can be explained by comparing these bands with those labeled by helical quantum numbers.

In fact, application of different sets of quantum numbers results in the same set of energies (eigenvalues of the same hamiltonian), but grouped into bands differently. Therefore, transition rules (4.8) between the representations of the two types directly relate these bands. Helical Brillouin zone is $\tilde{q}$ times larger than the linear one, and each helical *representation band* is subdivided into the linear ones (Sect. 4.1.4). The continuity of the helical *energy bands* implies that some of the linear bands must be connected at the edges of the reduced Brillouin zone [1]. Precisely, the band $\varepsilon^i_m(k)$ sticks to the band $\varepsilon^i_{m'}(k)$, with $m' \stackrel{\circ}{=} m + p$.

This is illustrated in Fig. 8.1, where we show the conduction bands of the (8,4) carbon nanotube given by both linear (k, m) and helical $(\tilde{k}, \tilde{m})$ quantum numbers. The $\tilde{m}$-bands can be thought of as consisting of the "unfolded" m-bands. There

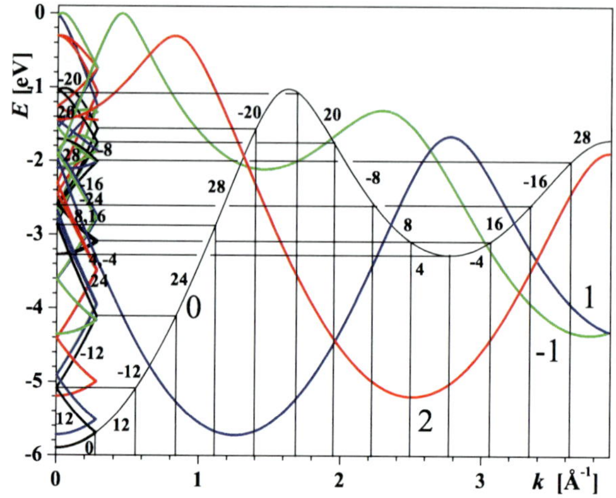

Fig. 8.1 Valence bands of the (8,4) carbon nanotube [2, 3]. Helical bands $\tilde{m} = 0, \pm 1, 2$ (large numbers) are over helical irreducible domain $\tilde{k} \in [0, 3.9\text{Å}^{-1}]$. In the *left* part are linear bands $m = 0, \pm 1, \ldots, \pm 27, 28$, over linear irreducible domain $k \in [0, 0.28\text{Å}^{-1}]$. In particular, *bold* linear bands (with m given by small numbers) yield helical band with $\tilde{m} = 0$

are $n = 4$ helical bands with $\tilde{k} \in (-14\pi/a, 14\pi/a]$ and $q = 56$ linear bands with $k \in (-\pi/a, \pi/a]$. Thus, the number of bands when represented in the helical quantum numbers is $\tilde{q} = 14$ times less than when represented in terms of the linear quantum numbers. However, the range of $\tilde{k}$ is increased by the same factor ($\tilde{q} = 14$) relative to the range of k.

Note that in the case of the groups with $\boldsymbol{Z} = \boldsymbol{T}$, these rules are ineffective, while for another achiral generalized translation group, $\boldsymbol{Z} = \boldsymbol{T}_{2n}$, they connect pairs of the bands. Further, considering only the first family groups, without additional symmetries, the obtained connectivity rules are applicable only to the point $k = \pi/a$.

8.1.2.2 Band Sticking and van Hove Singularities Induced by the Negative Parities

While vertical mirror and glide planes only increase the degeneracy of energy bands, not causing their additional sticking, the z-reversing elements give rise to the new connectivity rules. The easiest way to see this is to consider first the band structure assigned by the quantum numbers of the halving positive subgroup only, and a posteriori to analyze the influence of these symmetries. Although we discuss only linear quantum numbers, the same arguments and conclusions apply to the helical quantum numbers as well.

Thus, we begin with the bands $\varepsilon_m(k)$ assigned only by the momenta quantum numbers,[2] with k taking values over the entire Brillouin zone.

---

[2] Since possible appearance of several bands with the same assignation, as well as the parity of the vertical mirror/glide planes, is not important for further discussion, the corresponding superscripts are omitted.

The z-reversing symmetries, combining the quantum numbers k and $-k$ into single irreducible representation, imply that the band structure is symmetrical with respect to the $k = 0$ axis, i.e., that for each m there is m' such that $\varepsilon_m(-k) = \varepsilon_{m'}(k)$. Effectively, the assumed band structure is simply folded along the $k = 0$ axis and the bands are thus completely defined if their shape over only the positive half of the Brillouin zone is given (irreducible domain, Sect. 4.3.1). Therefore the band degeneracy is doubled within the interior of the irreducible domain (i.e., it is a sum of the degeneracies of $\varepsilon_m(k)$ and $\varepsilon_{m'}(k)$) and unchanged at the edges $k = 0, \pi/a$. However, if $\varepsilon_m(k)$ and $\varepsilon_{m'}(k)$ differ (i.e., $m \neq m'$), then these two bands stick together at $k = 0$, and the degeneracy at this point is the same as in the interior. Otherwise, when $\varepsilon_m(k) = \varepsilon_{m'}(k)$ (and $m = m'$) the degeneracy at $k = 0$ is halved, while the symmetry condition $\varepsilon_m(-k) = \varepsilon_m(k)$ implies that the band has extremum at $k = 0$; this is known as *van Hove singularity*. Quite analogously, the band structure is symmetrical with respect to the point $k = \pi/a$ (this follows from the described symmetry with respect to $k = 0$, and the periodicity of the band structure, meaning that the set of the eigenenergies at k and $k + 2\pi/a$ is the same).

The derived conclusions can be efficiently applied through the *compatibility relations* of Table 4.14; namely, in this table are listed all the band representations $_kD$ which become reducible when an edge point value $K = 0, \pi/a$ is substituted for k. In such cases the eigenstates at K transform according to one of the irreducible components (depending on the hamiltonian considered), meaning that the degeneracy $|_kD|$ of the energy band is halved at this edge. Consequently, there is no band sticking, but van Hove singularity appears. All other band representations remain irreducible at K (hence they are not listed in Table 4.14), and the K point degeneracy of the energy band is the same as in the interior of irreducible domain. Accordingly, there is another band $\varepsilon_{m'}(k)$ which sticks together with $\varepsilon_m(k)$ at K; its band representation $_kD'$ is the unique one satisfying $_KD = {}_KD'$. This is illustrated in Fig. 8.2.

It should be mentioned that even for the positive groups, time reversal symmetry may infer topological rules similar to those of the negative parities. In general, for the systems with ordered spins, the magnetic line groups and their co-representations with the corresponding compatibility relations should be used instead of the ordinary line groups. On the other hand, when the hamiltonian is invariant under the time reversal, the gray group and its co-representations are relevant. However, in this case the degeneracy and topology of the bands follow from the kind of the representations involved. Actually, for the band representations and the edge points one should determine the kind of the representations involved (captions of Tables 4.1–4.13). Then, the additional band sticking may appear when the edge point components combine into the single third kind representation, while van Hove singularities comes also in the cases when the band representation is of the second or third kind (thus with doubled degeneracy) while the edge point representations are not: then the edge point co-representation is reducible, although the ordinary one is not.

Finally, let us mention here that the derived topological rules may be easily used to derive various types of band shapes in polymers and nanotubes [4].

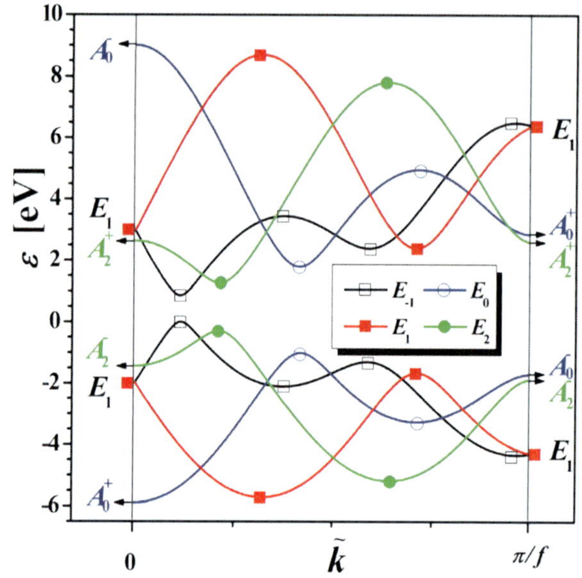

Fig. 8.2 Helical electron bands of the (8,4) carbon nanotube (Fig. 8.1). Assignation of the bands by the irreducible representations enables to compare the connectivity of the bands (*points*) and van Hove singularities (*arrows*) with the compatibility relations of the corresponding line group $\boldsymbol{T}^9_{56}(0.8\text{Å})\boldsymbol{C}_4$, $n = 4$, $\tilde{q} = 14$, and $p = 44$

8.2 Symmetry Breaking and Epikernels

System may reduce its symmetry $\boldsymbol{L}$ to a subgroup $\boldsymbol{L}'$ for many reasons: influence of the external fields, vibrational instability, continuous phase transition, etc. The state space $\mathcal{S}$ of the system remains the same, while the action of the new symmetry is realized by the subduced (restricted) representation $D(\boldsymbol{L}')$, being a subset of $D(\boldsymbol{L})$. Typically, an *order parameter* is related to the symmetry breaking. It is the invariant of the new symmetry, but not of the initial one. Order parameter is usually a component of the applied field, particular displacement, or a quantity getting a nonzero value during a phase transition. Being either single component object or a component of a vector or higher rank tensor, the order parameter A transforms according to the nonidentical representation $D^A(\boldsymbol{L})$, which after the symmetry breaking subduces a reducible representation of the broken group, with at least one identity irreducible component.

Therefore, the broken symmetry $\boldsymbol{L}'$ is an *epikernel* of $D^A(\boldsymbol{L})$; namely, epikernel of $D^A(\boldsymbol{L})$ is any subgroup of $\boldsymbol{L}$ such that the subduced representation $D^A(\boldsymbol{L}')$ contains identical representation $D^{\text{id}}(\boldsymbol{L}')$. In other words, in the decomposition $D^A(\boldsymbol{L}') = f^{\text{id}} D^{\text{id}}(\boldsymbol{L}') + \cdots$, the frequency number f^{id} does not vanish. Obviously, the order parameter is linear combination of the standard vectors of $D^A(\boldsymbol{L}')$. Note that *kernel* of the representation is a special epikernel, for which the whole space of $D^A(\boldsymbol{L})$ is transformed according to identity representation of $\boldsymbol{L}'$. Here we derive so-called epikernels of the irreducible representations of the line groups, i.e., we consider the cases $D^A(\boldsymbol{L}) = {}_kD^{\Pi}_m(\boldsymbol{L})$. This is the most important situation, and the results can be used to treat also the cases of reducible $D^A(\boldsymbol{L})$.

8.2.1 Epikernels of the First Family Groups

First, we briefly derive epikernels of the first family groups for arbitrary representations. To this end it is convenient to use helical quantum numbers. As the representations are one-dimensional, these are also their kernels. Therefore, the epikernel $\boldsymbol{T}_Q^{(\tilde{k},\tilde{m})}(f)$ is the set of elements ℓ_{ts} satisfying ${}_{\tilde{k}}A_{\tilde{m}}((C_Q|f)^t C_n^s) = 1$, which leads to equation $\frac{ft\tilde{k}}{2\pi} + \frac{s\tilde{m}}{n} \in \mathbb{Z}$ in t and s.

For irrational $\frac{f\tilde{k}}{2\pi}$, the only solution in t is $t = 0$. Then for s we obtain $s = s'n/\mathrm{GCD}(n, \tilde{m})$, i.e., the epikernel is pure rotational subgroup

$$\boldsymbol{C}_{\underline{n,\tilde{m}}}. \tag{8.3a}$$

In the opposite case $\tilde{k} = 2\pi\overline{k}/f\underline{k}$ the solutions (C.9) are two-parameter series of pairs $(t_{ij}, s_{ij}) = (t_i + t_j, s_i + s_j)$ corresponding to product of two screw-axes. The helices are obviously commensurate (as both fractional parts are multiples of f), and according to Theorem 3, the product is a line group of the first family. However, its rotational part must be a subgroup of $\boldsymbol{C}_n$, precisely $\boldsymbol{C}_{\underline{n,\tilde{m}}}$. Then it directly follows that $t = t'\underline{k}$ ($t' = 0, \pm 1, \dots$), i.e., that helical factor is $\boldsymbol{T}_{Q/\underline{k}}(\underline{k}f)$. Finally, the epikernel is

$$\boldsymbol{T}_Q^{(\tilde{k}=2\pi\overline{k}/f\underline{k},\tilde{m})}(f) = \boldsymbol{T}_{Q/\underline{k}}(\underline{k}f)\boldsymbol{C}_{\underline{n,\tilde{m}}}. \tag{8.3b}$$

Of course, the obtained Q should be additionally standardized in accordance with previously introduced convention (2.11). Note that commensurability of the rational epikernels is the same as for the initial group.

Using this result the epikernels of the other families can be easily found. Indeed, the first family subgroup of the epikernel is to be found according to (8.3), while the parities are easily handled afterward.

8.2.2 Equitranslational Epikernels

Here we restrict our considerations to the case of the commensurate line groups and to their irreducible representations with vanishing quasi-linear momentum, $k = 0$. This is the most frequent situation, corresponding to the translationally invariant order parameters, like Euclidean vectors or tensors. As at $k = 0$ all the translations (including the fractional ones) are represented by the identity operators, the set of matrices $D_m^{\Pi}(X) = {}_0D_m^{\Pi}(X|f)$ is effectively irreducible representation of the isogonal point group $\boldsymbol{P}_{\mathrm{I}}$. Therefore, we first look for the epikernels $\boldsymbol{P}'$ of the irreducible representations $D_m^{\Pi}(\boldsymbol{P}_{\mathrm{I}})$ of the axial point groups. Then to each element X of $\boldsymbol{P}'$ the same fractional translation f as in $\boldsymbol{L}$ is added to get the coset representative $(X|f)$. As the translational subgroup remains the same, the line group epikernel $\boldsymbol{L}'$

is obtained retaining in the decomposition (2.4) only the cosets corresponding to $\boldsymbol{P}'$ (this is called *extension* of $\boldsymbol{P}'$ to $\boldsymbol{L}'$):

$$\boldsymbol{L}_1 = \sum_{X \in \boldsymbol{P}_1} (X|f_X)\boldsymbol{T}. \tag{8.4}$$

To obtain the epikernels of the axial point groups we use the theorem [5] that if $\boldsymbol{P}'$ is an epikernel of $\boldsymbol{P}_\mathrm{I}$ for the irreducible representation $D_m^\Pi(P_I)$ with kernel $\boldsymbol{K}_m^\Pi = \{X \in \boldsymbol{P}_\mathrm{I} | D_m^\Pi(X) = I\}$, then $\boldsymbol{P}'/\boldsymbol{K}_m^\Pi$ is epikernel of a faithful irreducible representation of the factor group $\boldsymbol{P}_\mathrm{I}/\boldsymbol{K}_m^\Pi$. In addition, all the epikernels of $D_m^\Pi(P_I)$ are biuniquely related to the epikernels of $D_m^\Pi(\boldsymbol{P}_\mathrm{I}/\boldsymbol{K})$. As the factor groups of the axial point groups are again axial point groups, and the isomorphic groups have the same irreducible representations with isomorphic epikernels, only the faithful irreducible representations of the four nonisomorphic axial point groups are to be considered.

As usual, the main difficulty is to find epikernels of the first family line groups. Their isogonal point groups are $\boldsymbol{C}_q$, with irreducible representations $A_m(C_q^j) = \mathrm{e}^{\mathrm{i}mj2\pi/n}$. Then the unique epikernel is the kernel, determined by the equation $\mathrm{e}^{\mathrm{i}mj2\pi/q} = 1$. This yields Diophantine equation (C.6) in j: $mj \stackrel{q}{=} 0$, which is solved by j being multiple of q/q', with $q' = \mathrm{GCD}(m, q)$. Therefore, the searched for epikernel is the subgroup $\boldsymbol{C}_{q'}$ generated by $C_q^{q/q'}$. Corresponding cosets are easy to find from (2.17): $\ell(t, j'q/q') = (C_{q'}^{j'}|(t + \{\frac{j'p}{q'}\})a)$, implying that $p' = q'\{\frac{p}{q'}\}$. Particularly, for the achiral groups, when $p = 0$ also $p' = 0$ (i.e., epikernels of the symmorphic first family groups are symmorphic themselves), and when $q = 2p = 2n$, if $m/\mathrm{GCD}(n, m)$ is odd or even, the epikernels are, respectively, symmorphic or non-symmorphic: as $q' = \mathrm{GCD}(2, m/\mathrm{GCD}(n, m))\mathrm{GCD}(n, m)$, then for odd $m/\mathrm{GCD}(n, m)$ one gets $\{\frac{n}{\mathrm{GCD}(n,m)}\} = 0$, implying $p' = 0$, while in the opposite case $n/\mathrm{GCD}(n, m)$ must be odd, meaning that $q' = 2\mathrm{GCD}(n, \frac{m}{2})$ and $\{\frac{1}{2}\frac{n}{\mathrm{GCD}(n,m)}\} = \frac{1}{2}$, i.e., $p' = \mathrm{GCD}(n, \frac{m}{2}) = q'/2$. This also gives the procedure to determine the first family subgroup of the epikernel of the other families. It follows that for all achiral families epikernels are achiral, and for symmorphic families in particular the epikernels are symmorphic ($p = 0$). Thus only parities are to be additionally analyzed. It is easy to see that they pertain to the epikernels of the even representations only.

The equitranslational epikernels of the line groups are presented in Table 8.1, gathered according to the isogonal axial point groups.

Obviously, apart from the kernel, one-dimensional representations do not have other epikernels. Two-dimensional representations, however, besides the kernel (for which all the vectors of the irreducible subspace are invariant), have epikernels being supergroups of the kernel and subgroups of the whole group. In these cases the frequency number f^{id} of the identity representation is one, meaning that the order parameter is along a uniquely defined direction. However, such epikernels may not be invariant subgroups (to differ from the kernel), and their conjugated subgroups are also epikernels (leaving some other subspaces invariant).

Table 8.1 Equitranslational epikernels of the line groups. In the first column are isogonal axial point groups $\boldsymbol{P}_{\mathrm{I}}$ (underlined) above their epikernels $\boldsymbol{P}'_I$ for the irreducible representations $D^{\Pi}_{m}(\boldsymbol{P}_{\mathrm{I}})$ (column D^{Π}_{m}). The principle axis order of $\boldsymbol{P}_{\mathrm{I}}$ and $\boldsymbol{P}'_I$ is, respectively, q or q'' (when q/q'' must be odd) and q' (any divisor of q); when convenient, we separate q odd and even, denoted then by $\hat{q}$ and $2q$. In the next (one, two or three) columns F/F' follow the line groups $\boldsymbol{L}^{(F)}$ having isogonal point group $\boldsymbol{P}_{\mathrm{I}}$, above their equitranslational epikernels $\boldsymbol{L}^{(F')}$ for the representations ${}_0D^{\Pi}_{m}(\boldsymbol{L}^{(F)})$. The next column specifies parameters of $\boldsymbol{L}^{(F')}$. The different types of epikernels for the same (two-dimensional) representation D are in the consecutive rows, starting with the largest one; N is the number of different conjugated epikernels

$\boldsymbol{P}_{\mathrm{I}}$ / $\boldsymbol{P}'_I$	D^{Π}_{m}	F/F'	F/F'	F/F'	$q', \hat{q}', q'', p'$	N
$\underline{\boldsymbol{C}_q}$	A_0	1			$p' = p$	
$\boldsymbol{C}_{q'}$	A_m	1			$q'=\mathrm{GCD}(q,m)$ $p'=q'\{p/q'\}$	1
$\underline{\boldsymbol{S}_{2q}}$	A_0^+	2				
$\boldsymbol{S}_{2q''}$	A_m^+	2			$\mathrm{GCD}(2q,m) = 2q''$	1
$\boldsymbol{C}_{q'}$	A_m^+	1			$\mathrm{GCD}(2q,m)=q'$ $p'=0$	1
$\boldsymbol{S}_{2q''}$	A_m^-	2			$\mathrm{GCD}(2q,m\pm q) = 2q''$	1
$\boldsymbol{C}_{q'}$	A_m^-	1			$\mathrm{GCD}(2q,m\pm q)=q'$ $p'=0$	1
$\underline{\boldsymbol{C}_{\hat{q}\mathrm{h}}}$	A_0^+	3				
$\boldsymbol{C}_{\hat{q}'\mathrm{h}}$	A_m^+	3			$\mathrm{GCD}(\hat{q},m) = \hat{q}'$	1
$\boldsymbol{C}_{\hat{q}'}$	A_m^-	1			$\mathrm{GCD}(\hat{q},m)=\hat{q}'$ $p'=0$	1
$\underline{\boldsymbol{C}_{2q\mathrm{h}}}$	A_0^+	3	4			
$\boldsymbol{C}_{2q''\mathrm{h}}$	A_m^+	3	4		$\mathrm{GCD}(2q,m) = 2q''$	1
$\boldsymbol{C}_{q'\mathrm{h}}$		3	3		$\mathrm{GCD}(2q,m) = q'$	1
$\boldsymbol{S}_{2q'}$	A_m^-	2	2		$\mathrm{GCD}(2q,m) = q'$	1
$\boldsymbol{C}_{2q''}$		1	1		$\mathrm{GCD}(2q,m)=2q''$ $p'=pq''/q$	1
$\underline{\boldsymbol{D}_{\hat{q}}}$	A_0^+	5			$p' = p$	1
$\boldsymbol{C}_{\hat{q}}$	A_0^-	1			$p' = p$	1
$\boldsymbol{D}_{\hat{q}'}$	E_m	5			$\hat{q}'=\mathrm{GCD}(\hat{q},m)$ $p'=\hat{q}'\{p/\hat{q}'\}$	$\hat{q}/\hat{q}'$
$\boldsymbol{C}_{\hat{q}'}$		1				1
$\underline{\boldsymbol{D}_{2q}}$	A_0^+	5			$p' = p$	1
$\boldsymbol{C}_{2q}$	A_0^-	1			$p' = p$	1
$\boldsymbol{D}_q$	A_q^+	5			$p' = q\{p/q\}$	1
$\boldsymbol{C}_q$	A_q^-	1			$p' = q\{p/q\}$	1
$\boldsymbol{D}_{2q''}$	E_m	5			$2q''=\mathrm{GCD}(2q,m)$ $p''=2q''\{p/2q''\}$	q/q''
$\boldsymbol{C}_{2q''}$		1				1
$\boldsymbol{D}_{q'}$	E_m	5			$2q'=\mathrm{GCD}(2q,m)$ $p'=q'\{p/q'\}$	q/q'
$\boldsymbol{D}_{q'}$		5				q/q'
$\boldsymbol{C}_{q'}$		1				1
$\underline{\boldsymbol{C}_{\hat{q}\mathrm{v}}}$	A_0	6	7			1
$\boldsymbol{C}_{\hat{q}}$	B_0	1	1			1
$\boldsymbol{C}_{\hat{q}'\mathrm{v}}$	E_m	6	7		$\hat{q}' = \mathrm{GCD}(\hat{q},m)$	$\hat{q}/\hat{q}'$
$\boldsymbol{C}_{\hat{q}'}$		1	1			1
$\underline{\boldsymbol{C}_{2q\mathrm{v}}}$	A_0	6	7	8	$p' = p$	1
$\boldsymbol{C}_{2q}$	B_0	1	1	1	$p' = p$	1
$\boldsymbol{C}_{q\mathrm{v}}$	A_q	6	7	7	$p' = q\{p/q\}$	1
$\boldsymbol{C}_q$	B_q	1	1	1	$p' = q\{p/q\}$	1
$\boldsymbol{C}_{2q''\mathrm{v}}$	E_m	6	7	8	$2q''=\mathrm{GCD}(2q,m)$ $p''=2q''\{p/2q''\}$	q/q''
$\boldsymbol{C}_{2q''}$		1	1	1		1
$\boldsymbol{C}_{q'\mathrm{v}}$	E_m	6	7	6	$2q'=\mathrm{GCD}(2q,m)$ $p'=q'\{p/q'\}$	q/q'
$\boldsymbol{C}_{q'\mathrm{v}}$		6	7	7		q/q'
$\boldsymbol{C}_{q'}$		1	1	1		1
$\underline{\boldsymbol{D}_{q\mathrm{d}}}$	A_0^+	9	10			1
$\boldsymbol{S}_{2q}$	B_0^-	2	2			1
$\boldsymbol{D}_q$	B_0^+	5	5		$p' = 0$	1
$\boldsymbol{C}_{q\mathrm{v}}$	A_0^-	6	7			1
$\boldsymbol{D}_{q''\mathrm{d}}$	E_m^+	6	7		$\mathrm{GCD}(2q,m) = 2q''$	q/q''
$\boldsymbol{S}_{2q''}$		2	2			1
$\boldsymbol{D}_{q''\mathrm{d}}$	E_m^-	6	7		$\mathrm{GCD}(2q,m\pm q) = 2q''$	q/q''
$\boldsymbol{S}_{2q''}$		2	2			1
$\boldsymbol{D}_{q'}$	$E_{q/2}$	5	5		$q' = q/2,\ p' = 0$	2
$\boldsymbol{C}_{q'\mathrm{v}}$		6	7			2
$\boldsymbol{C}_{q'}$		1	1			1
$\boldsymbol{D}_{q'}$	E_m^-	5	5		$\mathrm{GCD}(2q,m\pm q)=q'$ $p'=0$	q/q'
$\boldsymbol{C}_{q'\mathrm{v}}$		6	7			q/q'
$\boldsymbol{C}_{q'}$		1	1			1
$\underline{\boldsymbol{D}_{\hat{q}\mathrm{h}}}$	A_0^+	11	12			1
$\boldsymbol{C}_{\hat{q}\mathrm{h}}$	B_0^+	3	3			1
$\boldsymbol{D}_{\hat{q}}$	B_0^-	5	5		$p' = 0$	1
$\boldsymbol{C}_{\hat{q}\mathrm{v}}$	A_0^-	6	7			1
$\boldsymbol{D}_{\hat{q}''\mathrm{h}}$	E_m^+	11	12		$\mathrm{GCD}(\hat{q},m) = \hat{q}''$	$\hat{q}/\hat{q}''$
$\boldsymbol{C}_{\hat{q}''\mathrm{h}}$		3	3			1
$\boldsymbol{D}_{\hat{q}'}$	E_m^-	5	5		$\mathrm{GCD}(\hat{q},m)=\hat{q}'$ $p'=0$	$\hat{q}/\hat{q}'$
$\boldsymbol{C}_{\hat{q}'\mathrm{v}}$		6	7			$\hat{q}/\hat{q}'$
$\boldsymbol{C}_{\hat{q}'}$		1	1			1
$\underline{\boldsymbol{D}_{2q\mathrm{h}}}$	A_0^+	11	12	13		1
$\boldsymbol{C}_{2q\mathrm{h}}$	B_0^+	3	3	4		1
$\boldsymbol{D}_{q\mathrm{h}}$	A_q^+	11	12	11		1
$\boldsymbol{D}_{q\mathrm{h}}$	B_q^+	11	12	12		1
$\boldsymbol{D}_{q\mathrm{d}}$	A_q^-	9	10	9	$p' = p$	1
$\boldsymbol{C}_{q\mathrm{d}}$	B_q^-	9	10	10		1
$\boldsymbol{D}_{2q}$	B_0^-	5	5	5	$p' = p$	1
$\boldsymbol{C}_{2q\mathrm{v}}$	A_0^-	6	7	8		1
$\boldsymbol{D}_{2q''\mathrm{h}}$	E_m^+	11	12	13	$\mathrm{GCD}(2q,m) = 2q''$	q/q''
$\boldsymbol{C}_{2q''\mathrm{h}}$		3	3	4		1
$\boldsymbol{D}_{q'\mathrm{h}}$	E_m^+	11	12	11	$\mathrm{GCD}(2q,m) = q'$	q/q''
$\boldsymbol{D}_{q'\mathrm{h}}$		11	12	12		q/q''
$\boldsymbol{C}_{q'\mathrm{h}}$		3	3	3		1
$\boldsymbol{D}_{q'\mathrm{d}}$	E_m^-	9	10	9	$\mathrm{GCD}(2q,m) = q'$	q/q''
$\boldsymbol{D}_{q'\mathrm{d}}$		9	10	10		q/q''
$\boldsymbol{S}_{2q'}$		2	2	2		1
$\boldsymbol{D}_{2q''}$	E_m^-	5	5	5	$\mathrm{GCD}(2q,m)=2q''$ $p'=pq''/q$	q/q''
$\boldsymbol{C}_{2q''\mathrm{v}}$		6	7	8		$2q/2q'$
$\boldsymbol{C}_{2q'}$		1	1	1		1

8.3 Optical and Vibrational Activity

Within quantum mechanics, physical processes induced by an external field are described by perturbation theory. Apart from the hamiltonian H_0 of the isolated system, total hamiltonian contains also the term Q describing interaction with the external field. This leads to the concept of quantum transitions between the symmetry-adapted eigenstates (5.37) of H_0. The response functions depend on probabilities of the particular transitions, which are determined by the matrix elements (5.38). For a qualitative analysis it is usually sufficient to find the matrix elements which do not vanish, i.e., to single out transitions contributing to the particular process.

Selection rules are clue to this problem. They simply reflect the conservation laws and selection rules for the line groups are given by (5.39). However, for precise calculation *Clebsch–Gordan coefficients* (Sect. 5.5) are needed. They manifest conservation of helical, linear, and angular momenta, as well as of the parities (when these are present in the symmetry group of the system) in the most detailed form.

8.3.1 Optical Transitions

When a system is exposed to the electromagnetic field, Hamiltonian of an electron is

$$H = \frac{2}{2m}(\boldsymbol{p} - e\boldsymbol{A})^2 + V + e\varphi, \tag{8.5}$$

where, $\boldsymbol{p}$ is electronic (generalized) momentum, V potential of the electron (without electromagnetic field), and φ and $\boldsymbol{A}$ the scalar and vector potentials of the electromagnetic field.

Within the Coulomb gauge, $\nabla \cdot \boldsymbol{A} = 0$, the monochromatic electromagnetic wave propagating with the wave vector $\boldsymbol{q}$ and frequency $\omega = cq$ is described by $\varphi = 0$ and

$$\boldsymbol{A}(\boldsymbol{q}) = \boldsymbol{a}(\boldsymbol{q})\mathrm{e}^{\mathrm{i}(\boldsymbol{q}\cdot\boldsymbol{r}-\omega t)} + \text{c.c.} = \mathrm{e}^{\mathrm{i}(\boldsymbol{q}\cdot\boldsymbol{r}-\omega t+\alpha)}\boldsymbol{e} + \text{c.c.} \tag{8.6}$$

where *polarization phase* α is defined by $a_1^2 + a_2^2 = |a_1^2 + a_2^2|\mathrm{e}^{2\mathrm{i}\alpha}$ (components in the basis transversal to $\boldsymbol{q}$), and (complex) *polarization vector* $\boldsymbol{e} = \boldsymbol{a}\mathrm{e}^{-\mathrm{i}\alpha} = \boldsymbol{e}_1 + \mathrm{i}\boldsymbol{e}_2$ determines mutually orthogonal (with real coordinates) principle axes of polarization. In the dipole approximation [6], valid within the optical domain, i.e., when $\boldsymbol{q}\cdot\boldsymbol{r} \ll 1$, the terms linear in the potential are dominate, which gives

$$H = H_0 + Q, \quad Q = -\frac{e}{m}\boldsymbol{A}\cdot\boldsymbol{p} \approx -\frac{e}{m}\left(\mathrm{e}^{\mathrm{i}(-\omega t+\alpha)}\boldsymbol{e} + \mathrm{e}^{\mathrm{i}(\omega t-\alpha)}\boldsymbol{e}^*\right)\cdot\boldsymbol{p}. \tag{8.7}$$

This form enables perturbative approach, namely, $H_0 = \frac{1}{2m}\boldsymbol{p}^2 + V$ is hamiltonian of the isolated system, and Q is the perturbation introduced by the electromagnetic wave. The first term of Q corresponds to the *absorption* and the second one to the *emission* of the energy quantum $\hbar\omega$.

As the absorption and emission terms for the monochromatic wave are proportional to the $\boldsymbol{e}\cdot\boldsymbol{p}$ and $\boldsymbol{e}^*\cdot\boldsymbol{p}$ components of the linear momentum, they transform like momentum operator, being a polar vector (Sect. 5.3). Its standard components, $p_0 = p_z$ and $p_\pm = p_x \mp \mathrm{i}p_y$, correspond to the *linear* $\boldsymbol{e}_0 = \boldsymbol{e}_z$ and *circular polarizations* $\boldsymbol{e}_\pm = \boldsymbol{e}_x \mp \mathrm{i}\boldsymbol{e}_y$ of the wave. According to the resulting electric field, we call them also *parallel* ($\|$) and *perpendicular polarizations* ($\perp$), while in these cases the wave propagates perpendicularly to the z-axis and along the z-axis, respectively.

Thus, optical properties of the system are determined by the matrix elements $\langle k_\mathrm{f} m_\mathrm{f} \Pi_\mathrm{f} \mid p_{0,\pm} \mid k_\mathrm{i} m_\mathrm{i} \Pi_\mathrm{i}\rangle$. They vanish unless the *optical transition* between initial and final state is allowed by the selection rules. Substituting quantum numbers of the polar vector (Table 5.3 and (5.31)) into the general expression (5.39), we find (for the order of the principle axis of the isogonal group larger than two) selection rules for the polarized optical transitions:

$$\| : \Delta k = \Delta\tilde{k} = 0,\ \Delta m = \Delta\tilde{m} = 0,\ \Pi_{U\mathrm{f}}\Pi_{U\mathrm{i}} \neq 1,\ \Pi_{\mathrm{vf}}\Pi_{\mathrm{vi}} \neq -1,\ \Pi_{\mathrm{hf}}\Pi_{\mathrm{hi}} \neq 1; \tag{8.8a}$$

$$\perp : \quad \Delta k = 0,\ \Delta\tilde{k} = \kappa,\ \Delta m = \Delta\tilde{m} = \pm 1,\ \Pi_{\mathrm{hf}}\Pi_{\mathrm{hi}} \neq -1. \tag{8.8b}$$

Note that for the perpendicular polarization there is no restriction regarding U and σ_v parities, as circular polarization is neither even nor odd under these transformations.

Optical properties of the system are expressed through several *optical response functions*. All of them are determined by the real part of the tensor of the *optical conductivity*:

$$\mathrm{Re}[\sigma_{lj}(\omega)] = \frac{2\pi e^2}{m_e^2\omega}\sum_{f,i}\langle\psi_f\mid p_l\mid\psi_i\rangle\langle\psi_i\mid p_j\mid\psi_f\rangle n_f(1-n_i)\delta(E_i-E_f-\hbar\omega), \tag{8.9}$$

where e and m_e are electron mass and charge, while $n_k = (\mathrm{e}^{(E_k-\mu)/kT}+1)^{-1}$ is the temperature-dependent probability of the occupancy of the state $\mid\psi_k\rangle$. In particular, for the temperatures for which $k_\mathrm{B}T/2$ (k_B is Boltzman constant) is less than a *gap* of a semiconductor, $n_f(1-n_f)$ is equal to 1 if $\mid\psi_i\rangle$ is bonding and $\mid\psi_f\rangle$ valence state, while otherwise vanishes. Note that $\mathrm{Re}[\sigma_{lj}(\omega)]$ is symmetric second-rank tensor. Further, as the angular momentum selection rules show that for fixed initial and final symmetry-adapted states $\langle\psi_f \mid p_0 \mid \psi_i\rangle$ and $\langle\psi_f \mid p_\pm \mid \psi_i\rangle$ cannot both be nonzero, it follows that the only nonzero standard components of conductivity tensor are $\mathrm{Re}[\sigma_\|(\omega)] = \mathrm{Re}[\sigma_{00}(\omega)]$ and $\mathrm{Re}[\sigma_\perp(\omega)] = \mathrm{Re}[\sigma_{++}(\omega)] = \mathrm{Re}[\sigma_{--}(\omega)]$, in accordance with the general form diag[$\mathrm{Re}[\sigma_\perp]$, $\mathrm{Re}[\sigma_\perp]$, $\mathrm{Re}[\sigma_\|]$], of the invariant symmetric tensors (for $q > 2$) given in Table 5.6.

8.3.2 Infrared Active Modes

Long wavelength light may be absorbed by ions, which are then excited and vibrate. Quantum mechanical formalism for such a perturbation is analogous to the electron–light interaction causing optical transitions, with the appropriate changes in the hamiltonian (8.5):

$$H = \frac{2}{2m}(\boldsymbol{P} + Ze\boldsymbol{A})^2 + V - Ze\varphi, \tag{8.10}$$

where $\boldsymbol{P}$ is momentum of the ion with the (positive) charge Ze, m is ion mass while V is the field made by the other ions or electrons of the isolated system. Therefore, the perturbation is again a *polar vector*, with the symmetry-adapted components discussed in Sect. 5.4. Such a transition is described as a creation of a *phonon*, being an elementary quant of ionic normal vibrational mode. Accordingly, the selection rules refer to the phonons which may be created (annihilated) by the absorption (emission) of light, and these phonons carry the quantum numbers given by (5.31) and Table 5.4. We conclude that only normal modes (phonons) transforming according to the polar-vector irreducible components may be infrared active and visible in the infrared spectra. More precise information about phonon activity in the polarized light may be derived from the same table, by extracting only the standard components corresponding to the polarization direction of the light.

The number of infrared active modes is obtained by adding numbers of such modes for all the orbits of the system. According to (7.2), this means that for the system $\boldsymbol{L}[N_1X_1,\ldots,N_KX_K]$ (*symmetry notation*, Sect. 3.5) the total number of the active modes is

$$N^{\mathrm{ir}} = \sum_{\lambda} |\lambda| \sum_{i=1}^{K} N_i f_i^{\lambda} - N_{\mathrm{ac}}^{\mathrm{ir}}. \tag{8.11}$$

Here the first sum is taken over all irreducible components $D^{(\lambda)}(\boldsymbol{L})$ of the polar vector (Table 5.4), f_i^{λ} is the frequency number of this representation in the dynamical representation of the orbit X_i, and $N_{\mathrm{ac}}^{\mathrm{ir}}$ is the number of acoustic modes counted by the first part of the expression; namely, the three *translational acoustic* modes have the quantum numbers of polar vector, and in the families 1 and 5, also the *twisting* mode has the quantum number of the z-component of the vector. Thus, the sum includes $N_{\mathrm{ac}}^{\mathrm{ir}} = 4$ for the families 1 and 5 and $N_{\mathrm{ac}}^{\mathrm{ir}} = 3$ for the other families of such modes. However, having zero frequency, they cannot be infrared active, and the result must be diminished by $N_{\mathrm{ac}}^{\mathrm{ir}}$.

Finally, recall that in Sect. 5.5 it is discussed that the matrix element allowed by symmetry may still vanish if the reduced matrix element is zero. In the present context this means that also the modes which are active by symmetry may actually be of low intensity or totally inactive. Hence, relation (8.11) gives complete set of potentially infrared active modes.

8.3.3 Raman Active Modes

Raman scattering is in the simplest way described as a three step process. It starts by the electronic excitation due to the absorption of the incident light. This changes spatial distribution of the electronic density, and previously stable ions are exposed to the electrostatic forces (within adiabatic approximation), which in the next step cause ionic vibration. Finally, the rest of the electronic excitation energy is emitted as the scattered light. Obviously, the first and the last steps are *optical transitions*, while the second one is activation of a normal mode, i.e., creation of a phonon. Rigorous quantum mechanical treatment reveals that ions may also be deexcited in the second step, causing increase of the electronic energy. Therefore *Raman shift*, i.e., energy difference of the incident and scattered light, can be either positive or negative (either phonon is created or annihilated). However, both schemes are described by the two electronic transitions and creation or annihilation of one phonon. Recall that in the dipole approximation, the optical transitions related to the incident and scattered electromagnetic wave polarized along $\boldsymbol{e}_{\mathrm{i}}$ and $\boldsymbol{e}_{\mathrm{f}}$ are described by $p_{\mathrm{i}} = \boldsymbol{e}_{\mathrm{i}} \cdot \boldsymbol{p}$ and $p_{\mathrm{f}} = \boldsymbol{e}_{\mathrm{f}} \cdot \boldsymbol{p}$. Hence, the total perturbation is $Q = p_{\mathrm{f}} \mathsf{R}^{\mathrm{fi}}(\boldsymbol{R}) p_{\mathrm{i}}$, where $\mathsf{R}^{\mathrm{fi}}(\boldsymbol{R})$ are components of the *Raman tensor*, essentially related to the electron–phonon interaction ($\boldsymbol{R}$ denotes all the ionic coordinates). As far as the transformations in the electronic space are considered, Q is axial second-rank tensor. Usually, for quasi-one-dimensional systems, it is given by the components describing the following polarizations of the incident and scattered light: parallel $\mathsf{R}^{\|\|}$, perpendicular $\mathsf{R}^{\perp\perp}$, and two crossed $\mathsf{R}^{\|\perp}$ and $\mathsf{R}^{\perp\|}$. The initial and final electronic states are identical so the total matrix element of the perturbation is

$$\langle k_{\mathrm{i}}^{\mathrm{el}} m_{\mathrm{i}}^{\mathrm{el}} \Pi_{\mathrm{i}}^{\mathrm{el}} | \left(\langle k_{\mathrm{f}}^{\mathrm{ph}} m_{\mathrm{f}}^{\mathrm{ph}} \Pi_{\mathrm{f}}^{\mathrm{ph}} | \; Q \; | k_{\mathrm{i}}^{\mathrm{ph}} m_{\mathrm{i}}^{\mathrm{ph}} \Pi_{\mathrm{i}}^{\mathrm{ph}} \rangle \right) | k_{\mathrm{i}}^{\mathrm{el}} m_{\mathrm{i}}^{\mathrm{el}} \Pi_{\mathrm{i}}^{\mathrm{el}} \rangle .$$

Therefore, the braced ionic (i.e., phonon) part must be invariant. On the other side, the selection rules (5.39) in the ionic space allow creation (annihilation) of the normal vibration mode (phonon) with the quantum numbers k^{ph}, m^{ph}, and Π^{ph}, relating initial and final ionic states with quantum numbers $k_{\mathrm{i}}^{\mathrm{ph}}$, $m_{\mathrm{i}}^{\mathrm{ph}}$, $\Pi_{\mathrm{i}}^{\mathrm{ph}}$ and $k_{\mathrm{f}}^{\mathrm{ph}}$, $m_{\mathrm{f}}^{\mathrm{ph}}$, $\Pi_{\mathrm{f}}^{\mathrm{ph}}$, respectively (analogously for the helical quantum numbers). Finally, as the perturbation is axial second-rank tensor, we see that $D^{\mathsf{P}}(\boldsymbol{L}) \otimes D^{\mathsf{P}}(\boldsymbol{L}) \otimes D^{(k_{\mathrm{i}}^{\mathrm{ph}} m_{\mathrm{i}}^{\mathrm{ph}} \Pi_{\mathrm{i}}^{\mathrm{ph}})}(\boldsymbol{L})$ must contain identical representation, i.e., the representation of the activated normal mode must be among the components of $D^{\mathsf{A}}(\boldsymbol{L})$ (Table 5.5). In conclusion, among the normal modes with the symmetry defined by the irreducible representations given in Table 5.5 are those which can be activated by Raman scattering. They are finally selected by the nonvanishing Clebsch–Gordan coefficients in (5.38).

As well as for the infrared active modes, one has to bear in mind that the reduced matrix element (including all the specific physical quantities) may annihilate the total matrix element, in which case the mode will be inactive, although allowed by symmetry. It turns out that for non-chiral and non-polar systems, the modes corresponding to the antisymmetric part of the axial tensor (of perturbation) are inactive.

The number of Raman active modes can be found as described for the infrared activity. For the system $\boldsymbol{L}(N_1 S_1, \ldots, N_K S_K)$ it is

$$N^{\mathrm{R}} = \sum_{\lambda} |\lambda| \sum_{i=1}^{K} N_i f_i^{\lambda} - N_{\mathrm{ac}}^{\mathrm{R}}. \tag{8.12}$$

Like in (8.11), the first sum is taken over all irreducible components of the axial tensor (or its symmetric part, if only such modes are searched for), listed in Table 5.5, while $N_{\mathrm{ac}}^{\mathrm{R}}$ is the number of the *acoustic* modes included in the first term.

8.3.4 Vibronic Activity: Jahn–Teller Theorem

The *Jahn–Teller theorem* asserts that for any degenerate electronic state, some normal modes of the phonon spectrum will be activated due to the vibronic coupling to produce nonsymmetrical distortion of the ions. The exceptions are linear ion configurations as well as the electronic Krammer's degeneracy. The theorem is verified [7] in 1937 for molecules, i.e., point groups, and in 1993 for quasi-one-dimensional crystals [8]. However, there are exceptions [9] for some highly symmetric diperiodic structures (in particular for the CuO layers relevant for high temperature superconductivity), while the systematic investigation of three-dimensional crystals (space groups) has not been performed yet. Here we present outline of the proof of the theorem for the line groups.

From the group theoretical point of view, electronic orbital degeneracy means that the state $|E\rangle$ of the electronic subsystem belongs to the multidimensional real (Sect. 4.3.4) irreducible representation $E(\boldsymbol{L})$. Within the linear vibronic coupling, the mean value $\langle E| H |E\rangle$ of the hermitian real operator, transforming according to the irreducible component $D^H(\boldsymbol{L})$ of the dynamical representation $D^{\mathrm{dyn}}(\boldsymbol{L})$ of the system, multiplies the normal mode Q^H associated to this representation. Hence, when $< E|H|E > \neq 0$, Q^H is the *active mode*: its nonzero mean displacement is a distortion of the configuration. If it is not totally symmetric or *acoustic* (*translational* or *rotational*) mode, the symmetry $\boldsymbol{L}$ of the system is broken. The Jahn–Teller theorem points out that such active modes always exist, i.e., that for each $E(\boldsymbol{L})$, the symmetrized square $[E^2(\boldsymbol{L})]$ contains at least one irreducible component which also pertains to $D^{\mathrm{dyn}}(\boldsymbol{L})$ and which does not correspond to symmetrical, translational, or rotational modes.

As it has been stressed out (Sect. 3.4), the set of orbits S_i of a system S with symmetry $\boldsymbol{L}$ contains at least one of the symmetry fixing sets (Table 3.5) of $\boldsymbol{L}$ (otherwise not $\boldsymbol{L}$ but a supergroup of $\boldsymbol{L}$ would be the symmetry). Further, as the dynamical representation is additive over orbits, it is easily found by use of (7.1). Irreducible components for each orbit are given in Tables 7.1–7.13.

For the present task, it is sufficient to check the conditions of the Jahn–Teller theorem only for the symmetry fixing sets. As irreducible components of the symmetrized squares of the real representations of the line groups are found [10–12], it remains to construct the dynamical representations $D_S^{\mathrm{dyn}}(\boldsymbol{L})$ of the symmetry fixing sets listed in Table 3.5 and subtract from them the *translational* and *rotational acoustic* modes (Sect. 7.2), in order to get the vibrational representation. Finally, comparing this representations and symmetrized squares, we straightforwardly verify Jahn–Teller theorem for the line groups, proving vibronic instability of the stereo-regular polymers, nanotubes, and other systems with line group symmetry.

8.3.4.1 Example: Vibronic Instability of *trans*-Polyacetylene

One of the most frequently discussed [13] examples of the vibronic instability is distortion of the *trans*-polyacetylene with uniform bond length, yielding *trans-transoid* isomer with alternating bond lengths (Fig. 3.4). Here we briefly repeat the relevant symmetry arguments. At first, we see from Table 4.13 of the representations of the group $\boldsymbol{L}^{(13)} = \boldsymbol{T}_2^1\boldsymbol{D}_{1\mathrm{h}}$ that there can be at most four types of the π-electron energy bands, ${}_kE_m^{\Pi_\mathrm{v}}$ ($m = 0, 1, \Pi_\mathrm{v} = A, B$), all of them double degenerate. As at $k = \pi/a$ there is no one-dimensional irreducible representation of the line groups from the 13 family, compatibility relations (Table 4.14) at this point give no dimensionality reduction; in addition, since $n = 1$, at $k = \pi/a$ there are only two double degenerate representations ${}_\pi E_0^A$ and ${}_\pi E_0^B$. Compatibility relations ${}_kE_m^{\Pi_\mathrm{v}} \to {}_\pi E_0^{\Pi_\mathrm{v}}$ stick together bands ${}_kE_0^A$ and ${}_kE_0^A$ into ${}_\pi E_0^A$, as well as the bands ${}_kE_0^B$ and ${}_kE_1^B$ into ${}_\pi E_0^B$ (Fig. 8.3). There is one electron per monomer, i.e., two electrons per translational unit cell, and the Fermi radius equals π.

At first, we consider carbon atoms only, within the Hückel's approximation with π-electrons. Taking into account p_y-orbital per atom, the C —C bonding π_u orbital [13] is assigned (as y-component of polar vector, Table 5.4) by B_0^+ representation of the stabilizer group $\boldsymbol{D}_{1\mathrm{h}}$ (or B_1^+ of the isogonal group $\boldsymbol{D}_{2h}$, i.e., ${}_0B_1^+$ of

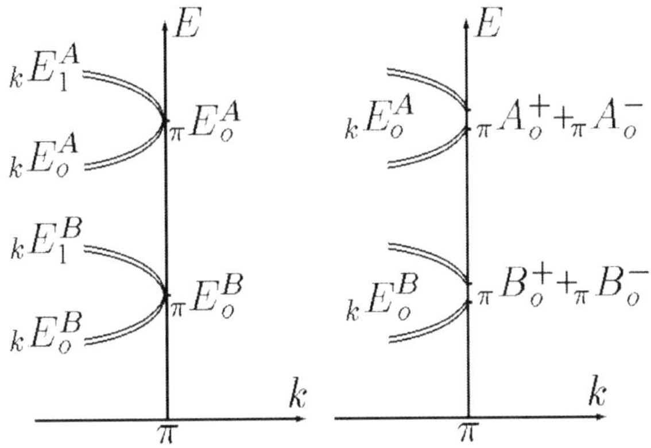

Fig. 8.3 Band structure at the end of the Brillouin zone (reduced representation is used) of *trans*-polyacetylene and "dimerized" *trans*-polyacetylene

$L^{(13)}$). All these orbitals of the orbit e_1 give the induced representation of the line group:

$$B_0^+(D_{1h} \uparrow L^{(13)}) = e_1(L^{(13)}) \otimes {}_0B_1^+ = {}_0B_0^+ + {}_0B_1^+ + {}_kE_0^B + {}_kE_1^B + {}_\pi E_0^B$$

($e_1(L^{(13)})$ is the permutational representation of the orbit e_1). This implies that the electronic state at Fermi level is ${}_\pi E_0^B$, connecting the bands ${}_kE_0^B$ and ${}_kE_1^B$ (Fig. 8.3, left panel). So, the electronic state space is half-filled and *trans*-polyacetylene is an intrinsic metal with the degenerate Fermi level. This is a classical situation of the cooperative Jahn–Teller vibronic instability. The candidates for the soft mode are those transforming according to the irreducible components of the symmetrized square $[{}_\pi E_0^B]^2 = {}_0A_0^+ + {}_0A_1^+ + {}_0A_1^-$. All of them are contained in the dynamical representation of the orbit of C-ions (Table 7.13):

$$D_{e_1}^{\text{dyn}} = {}_0A_0^+ + {}_0A_0^- + {}_0A_1^+ + {}_0A_1^- + {}_0B_0^+ + {}_0B_1^+ + \\ + \sum_k [{}_kE_0^B + {}_kE_1^B + 2({}_kE_0^A + 2{}_kE_1^A)] + 2{}_\pi E_0^A + {}_\pi E_0^B.$$

After neglecting the totally symmetrical representation and x-component of the translations, in $[{}_\pi E_0^B]^2$ remains ${}_0A_1^-$ (Fig. 7.1), satisfying all the requirements for the Jahn–Teller effect.

If one takes into account both e_1-type orbits the situation is a little bit different. The dynamical representation now becomes $2D_{e_1}^{\text{dyn}}$ and contains ${}_0A_0^-$ twice. The vibrations of the polymer are described by the linear combinations of the independent normal displacements (Fig. 7.1): atoms from the different orbits can oscillate out-of-phase (C —H bending) and in-phase (C —C stretching). For the phase transition the candidate is only the last mode (C —H bonds are tighter and hydrogen atoms follow neighboring C-ions), yielding the Peierls dimerization. This is equitranslational structural phase transition, with the soft mode ${}_0A_1^-$ being the order parameter. The symmetry predicts its epikernel $L^{(9)} = \mathbf{L}\bar{1}m$ for the symmetry group of new configuration (Table 8.1), i.e., *trans*-transoid isomer (Fig. 3.4). The reflection σ_h is representative of the "lost" coset ($L2_1/mcm = L\bar{1}m + \sigma_\text{h}L\bar{1}m$), thus restoring the initial symmetry via two possible domains with the soliton [14] in the role of the Goldstone mode.

Both representations ${}_kE_0^B$ and ${}_kE_1^B$ of $L2_1/mcm$ subduce into ${}_kE_0^B$ of $L\bar{1}m$, without any requirement on their connection at $k = \pi/a$; moreover, at $k = \pi/a$ this representation reduces into ${}_\pi B_0^+ + {}_\pi B_0^-$. Hence, the energy gains a *gap* at Fermi level, and the distorted isomer is an intrinsic insulator, with nondegenerate ground state (Fig. 8.3, right panel).

In the previous example the order parameter was one-dimensional, which is the simplest case. Two- or four-dimensional active modes are possible also (e.g., the representation ${}_\pi G_2$ occurs in the orbit e_1 of the group $L16_8/mcm$, as well as in the symmetrized square $[{}_{\pi/2}G_1]^2$), giving rise to more complicated, and possibly physically more interesting phenomena [10–12].

8.4 Diffraction

Diffraction measurements reveal *scattering intensity* $S(\boldsymbol{k})$ along cross sections of $\boldsymbol{k}$ space. The *scattering vector* $2\pi\boldsymbol{k}$ is the difference $\boldsymbol{k}_{\rm sc}-\boldsymbol{k}_{\rm in}$ of the scattered and incident wave vectors. Within kinematical diffraction theory, $S(\boldsymbol{k})$ is determined by the *scattering amplitude* averaged per atom $F(\boldsymbol{k})$:

$$S(\boldsymbol{k})=|F(\boldsymbol{k})|^2,\quad F(\boldsymbol{k})=\tfrac{1}{N}\sum_i f_i(\boldsymbol{k})\mathrm{e}^{2\pi\mathrm{i}\boldsymbol{k}\cdot\boldsymbol{r}_i};$$

here, $f_i(\boldsymbol{k})$ is the scattering amplitude of the ith atom ($i=1,\ldots,N$) positioned at $\boldsymbol{r}_i$.

8.4.1 Symmetry and Orbit Amplitudes

For the system with symmetry group $\boldsymbol{L}$, the sum over atoms splits into the sums over symcell atoms A and over their orbits; the later have $|\boldsymbol{Y}_A|=|\boldsymbol{L}|/|\boldsymbol{S}_A|$ atoms, where $\boldsymbol{Y}_A$ is the transversal and $\boldsymbol{S}_A$ is the stabilizer of the symcell atom A. Thus, the total amplitude $F(\boldsymbol{k})$ is expressed in terms of the orbit amplitudes $F_A(\boldsymbol{k})$:

$$F(\boldsymbol{k})=\frac{\sum_{A\in S}|\boldsymbol{Y}_A|F_A(\boldsymbol{k})}{\sum_{B\in S}|\boldsymbol{Y}_B|}=\frac{\sum_{A\in S}\frac{1}{|\boldsymbol{S}_A|}F_A(\boldsymbol{k})}{\sum_{B\in S}\frac{1}{|\boldsymbol{S}_B|}},\quad F_A(\boldsymbol{k})=f_A(\boldsymbol{k})\frac{\sum_{y_A}\mathrm{e}^{2\pi\mathrm{i}\boldsymbol{k}\cdot\boldsymbol{r}_{y_AA}}}{|\boldsymbol{Y}_A|},\tag{8.13}$$

where $\boldsymbol{r}_A$ is position of the atom A while the atom y_AA (i.e., A moved by the transversal element y_A) is at $\boldsymbol{r}_{y_AA}$. Therefore, we calculate scattering amplitudes for orbits, and then superpose them according to (8.13). As only one type of atoms is involved, the scattering amplitude of an orbit is factorized:

$$F_A(\boldsymbol{k})=f_A(\boldsymbol{k})G^{\boldsymbol{Y}_A}(\boldsymbol{k}),\quad G^{\boldsymbol{Y}_A}(\boldsymbol{k})=\frac{1}{|\boldsymbol{Y}_A|}\sum_{y_A\in\boldsymbol{Y}_A}\mathrm{e}^{2\pi\mathrm{i}\boldsymbol{k}\cdot\boldsymbol{r}_{y_AA}}.\tag{8.14}$$

The first factor, $f_A(\boldsymbol{k})$, is input which comprises relevant physical information on the diffraction power of the atoms. The second, *transversal factor* $G^{\boldsymbol{Y}_A}(\boldsymbol{k})$ is purely geometrical determined only by the *conformation class* (Sect. 3.1.4).

Therefore, geometrical factor can be calculated a priori, for each transversal $\boldsymbol{Y}$ of any line group. To this end we use cylindrical coordinates $\boldsymbol{k}=(k_\perp,\Phi,k_z)$ and $\boldsymbol{r}_{yA}=(D_A/2,\varphi_{yA},z_{yA})$ of the scattering vector and atom yA. Substituting scalar product $\boldsymbol{k}\cdot\boldsymbol{r}_{yA}=k_\perp\rho_A\cos(\varphi_{yA}-\Phi)+k_zz_{yA}$ in the *Jacobi–Anger expansion* $\mathrm{e}^{\mathrm{i}x\cos\varphi}=\sum_{l=-\infty}^{+\infty}\mathrm{i}^lJ_l(x)\mathrm{e}^{\mathrm{i}l\varphi}$ over the Bessel functions $J_l(x)$, (8.14) becomes

$$G_A^{\boldsymbol{Y}}(\boldsymbol{k})=\sum_l\mathrm{i}^lJ_l(\pi D_Ak_\perp)\mathrm{e}^{-\mathrm{i}l\Phi}\frac{1}{|\boldsymbol{Y}|}\sum_z\mathrm{e}^{\mathrm{i}l\varphi_{yA}+\mathrm{i}2\pi z_{yA}k_z}.\tag{8.15}$$

This general expression will be specified for each of the 15 conformation classes of the systems with line group symmetry defined by the different transversals $\boldsymbol{Y}^{(i)}$ (Table 3.4). As usual we start with the first family groups (i.e., with the transversal $\boldsymbol{Y}^{(1)}$), and after that the remaining results we will find in two steps using the expressions found for the halving subgroups [15].

8.4.2 Geometrical Factors of the Line Group Orbits

For the fist family groups the general element is $\ell_{ts} = (C_Q|f)^t C_n^s$. Taking it as y in (8.15) one gets

$$G_A^{\boldsymbol{T}_Q \boldsymbol{C}_n}(\boldsymbol{k}) = \sum_l \mathrm{i}^l J_l(D\pi k_\perp)\mathrm{e}^{-\mathrm{i}l(\Phi-\varphi_A)}\mathrm{e}^{2\pi\mathrm{i}k_z z_A}\left(\frac{1}{n}\sum_s \mathrm{e}^{\mathrm{i}l\frac{2\pi}{n}s}\right)\left(\frac{1}{|\boldsymbol{T}_Q|}\sum_t \mathrm{e}^{2\pi\mathrm{i}(k_z f+\frac{l}{Q})t}\right).$$

The sum over s vanishes unless $l = \tilde{M}n$ for any integer $\tilde{M}$. Also, when $l = \tilde{M}n$ the sum over t is nonzero only when the bracket in the exponent is an integer, $\tilde{K}$. This implies that the amplitudes are distributed within the countable set of planes with $k_z = (\frac{n\tilde{M}}{Q} + \tilde{K})/f$, which are called *layer lines*. For incommensurate groups the layer lines are quasi-continually distributed: each pair of integers $\tilde{K}$ and $\tilde{M}$ defines a layer line with vectors $\boldsymbol{k}$ constrained by $\boldsymbol{k}_{\tilde{K}}^{\tilde{M}} = (\frac{D}{2}, \Phi, \frac{-n\tilde{M}+Q\tilde{K}}{Qf})$; geometrical factor along it is given in the row 1' of Table 8.2. However, for commensurate groups $\boldsymbol{T}_q^r(f)\boldsymbol{C}_n = \boldsymbol{L}q_p(a = \tilde{q}f)$ (here $Q = q/r$), all the different pairs of $\tilde{M}$ and $\tilde{K}$ satisfying Diophantine equation $-r\tilde{M} + \tilde{K}\tilde{q} = K$ correspond to the same layer line $k_z = K/a$. As $\tilde{q}$ and r are co-primes, according to (C.8), the solutions are $\tilde{M} = -K\tilde{p} + M\tilde{q}$ (here, $\tilde{p} = p/n$) and $\tilde{K} = K\frac{1-r\tilde{p}}{\tilde{q}} - Mr$ for any integer M. Thus, summing over M for the fixed layer line defined by $\boldsymbol{k}_K = (\frac{D}{2}, \Phi, \frac{n\tilde{M}+Q\tilde{K}}{Qf})$, we find the geometrical factor of the row 1 of the table within discrete layer lines spaced by $1/a$.

Especially for the achiral first family groups, being important as subgroups of several other families, we find

$$G_A^{\boldsymbol{T}\boldsymbol{C}_n}(\boldsymbol{k}_K) = \sum_M \mathrm{i}^{Mn} J_{Mn}(D\pi k_\perp)\mathrm{e}^{\mathrm{i}Mn(\varphi_A-\Phi)}\mathrm{e}^{2\pi\mathrm{i}K\frac{z_A}{a}}, \tag{8.16a}$$

$$G_A^{\boldsymbol{T}_{2n}^1\boldsymbol{C}_n}(\boldsymbol{k}_K) = \sum_M \mathrm{i}^{(2M-K)n} J_{(2M-K)n}(D\pi k_\perp)\mathrm{e}^{\mathrm{i}(2M-K)n(\varphi_A-\Phi)}\mathrm{e}^{2\pi\mathrm{i}K\frac{z_A}{a}}. \tag{8.16b}$$

For the remaining orbit types the results from Table 8.2 are obtained by the usual two step procedure. In fact, according to (8.13), if $\boldsymbol{Y} = \boldsymbol{Y}' + z\boldsymbol{Y}'$, then $G_A^{\boldsymbol{Y}}(\boldsymbol{k}_K) = \frac{1}{2}\left(G_A^{\boldsymbol{Y}'}(\boldsymbol{k}_K) + G_{zA}^{\boldsymbol{Y}'}(\boldsymbol{k}_K)\right)$. Thus, for the transversals having the first

Table 8.2 Geometrical factors for conformation classes. For each class (Column CC, primes single out the incommensurate cases), the diffraction space symmetry group $\tilde{\boldsymbol{P}}_I^{\mathcal{I}} = \tilde{\boldsymbol{Y}}_I + \mathcal{I}\tilde{\boldsymbol{Y}}_I$ is in the next column (when it depends on the parity of n, the odd case is up). The last column gives geometrical factors G_A^Y in the layer lines specified by $\boldsymbol{k}_K = (k_\perp, \Phi, \frac{K}{a})$ for commensurate and $\boldsymbol{k}_{\tilde{K}}^{\tilde{M}} = (k_\perp, \Phi, \frac{Q\tilde{K}-n\tilde{M}}{Qf})$ for incommensurate groups; outside them geometrical factors vanish. Here q, n, p, f, and a are line group parameters, while $d = D_A \pi k_\perp$

CC	$\tilde{\boldsymbol{P}}_I^{\mathcal{I}}$	$G_A^Y(\boldsymbol{k}_K)$ or $G_A^Y(\boldsymbol{k}_{\tilde{K}}^{\tilde{M}})$
1	$\boldsymbol{D}_{q\mathrm{d}}$ $\boldsymbol{D}_{q\mathrm{h}}$	$\sum_M \mathrm{i}^{Mq-Kp} J_{Mq-Kp}(d) \mathrm{e}^{\mathrm{i}(Kp-Mq)(\Phi-\varphi_A)} \mathrm{e}^{2\pi \mathrm{i} K \frac{z_A}{a}}$
1_1	$\boldsymbol{D}_{2n\mathrm{h}}$ $\boldsymbol{D}_{n\mathrm{h}}$	$\sum_M \mathrm{i}^{Mn} J_{Mn}(d) \mathrm{e}^{-\mathrm{i}Mn(\Phi-\varphi_A)} \mathrm{e}^{2\pi \mathrm{i} K \frac{z_A}{a}}$
1_2	$\boldsymbol{D}_{2n\mathrm{h}}$	$\sum_M \mathrm{i}^{(2M-K)n} J_{(2M-K)n}(d) \mathrm{e}^{\mathrm{i}(K-2M)n(\Phi-\varphi_A)} \mathrm{e}^{2\pi \mathrm{i} K \frac{z_A}{a}}$
$1'$	$\boldsymbol{D}_{\infty\mathrm{h}}$	$\mathrm{i}^{n\tilde{M}} J_{n\tilde{M}}(d) \mathrm{e}^{-\mathrm{i}n\tilde{M}(\Phi-\varphi_A)} \mathrm{e}^{2\pi \mathrm{i} \frac{Q\tilde{K}-n\tilde{M}}{Q} \frac{z_A}{f}}$
2	$\boldsymbol{D}_{n\mathrm{d}}$ $\boldsymbol{D}_{2n\mathrm{h}}$	$\sum_M \mathrm{i}^{M(n+1)} J_{Mn}(d) \mathrm{e}^{-\mathrm{i}Mn(\Phi-\varphi_A)} \cos(2\pi K \frac{z_A}{a} - \frac{M\pi}{2})$
3	$\boldsymbol{D}_{2n\mathrm{h}}$ $\boldsymbol{D}_{n\mathrm{h}}$	$\sum_M \mathrm{i}^{Mn} J_{Mn}(d) \mathrm{e}^{-\mathrm{i}Mn(\Phi-\varphi_A)} \cos 2\pi K \frac{z_A}{a}$
4	$\boldsymbol{D}_{2n\mathrm{h}}$	$\sum_M \mathrm{i}^{(2M-K)n} J_{(2M-K)n}(d) \mathrm{e}^{\mathrm{i}(K-2M)n(\Phi-\varphi_A)} \cos 2\pi K \frac{z_A}{a}$
5	$\boldsymbol{D}_{q\mathrm{d}}$ $\boldsymbol{D}_{q\mathrm{h}}$	$\sum_M \mathrm{i}^{Mq-Kp} J_{Mq-Kp}(d) \mathrm{e}^{\mathrm{i}(Kp-Mq)\Phi} \cos[(Mq-Kp)\varphi_A + 2\pi K \frac{z_A}{a}]$
$5'$	$\boldsymbol{D}_{\infty\mathrm{h}}$	$\mathrm{i}^{n\tilde{M}} J_{n\tilde{M}}(d) \mathrm{e}^{-\mathrm{i}n\tilde{M}\Phi} \cos(2\pi \frac{Q\tilde{K}-n\tilde{M}}{Q} \frac{z_A}{f} - n\tilde{M}\varphi_A)$
6	$\boldsymbol{D}_{2n\mathrm{h}}$ $\boldsymbol{D}_{n\mathrm{h}}$	$\sum_M \mathrm{i}^{Mn} J_{Mn}(d) \mathrm{e}^{-\mathrm{i}Mn\Phi} \mathrm{e}^{\mathrm{i}2\pi K \frac{z_A}{a}} \cos Mn\varphi_A$
7	$\boldsymbol{D}_{2n\mathrm{h}}$ $\boldsymbol{D}_{n\mathrm{h}}$	$\sum_M \mathrm{i}^{Mn+K} J_{Mn}(d) \mathrm{e}^{-\mathrm{i}Mn\Phi} \mathrm{e}^{2\pi \mathrm{i} K \frac{z_A}{a}} \cos(Mn\varphi_A - \frac{K\pi}{2})$
8	$\boldsymbol{D}_{2n\mathrm{h}}$	$\sum_M \mathrm{i}^{(2M-K)n} J_{(2M-K)n}(d) \mathrm{e}^{\mathrm{i}2\pi K \frac{z_A}{a}} \mathrm{e}^{\mathrm{i}(K-2M)n\Phi} \cos(K-2M)n\varphi_A$
9	$\boldsymbol{D}_{n\mathrm{d}}$ $\boldsymbol{D}_{2n\mathrm{h}}$	$\sum_M \mathrm{i}^{M(n+1)} J_{Mn}(d) \mathrm{e}^{-\mathrm{i}Mn\Phi} \cos(2\pi K \frac{z_A}{a} - \frac{M\pi}{2}) \cos Mn\varphi_A$
10	$\boldsymbol{D}_{n\mathrm{d}}$ $\boldsymbol{D}_{2n\mathrm{h}}$	$\sum_M \mathrm{i}^{M(n+1)+K} J_{Mn}(d) \mathrm{e}^{-\mathrm{i}Mn\Phi} \cos(2\pi K \frac{z_A}{a} - \frac{M\pi}{2}) \cos(Mn\varphi_A - \frac{K\pi}{2})$
11	$\boldsymbol{D}_{2n\mathrm{h}}$ $\boldsymbol{D}_{n\mathrm{h}}$	$\sum_M \mathrm{i}^{Mn} J_{Mn}(d) \mathrm{e}^{-\mathrm{i}Mn\Phi} \cos 2\pi K \frac{z_A}{a} \cos Mn\varphi_A$
12	$\boldsymbol{D}_{2n\mathrm{h}}$ $\boldsymbol{D}_{n\mathrm{h}}$	$\sum_M \mathrm{i}^{Mn+K} J_{Mn}(d) \mathrm{e}^{-\mathrm{i}Mn\Phi} \cos 2\pi K \frac{z_A}{a} \cos(Mn\varphi_A - \frac{K\pi}{2})$
13	$\boldsymbol{D}_{2n\mathrm{h}}$	$\sum_M \mathrm{i}^{(2M-K)n} J_{(2M-K)n}(d) \mathrm{e}^{\mathrm{i}(K-2M)n\Phi} \cos 2\pi K \frac{z_A}{a} \cos(K-2M)n\varphi_A$
14	$\boldsymbol{D}_{\infty\mathrm{h}}$	$\mathrm{e}^{\mathrm{i}2\pi K \frac{z_A}{a}}$
15	$\boldsymbol{D}_{\infty\mathrm{h}}$	$\cos(\pi K \frac{z_A}{a})$

family group as a halving subset, the geometrical factors are immediately found from the above-derived ones. Further, such transversals are halving subsets for the remaining ones, and the procedure can be applied once again.

8.4.3 *Characteristics of the Diffraction Patterns*

8.4.3.1 Symmetry of the Intensity Distribution

To find the symmetry of the diffraction space we consider the action of the geometrical transformations on the intensity distribution $S(\boldsymbol{k}) = \sum_{ij} f_i(\boldsymbol{k}) f_j(\boldsymbol{k}) \mathrm{e}^{\mathrm{i}2\pi \boldsymbol{k}\cdot(\boldsymbol{r}_i - \boldsymbol{r}_j)}$, being the only observable quantity in the diffraction experiments. As the translations leave $\boldsymbol{k}$ vectors invariant, the action of the transformation $\ell = (R|f)$ is

$$\ell S(\boldsymbol{k}) \stackrel{\mathrm{def}}{=} S(\ell^{-1}\boldsymbol{k}) = \sum_{ij} f_i(\boldsymbol{k}) f_j(\boldsymbol{k}) \mathrm{e}^{\mathrm{i}2\pi(R^{-1}\boldsymbol{k})\cdot(\boldsymbol{r}_i - \boldsymbol{r}_j)} = \sum_{ij} f_i(\boldsymbol{k}) f_j(\boldsymbol{k}) \mathrm{e}^{\mathrm{i}2\pi \boldsymbol{k}\cdot(R\boldsymbol{r}_i - R\boldsymbol{r}_j)}.$$

Due to the summation over the pairs of atoms, the elements of the isogonal group $\boldsymbol{P}_{\mathrm{I}}$ of $\boldsymbol{L}$ act simply as permutations of the terms, not affecting the sum. Further, only the difference $\boldsymbol{r}_i - \boldsymbol{r}_j$ of the atomic position vectors is involved; the inversion changes its sign and only intertwines the terms. Consequently, the symmetry of the intensity distribution is the isogonal point group of the symmetry of the atomic conformation, extended by the spatial inversion. For the conformation classes we combine the results on their symmetry, isogonal groups, and their extension (Tables 3.4, 2.2, and 2.1), to get the symmetry of the diffraction space $\tilde{\boldsymbol{P}}_I^{\mathcal{I}}$ given in Table 8.2.

8.4.3.2 Properties Related to the First Family Subgroup

Each of the geometrical factors contains products of Bessel functions of $k_\perp$ and trigonometric functions of Φ and k_z. The orders of the Bessel functions are determined by the first family subgroup: all of them are multiples of n (for the commensurate classes 1 and 5 recall that n divides both p and q) and which multiples are involved in Kth layer line is determined by the helical axis.

For incommensurate helical axes layer lines are densely distributed, and geometrical factor along a layer line is a single term. For commensurate axes there is a countable set of equally spaced (by $1/a$) layer lines; geometrical factor for Kth one is a sum of terms with Bessel functions of orders $qM - pK = n(\tilde{q}M - \tilde{p}K)$ ($M = 0, \pm 1, \dots$), differing by multiple of q, and repeating for every $\tilde{q}$th layer line. Due to this periodicity and spatial inversion of the diffraction space (making equal distributions along layer lines K and $-K$), only the layer lines $0 \leq K \leq \tilde{q}/2$ are different.

Several properties of Bessel functions are manifested in diffraction patterns. As $J_{-\alpha}(x) = (-1)^\alpha J_\alpha(x)$, the position $x_{\alpha,i}$ of ith nonnegative extreme of $J_\alpha(x)$ satisfies $x_{\alpha,i} = x_{-\alpha,i}$. The position $x_{\alpha,i}$ increases with $|\alpha|$; while $x_{0,1} = 0$, for $\alpha > 0$ the function $J_\alpha(x)$ is almost zero until the region close to $x_{\alpha,1}$. For Kth layer line, two orders with the least absolute values are denoted by β_K and β'_K; their values are $\beta_K \stackrel{\circ}{=} -Kp$ (equality modulo interval $(-q/2, q/2]$), and $\beta'_K = \beta_K - q$ if $\beta_K \geq 0$ and $\beta'_K = -\beta_K + q$ otherwise. Thus, $|\beta_K| + |\beta'_K| = q$, yielding that $|\beta_K| = |i_K|n$ (for essentially different layer lines $K = 0, \dots, [\tilde{q}/2]$) and $|\beta'_K| - |\beta_K| = q - 2|i_K|n$ for $i_K \stackrel{\circ}{=} -K\tilde{p}$ (equal modulo interval $(-\tilde{q}/2, \tilde{q}/2]$). Consequently, $\beta_K = 0$ only for $K = 0, \pm\tilde{q}, \pm 2\tilde{q}, \dots$, these layer lines are characterized by broad maximum at $k_\perp = 0$ produced by the term with J_0. In other layer lines there is *intensity gap*, a dark central region, as the first maximum is at a distance $k_\perp = x_{\beta,1}/\pi D > 0$. The layer lines with $K = \pm 1$, i.e., the closest to the equatorial one have the leading term $J_{\mp p}$. The narrowest intensity gap is for the minimal nonzero β_K; this is $\mp n$ and corresponds to the layer lines $K = \pm r + s\tilde{q}$ ($s = 0, \pm 1, \dots$).

Dependence of intensity distribution on Φ, included by the phase factors, yields the interference terms in the absolute square of geometrical factor. The angular period of all the terms in $S(\boldsymbol{k})$ is at most $2\pi/q$, as q is the principle axis order of the pattern (having symmetry group $\tilde{\boldsymbol{P}}_I^{\mathcal{I}}$). The most characteristic region is the central part of a pattern ($k_\perp < x_{\beta',1}/\pi D$), where all the terms except the leading one $J_\beta^2(\pi D k_\perp)$ are negligible. Consequently, several concentric seemingly solid (nonmodulated) circles are observed at radii corresponding to the extremes of $J_\beta(\pi D k_\perp)$. The number of such circles is less than (though close to) the number of zeros of $J_\beta(\pi D k_\perp)$ which are less than $x_{\beta',1}/\pi D$). Thus, the circles do not appear if and only if $|\beta'_K| = |\beta_K|$, which is the case only for even $\tilde{q}$ for the layer lines $K = \pm\tilde{q}/2, \pm 3\tilde{q}/2, \pm 5\tilde{q}/2, \ldots$; then $\tilde{p}$ is odd, $\tilde{p} = 2P + 1$, and substituting $M - P$ by M the geometrical factor reads

$$G_A^{(1)}(\boldsymbol{k}_K) = 2\mathrm{e}^{2\pi \mathrm{i} K \frac{z_A}{a}} \sum_{M \geq 0} \mathrm{i}^{Mq + \frac{q}{2}} \cos[(2M+1)(\Phi - \varphi_A)\frac{q}{2}] J_{Mq+\frac{q}{2}}(d).$$

The cosine factor introduces significant angular modulation and vanishes for

$$\Phi_j = \varphi_A + \pi(2j+1)/q, \; j = 0, \ldots, q-1. \tag{8.17a}$$

This leads to the *extinction* of these layer lines, i.e., they disappear in the diffraction patterns for normal incidence with wave along q directions Φ_j.

Thus, to reveal the first family subgroup $\boldsymbol{L}^{(1)}$ one first examines gapless layer lines; while in chiral systems they are rare, in achiral ones either all or one half lines are such. The distance between them is $1/f$. Numerical fit of radii of the first several intensity maxima to the first extremes of $J_0(D\pi k_\perp)$ suffices to find the diameter D.

Chiral systems ($Q/n > 2$) may be incommensurate or commensurate and belong either to the first or to the fifth conformation class (Figs. 8.4 and 8.5, left column). The incommensurate systems produce dense layer lines, each being completely axially symmetric. The least gap characterizes layer line with $\tilde{M} = \pm 1$ at $k_z = \mp n/Qf + \tilde{K}/f$, and radius $x_{n,1}/\pi D$ of the first extreme (or positions of the next peaks) enables to find Q/n and n. The equal distance $1/a = 1/\tilde{q}f$ between adjacent layer lines justifies that the system is commensurate and gives $\tilde{q}$; alternatively, it can be found as the ordinal of the first gapless line above the equatorial one. The height (i.e., the value of K) of the layer line with the narrowest intensity gap equals r, while its radius of the gap, equal to $x_{n,1}/\pi D$, gives n (and then q). Alternatively, q (and n) can be found as the order of the principle axis of diffraction space (i.e., of any layer line). Finally, to distinguish from $\boldsymbol{Y}^{(1)}$ and $\boldsymbol{Y}^{(1')}$, the intensities of layer lines for $\boldsymbol{Y}^{(5)}$ and $\boldsymbol{Y}^{(5')}$ are K dependent due to $\cos(2\pi \frac{Q\tilde{K} - n\tilde{M}}{Q} \frac{z_A}{f} - n\tilde{M}\varphi_A)$. However, the restriction of the area around $k_\perp = 0$ visible in real experiments seems to prevent experimental verification of the incommensurability of a system. At first, within this finite region layer lines of an incommensurate system are not dense (intensity gap of vast majority of layer lines is greater than this region), while commensurate system may produce axially symmetric patterns in this region (solid circles). Therefore, together with n and f one always finds rational q/r, being in incommensurate cases only an approximation of the irrational parameter Q. Possible

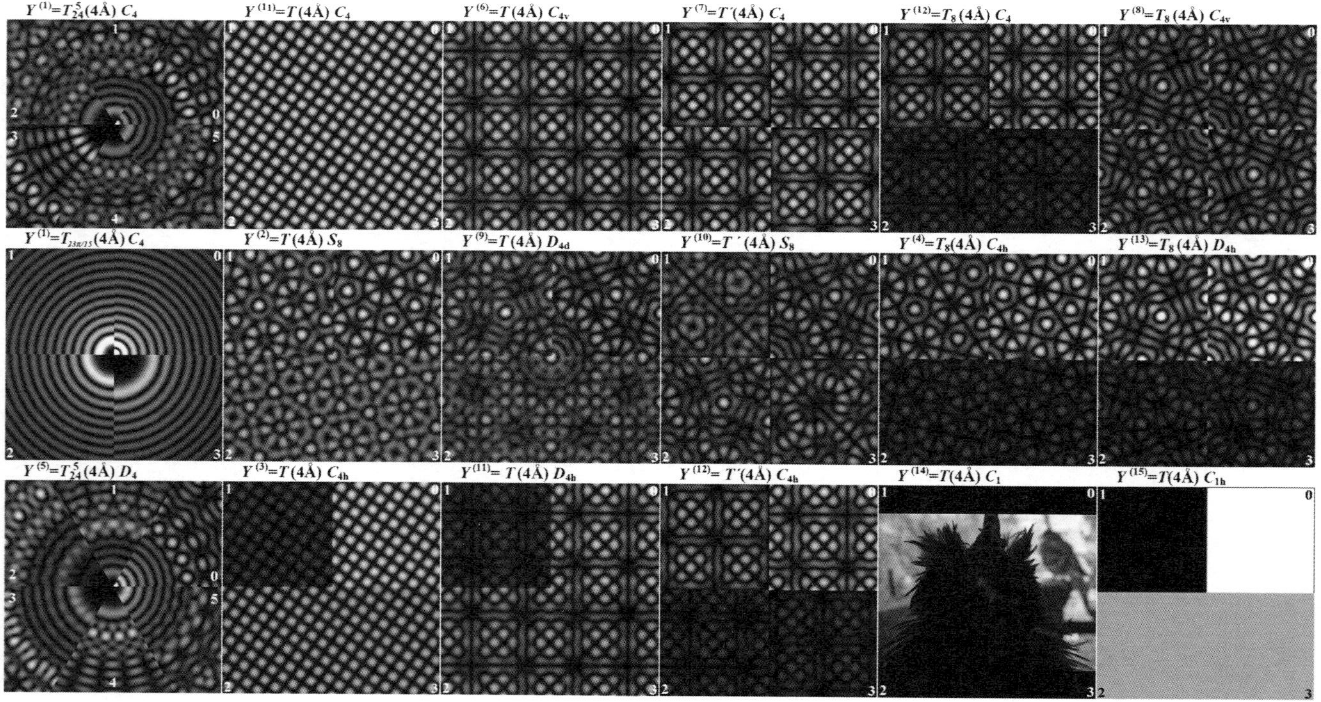

Fig. 8.4 Layer lines for conformation classes. Simulated diffraction patterns for mono-orbit systems (Fig. 3.2) generated from the orbit representative positioned at (cylindrical coordinates) $\boldsymbol{r} = (3Å, \frac{\pi}{18}, 0.8Å)$ (only for the classes 14 and 15 $\boldsymbol{r} = (0Å, 0, 0.8Å)$) by the transversals $\boldsymbol{Y}^{(F)}$ for $n = 4$ and $f = 4\,Å$. For achiral classes each layer line $K = 0, 1, 2, 3$ (indicated as a white number) is given in a (radial) quarter of the panel. For chiral commensurate classes (the first and the last panel of the *left* column) $\tilde{q} = 6$, and the layer lines $K = 0, \ldots, 5$ are plotted. For the incommensurate transversal (in the *middle* of the *left* column), the layer lines with $\tilde{M} = 0, 1, 2, 3$ (corresponding to $\tilde{K} = 0, 0, 1, 2$ at heights $k_z = 0, 0.21, 0.17, 0.12Å^{-1}$, respectively) are plotted. The plotted region is $|k_x|, |k_y| \leq 2\,Å$

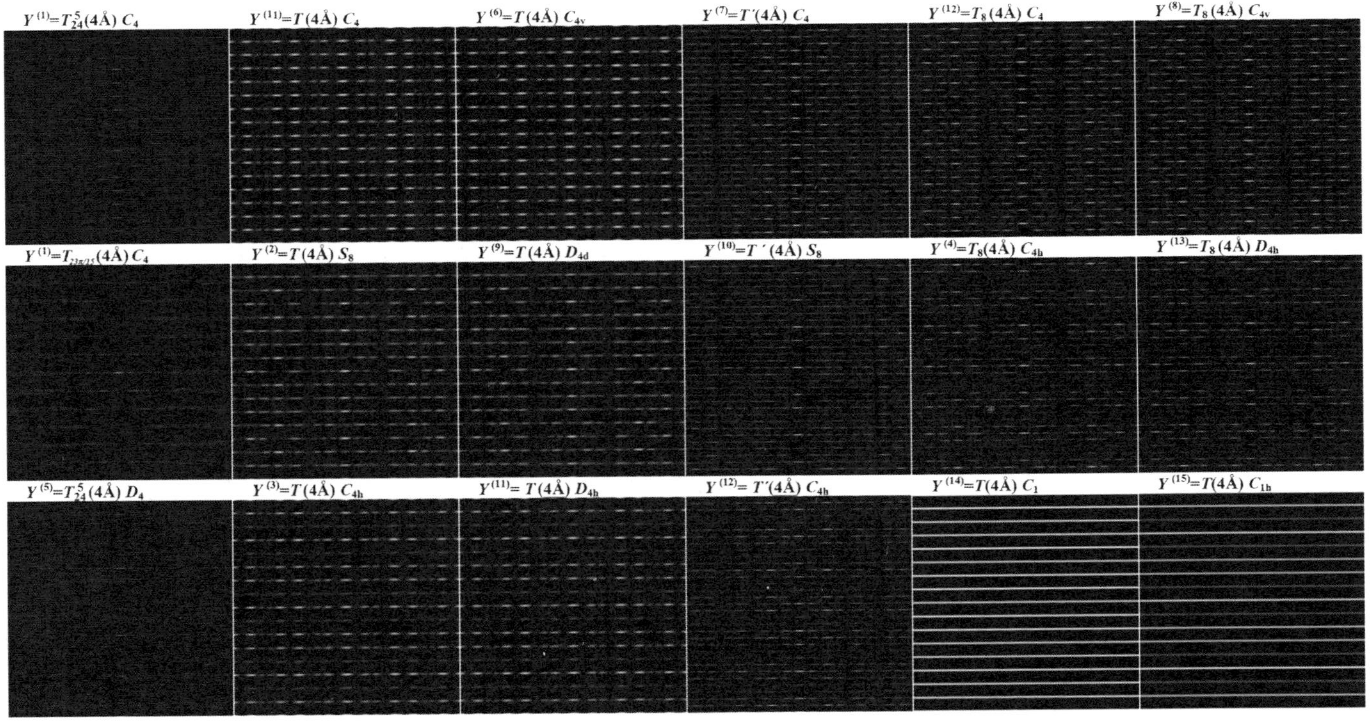

Fig. 8.5 Normal incidence patterns for conformation classes. For the same systems as in Fig. 8.4 the simulated diffraction intensity for normal incidence of the light along the direction $\Phi = 0$, except $\Phi = \pi/6$ for $\boldsymbol{Y}^{(7)}$, $\boldsymbol{Y}^{(10)}$, and $\boldsymbol{Y}^{(12)}$ (fourth column; odd layer lines extinct for $\Phi = 0$)

control is to compare if this way found parameter p agrees with the intensity gap in the closest to equatorial visible layer line, which in the commensurate case should correspond to J_p.

The achiral systems are commensurate with $\tilde{q}$ being 1 or 2. In the first case, $\boldsymbol{L}^{(1)} = \boldsymbol{T}(a)\boldsymbol{C}_n$ ($q = n$, $\tilde{q} = 1$ and $p = 0$) all the multiples nM of n are involved for each K, and all the layer lines are gapless, unless glide plane is present (classes $\boldsymbol{Y}^{(7)}$, $\boldsymbol{Y}^{(10)}$ and $\boldsymbol{Y}^{(12)}$ in the fourth column of Figs. 8.4 and 8.5). In fact, for odd layer lines $\sin Mn\varphi_A$ is a divisor of the Mth term of geometrical factor; besides appearing of *intensity gap* (due to vanishing of the term with zero order Bessel function), this yields *extinction* along $2n$ normal incidence directions

$$\Phi_j = \pi j/n, \; j = 0, \ldots, 2n-1. \tag{8.17b}$$

These characteristics are much alike those of the another achiral case $\boldsymbol{L}^{(1)} = \boldsymbol{T}^1_{2n}(a)\boldsymbol{C}_n$ (classes $\boldsymbol{Y}^{(1_2)}$, $\boldsymbol{Y}^{(4)}$, $\boldsymbol{Y}^{(8)}$, and $\boldsymbol{Y}^{(13)}$ in the last two right columns). Still, besides the angle of extinction of odd layer lines, the two cases differ in equatorial line: besides J_0, the number and radii of the circles can be numerically fitted either by J_{2n} for $\boldsymbol{L}^{(1)} = \boldsymbol{T}^1_{2n}(a/2)\boldsymbol{C}_n$, and J_n for $\boldsymbol{L}^{(1)} = \boldsymbol{T}(a)\boldsymbol{C}_n$. Generally, in the latter case the number of circles is less, while the number of white spots is greater. While for $\boldsymbol{Y}^{(1_2)}$, $\boldsymbol{Y}^{(4)}$ *extinction* directions are given by (8.17a), for $\boldsymbol{Y}^{(8)}$ and $\boldsymbol{Y}^{(13)}$ cosine factor disappears for φ_A independent directions

$$\Phi_j = \pi(2j+1)/2n, \; j = 0, \ldots, 2n-1. \tag{8.17c}$$

8.4.3.3 Parities

While for $\boldsymbol{Y}^{(1)}$ each Bessel function is multiplied by a Φ-dependent phase, coordinates of the orbit representative, and counters K and M. This multipliers equivalently ponder all the terms. On the contrary, parities in other conformation classes introduce cosine functions which may affect contribution of the terms: σ_h yields cosine dependence on K and z_A, σ_v on M and φ_A, glide plane on K, M, and φ_A, roto-reflectional plane on K, z_A, and M, and U-axis on the all four parameters. If depend on M, these cosine terms give additional radial modulation of the patterns in comparison to the $\boldsymbol{Y}^{(1)}$. Also, position of orbit representatives may have important effect on patterns (some layer lines may completely disappear).

Horizontal mirror plane infers factor $\cos^2(2\pi K z_a/a)$ in all the terms, making K-dependent intensities, but not changing the patterns of the layer lines. On the other hand, inferring φ_A and M depending factor, vertical mirror plane modulates relative contribution of the terms, thus changing patterns but not the relative intensity of the layer lines. Finally, U-axis and $\boldsymbol{S}_{2n}$ symmetry affect both the patterns and their relative intensities; these two are easily distinguished, as odd M terms disappear only in the latter cases.

8.4.4 Applications to the Multiorbit Systems

Above results for conformation classes, i.e., single orbit systems (like carbon nanotubes 9.2.1.8), give exhaustive characterization of the diffraction patterns of the quasi-one-dimensional crystals. For a multiorbit system with symmetry group $\boldsymbol{L}^{(F)}$ $(F = 1, \ldots, 13)$, all the orbits should be determined and then found in the column 2 of the Fth part of Table 3.4. For each of these orbits its conformation class is in the first column of the same row. Finally, we substitute scattering amplitudes of these conformation classes in (8.13) and obtain total scattering amplitude. The symmetry of the distribution is the intersection of the symmetries of the included conformation classes, and at least it is equal to $\boldsymbol{P}_{\mathcal{I}}^{(F)}$, being the isogonal group $\boldsymbol{P}^{(F)}$ of $\boldsymbol{L}^{(F)}$, extended by spacial inversion.

Finally, let us stress out that the presented results are obtained within kinematical approximation, meaning that their validity is restricted by the applicability of the model. Still, the symmetry of the diffraction space is model independent. In this context, it should be noted that the range of k is restricted in the real experiment, which leads to considerably fast fading of the patterns with increase of k. This may cause that some of the features discussed above are not easily visible. However, this varies with the type of the used beam; e.g., for X-ray diffraction the visible range of k is much greater than for electron diffraction.

8.5 Numerical Implementations of the Line Groups

As it has been discussed, symmetry group generates the whole structure from the *symcell*, i.e., a set containing one atom from each orbit (Sect. 2.1.4). Therefore, it is intuitively clear that symcell and symmetry group determine all the physical properties of the considered system. However, most of the numerical algorithms used to derive such properties do not use the full symmetry but just the translational invariance. This is satisfactory enough for two- and three-dimensional crystals, since in such systems translational group is a subgroup of low index of the total symmetry, making elementary cell at most several times larger than the symcell. However, in the systems with line group symmetry, the situation is frequently quite different. Incommensurate structure has no elementary cell, while symcell is well defined. As for commensurate systems, elementary cell is typically much larger than the symcell (e.g., for carbon nanotubes symcell consists of a single atom, while there are hundreds or thousands of them in elementary cell). In this context we briefly discuss numerical implementation of the full-line group symmetry. Generally, physical predictions are based on the known (electron and/or phonon) eigenenergies, causing that various numerical algorithms are devoted to spectral problem of effective hamiltonians obtained within different physical approximations. With respect to the adopted physical models, these algorithms can be roughly divided into two groups: tight-binding approximation and density functional theory. Their combinations are also developed. In both models full symmetry implementation effectively reduces calculations to the symcell with no further approximation, but in the conceptually different ways.

8.5.1 Tight-Binding Methods

In the tight-binding electronic models we empirically chose several relevant orbitals $| A; \psi\rangle$ ($\psi = 1, \dots, n_A$) from each atom A, building the total tight-binding state space $\mathcal{S}$ by linear combinations of all of them. As for phonons, $\mathcal{S}$ is the space of the atomic displacements, having the same structure, with three independent displacement $| A; \psi\rangle$ ($\psi = 1, 2, 3$) of each atom used instead of the atomic orbitals. This space is infinite dimensional, as the systems with line group symmetry are with infinitely many atoms. Hamiltonian is defined according to the model of interaction: an assumption on the mutually interacting atoms is made (e.g., the first or higher neighbors are considered, depending on the range of the interaction), and interaction for each such pair is defined by a number of parameters. These are essentially hamiltonian matrix elements, such as hopping integrals or force constants, but the model treats them as the empirical data suitable for further optimization; in the hybrid density functional tight-binding methods, these matrix elements are calculated by density functional procedure. Hamiltonian is built up by collecting all of them and it remains to apply an eigenvalue code.

At this stage symmetry may be efficiently implemented due to the special (so-called inductive) structure of the space and hamiltonian. The total space is obviously the direct sum of the orbit subspaces, which are built from the orbitals of atoms being on the same orbit. In fact, if A is symcell atom, then its orbit (Sect. 3.1) is generated by the action of the corresponding transversal $\boldsymbol{Y}_A$ on A. Symmetry of the system requires that the orbitals of all other atoms on this orbit are physically the same as for A, meaning that they are also generated by the action of the transversal elements y_A:

$$|y_A A; \psi\rangle = y_A\, |A; \psi\rangle. \tag{8.18}$$

Therefore, arbitrary state may be expanded in the form

$$|\psi\rangle = \sum_{A=1}^{S} \sum_{y_A} \sum_{\psi} c_{A y_A \psi}\, y_A\, |A; \psi\rangle. \tag{8.19}$$

Simultaneously, the matrix elements must obey the same symmetry requirement: interaction of pairs A-B and ℓA-ℓB is the same for any symmetry transformation ℓ. Therefore, it is intuitively clear that only the space $\mathcal{S}_S$ of the orbitals/displacements of the symcell atoms determines the whole space and total hamiltonian and that the eigenproblem can be reduced to this space. It is important that this is finite dimensional problem, as the symcell contains finite number S of atoms, and the dimension of $\mathcal{S}_S$ is $|\mathcal{S}_S| = \sum_{A=1}^{S} n_A$.

The described structure of the total space enables to implement full symmetry of the system and to reduce the eigenvalue problem to the finite dimensional one. The procedure, briefly reviewed in Appendix F, is known the *modified group projector technique*. Suppose that there is common transversal $\boldsymbol{Y}$ of all the symcell atoms, and that there is (finite) subgroup $\boldsymbol{P}$ of $\boldsymbol{L}$ such that $\boldsymbol{Y}$ is the set of its coset representatives: $\boldsymbol{L} = y_1\boldsymbol{P} + y_2\boldsymbol{P} + \dots$. In fact, this is always satisfied for mono-orbit systems or systems with the same type of orbits, when $\boldsymbol{P}$ is the stabilizer of the orbit. If this is not the case, the procedure may be further elaborated to use symcell only, but for simplicity we consider here a case when there is a common transversal: we consider larger set of atoms (i.e., S is larger than the number of symcell atoms) such that it generates the same structure by a transversal $\boldsymbol{Y}$ (being at most intersection of the original transversals) and use $\boldsymbol{P}$ which intertwines (at most) these atoms. Due to the product structure of the line groups, monomer has this property, with the transversal and $\boldsymbol{P}$ being generalized translations and point factor, respectively.

According to these assumptions, symmetry-adapted eigenvector is

$$|\lambda l; t_\lambda\rangle = \sum_{A=1}^{S}\sum_{y}\sum_{\psi} c_{Ay\psi}^{(\lambda l;t_\lambda)}\,|yA;\psi\rangle. \tag{8.20}$$

However, the expansion coefficients $c_{Ay\psi}^{(\lambda l;t_\lambda)}$ are related to those corresponding to the symcell atoms, $c_{A\psi}^{(\lambda l;t_\lambda)}$, because the orbitals within same orbit are related by the symmetry transformations (8.18), while the total state obeys the rule (8.1). When these rules are substituted in (8.20) we get

$$|\lambda l; t_\lambda\rangle = \frac{1}{\sqrt{|\boldsymbol{Y}|}}\sum_{y\in\boldsymbol{Y}}\sum_{l' A\psi_A} c_{A\psi_A}^{(\lambda l';t_\lambda)} D_{ll'}^{(\lambda)^*}(y)\,|y_t A, \psi_A\rangle. \tag{8.21}$$

It remains to find the coefficients $c_{A\psi_A}^{(\lambda l';t_\lambda)}$. To this end we construct *pulled down* hamiltonian matrix $H_{AB}^{\downarrow\lambda}$, with the submatrices corresponding to the atoms of symcell

$$H_{AB}^{\downarrow\lambda} = \sum_{y=1}^{N_B^A}\sum_{\psi_A\psi_B}\langle A, \psi_A|\,H\,|yB, \psi_B\rangle\,|B\psi_B\rangle\langle A\psi_A|\otimes D^{(\lambda)^T}(y). \tag{8.22}$$

Here, for the fixed atom A, we sum over the transversal elements y for which the pair A-yB is within the range of interaction. Solving the common eigenproblem of $H_{AB}^{\downarrow\lambda}$ and of the projector $L^{\lambda\downarrow} = \frac{|\lambda|}{|\boldsymbol{P}|}\sum_{\ell\in\boldsymbol{P}} D^{\downarrow}(\ell)\otimes D^{(\lambda)^*}(s)$. There are f^λ eigenvectors of $H_{AB}^{\downarrow\lambda}$ (with the eigenvalues $E_{\lambda t_\lambda}$) which are in the range of $L^{\lambda\downarrow}$. They have form $|0;\lambda t_\lambda\rangle = \sum_{lA\psi_A} c_{A\psi_A}^{(\lambda l;t_\lambda)}\,|A\psi_A\rangle\langle\lambda l\,|$, enabling to determine these coefficients, and therefore the symmetry-adapted eigenvectors (8.21) with the same eigenvalue $E_{\lambda t_\lambda}$ (degenerate $|\lambda|$ times).

To summarize, the energy bands of the systems with line group symmetry can be found by solving eigenvalue problem of the matrix $H^{\downarrow\lambda}$ for each irreducible representation. Therefore, assignation of the bands by the complete set of conserved quantum numbers is automatically obtained, together with the symmetry-adapted eigenbasis. The dimension of this matrix is equal to the dimension of the irreducible representation times number of orbitals included in the extension of symcell described above (monomer, at most). For mono-orbit systems only single atom of symcell is considered, which greatly simplifies problem, allowing in some cases (e.g., carbon nanotubes, Sect. 9.2.1.4) analytical solution.

8.5.2 Density Functional Relaxation

The most precise numerical methods to find stable configurations of structures are based on quantum mechanical density functional theory. Although the success of this theory stems from the better treatment of the electronic correlations than in other methods, density functional theory is still an approximation. Therefore, many diverse density functional algorithms are adopted in order to improve predictions of particular physical properties. Common characteristics of all the variants are that the configuration of the system is slightly varied in the vicinity of some initial position, and the configuration corresponding to the minimum of the total energy (including electronic and ionic) is considered as the stable one. This procedure is called *relaxation*. Concerning consumed computer time, the critical parameters are number of atoms of the studied structure and the dimension of the approximate quantum mechanical state space.

Number of atoms in the structure is important, as their coordinates are to be varied. Currently, the structures with not more than a few hundreds of atoms may be successfully relaxed. For the crystals, the methods implementing translational symmetry are elaborated to enable relaxation of elementary cell only. Still, as it has been stressed out, quasi-one-dimensional crystals frequently have large elementary cells, with sufficiently many atoms to prevent calculations, and for incommensurate systems such approach is completely inapplicable. However, application of full symmetry for the most of the studied structures enables acceptably rapid calculations. In fact, according to the topological theorem of Abud and Sartori [16], extremes of the total energy correspond to the special configurations where the symmetry is increased with respect to the vicinity. Further, a slight change of the position of an atom of the structure with line group symmetry, will reduce the total symmetry significantly, meaning that this is not a stable configuration. Therefore, it is sufficient to vary only those very special collective coordinates (combining coordinates of various atoms) which do not diminish the symmetry. Obviously, as the atoms within same orbit are all obtained by the action of the symmetry group from a single one, only one atom per orbit may be independently changed. This way, it is obvious that only the symcell atoms coordinates are to be varied. Note that in carbon nanotubes, despite large number of atoms in the elementary cell, symcell contains only one atom. Some symmetry operation may further forbid variation of some of the coordinates; in particular, atoms being in the mirror/glide plane must not change the perpendicular to the plane coordinate.

In addition to the symcell atom coordinates, there are also parameters of the symmetry group itself which can be varied without diminishing symmetry. To understand this, note that the line groups of the same family with same n are isomorphic, independent of Q and f. Therefore, change of these continual parameters do not diminish symmetry. Accordingly, in the numerical relaxation Q and f should also be varied. While f is responsible for spacing of atoms along z-axis, helicity of the system is governed by Q. Note that even when the initial configuration is commensurate, the relaxed one may not be, such although the numerical floating point calculations will give rational result for Q. In particular, such effect may be expected when externals fields or mechanical influence [17] like twisting is applied. The question, of commensurability of quasi-one-dimensional solids will be referred to in the context of carbon nanotubes in Sect. 9.2.

The number of probe functions, i.e., the dimensionality of the approximate quantum mechanical state space of electrons is a compromise between the memory and run-time on the one side and accuracy on the other. Usually a regular greed of several hundreds to thousand plane waves is used. However, the symmetry inspires a new scheme. It can be expected (as illustrated by carbon nanotubes in Sect. 9.2.1.9) that in the expansion of the ground state density (which is invariant function) over line group harmonics, only several lowest harmonics have significant contribution. Even in the cases of the most complex quasi-one-dimensional systems, the number of such harmonics is less than hundred. Therefore, using basis of the lowest harmonics, the dimension of the numerical problem is reduced and the comparative accuracy is improved; namely, the contribution of the neglected part of the Hilbert space is a priori known to be negligible.

References

1. M. Damnjanović, I. Milošević, T. Vuković, J. Maultzsch, J. Phys. A **36**, 5707 (2003)
2. M. Damnjanović, T. Vuković, I. Milošević, J. Phys. A **33**, 6561 (2000)
3. T. Vuković, I. Milošević, M. Damnjanović, Phys. Rev. B **65**, 045418 (2002)
4. I. Božović, J. Delhalle, Phys. Rev. B **29**, 4733 (1984)
5. E. Ascher, J. Phys. C: Solid State Phys. **10**, 1365 (1977)
6. L.D. Landau, E.M. Lifshitz, *The Classical Theory of Fields* (Elsevier, Burlington, 1980)
7. H.A. Jahn, E. Teller, Proc. Roy. Soc. A **161**, 220 (1937)
8. I. Milošević, M. Damnjanović, Phys. Rev. B **47**, 7805 (1993)
9. I. Milošević, B. Nikolić, M. Damnjanović, M. Krčmar, J. Phys. A **31**, 3625 (1998)
10. I. Božović, N. Božović, J.Phys. A: Math. Gen **22**, 145 (1989)
11. I. Božović, N. Božović, J.Phys. A: Math. Gen **23**, 2775 (1990)
12. I. Božović, N. Božović, J.Phys. A: Math. Gen **23**, 5131 (1990)
13. J.C.W. Chien, *Polyacetylene* (Academic Press, New York, 1984)
14. W.P. Su, J.R. Schrieffer, A.J. Hegger, Phys. Rev. Lett. **42**, 1698 (1979)
15. T. Vuković, I. Milošević, M. Damnjanović, Phys. Rev. B **79**, 165439 (2009)
16. H. Abud, G. Sartori, Ann. Phys. **150**, 307 (1983)
17. Y. Li, S.V. Rotkin, U. Ravaioli, Nano Lett. **3**, 183 (2003)

Chapter 9
Nanotubes

Abstract The first reports on synthesis of carbon [1] and transition metal chalcogenide [2] nanotubes triggered extensive research on both organic and inorganic nanostructures which proved to have potential of becoming a key nanotechnological material due to the outstanding physical properties. It was found that many compounds which crystallize in a bulk or/and in a layered form can grow into the cylindrical structures, under specific conditions. After the discoveries of nanotubes made of carbon, transition metal chalcogenides and oxides [3], boron nitride [4], silicon [5] and metal (e.g., Au [6]), recent discovery of the functional semiconducting oxide nanostructures [7] paved the way for synthesis of diverse nanosized forms of zinc oxide as well. Diameters of the synthesized nanotubes (or lateral dimensions of the other nanostructures) vary from few Angstroms to few micrometers. In this chapter we derive symmetry of arbitrary nanotubes and discuss their common symmetry-based properties. Then we focus on carbon nanotubes: deriving easily many of their famous properties, we show that symmetry is the most profound way of understanding them.

9.1 Symmetry of Nanotubes

The so-called layer folding offers the simplest description and parametrization of the structure of nanotube and also enables to derive their symmetry. Let us assume that the two-dimensional layer has lattice basis vectors $\boldsymbol{A}_1$ and $\boldsymbol{A}_2$, chosen such that $A_1 \geq A_2$ and the angle between them is $\alpha \in (0, \pi/2]$. Then we define dimensionless parameters X and Y:

$$X = \frac{A_1^2}{A_2^2} \geq 1, \quad Y = \frac{A_1}{A_2} \cos\alpha \geq 0. \tag{9.1}$$

The nanotube (n_1, n_2) is obtained by folding the layer in a way that the *chiral vector* $\boldsymbol{c} = (n_1, n_2) = n_1\boldsymbol{A}_1 + n_2\boldsymbol{A}_2$ becomes circumference of the tube (Fig. 9.1). Alternatively, nanotube is defined by length c (giving the tube's diameter $D = c/\pi$) and slope θ (called *chiral angle*) of the chiral vector $\boldsymbol{c}$:

Damnjanović, M., Milošević, I.: *Nanotubes*. Lect. Notes Phys. **801**, 143–169 (2010)
DOI 10.1007/978-3-642-11172-3_9

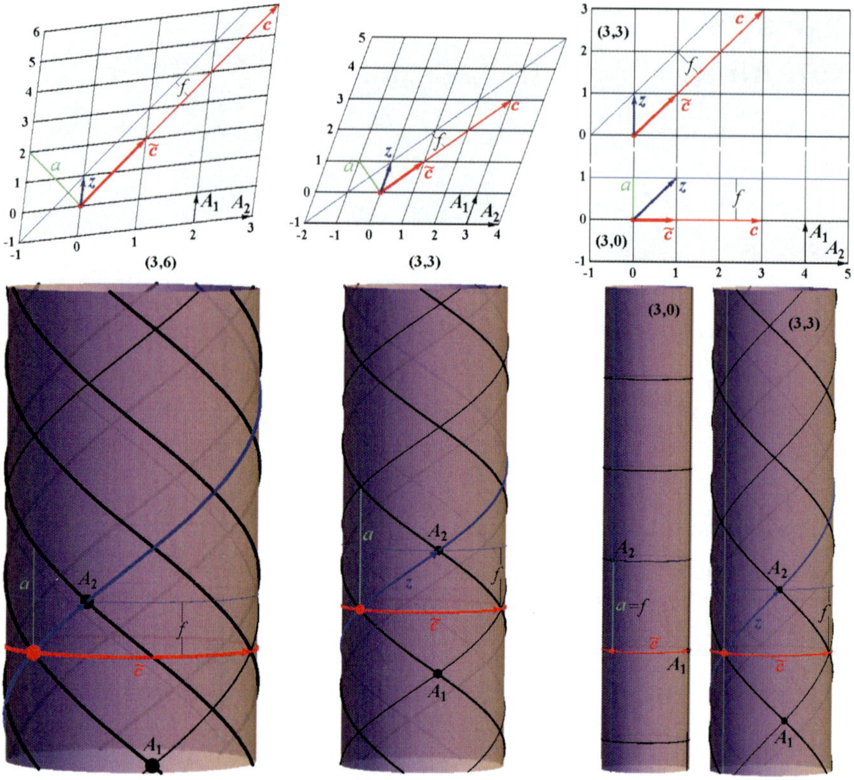

Fig. 9.1 Layer folding. Two dimensional lattices (*up*), defined by the basis $\boldsymbol{A}_1$ and $\boldsymbol{A}_2$, are rolled to the nanotubes (*below*) according to the chiral vectors $\boldsymbol{c}$ (*red*). The vectors $\tilde{\boldsymbol{c}}$ (*red*) and $\boldsymbol{z}$ (*blue*, line shows other possible choices $\boldsymbol{z}_s$) are roto-helical generators C_n and Z, while f if fractional translation, and for commensurate cases a is nanotube period (*green*). *Left*: layer $\boldsymbol{A}_1 = (12, \sqrt{2})$, $\boldsymbol{A}_2 = (\frac{1}{\sqrt{2}}, 6)$ $(X = w = 4,\ Y = J + X = 24\sqrt{2}/73,\ x = 0,\ y = 1)$; nanotube $\boldsymbol{c} = (3, 6)$ $(\boldsymbol{z} = (0, 1),\ \boldsymbol{L}^{(1)} = \boldsymbol{T}^1_{12}(f = \frac{a}{4} = 3\sqrt{2} - \frac{1}{2})\boldsymbol{C}_3,\ a = 4f)$. *Middle*: rhombic layer with $\alpha = 70°$ (Sect. 9.1.3); nanotube $\boldsymbol{c} = (3, 3)$ $(\boldsymbol{z} = (0, 1),\ \boldsymbol{L}^{(1)} = \boldsymbol{T}^1_6(f = a/2 \approx 0.57A_1)\boldsymbol{C}_3,\ a = 2f)$. *Right*: rectangular layer $\boldsymbol{A}_1 = (\pi, 0)$, $\boldsymbol{A}_2 = (0, 3)$ (Sect. 9.1.3); nanotubes $\boldsymbol{c} = (3, 0)$ $(\boldsymbol{z} = (1, 1),\ \boldsymbol{L}^{(1)} = \boldsymbol{T}^1_3(f = a = A_2)\boldsymbol{C}_3)$ and $\boldsymbol{c} = (3, 3)$ $(\boldsymbol{z} = (0, 1),\ \boldsymbol{L}^{(1)} = \boldsymbol{T}^1_{\frac{9+\pi^2}{3}}(f = \frac{3\pi}{\sqrt{9+\pi^2}})\boldsymbol{C}_3$, incommensurate tube)

$$c = A_2\sqrt{n_1^2 X + n_2^2 + 2n_1 n_2 Y}, \quad \sin\theta = n_2 A_2 / c. \tag{9.2}$$

It is enough to consider nanotubes with $n_2 \geq 0$, i.e., $0 \leq \theta < \pi$, as indices $(-n_1, -n_2)$ and (n_1, n_2) describe the same nanotube structure.

Nanotubes, being quasi-one-dimensional systems have line group symmetry. From the very beginning, it has been recognized [8, 9] that such a large symmetry substantially determines fundamental physical characteristics of carbon nanotubes [10, 11]. Additionally, it had proved to be important technically facilitating the calculations. For example, only due to the symmetry, the electronic bands of carbon nanotubes are analytically calculated in a good approximation, which has been the most important result for further development of the field.

9.1.1 Folded Translations: The First Family Subgroup

The translations of the layer become roto-helical operations on the tube, i.e., two-dimensional translational group is folded into the first family subgroup $\boldsymbol{L}^{(1)}$ of the nanotube line group. In particular, the minimal lattice vector collinear with $\boldsymbol{c}$ is the reduced chiral vector $\tilde{\boldsymbol{c}} = \boldsymbol{c}/n = \tilde{n}_1\boldsymbol{A}_1 + \tilde{n}_2\boldsymbol{A}_2$, where[1] $n = \mathrm{GCD}(n_1, n_2)$ (hence $\tilde{n}_1$ and $\tilde{n}_2$ are co-primes). It corresponds to the minimal pure rotation around the nanotube axis, generating $\boldsymbol{C}_n$ (the circle is closed after n successive rotations, $\boldsymbol{c} = n\tilde{\boldsymbol{c}}$). Further, $\boldsymbol{L}^{(1)}$ contains combined transformations from $\boldsymbol{Z}$ and $\boldsymbol{C}_n$, which means that the $(C_Q|f)$, generator of $\boldsymbol{Z}$, must correspond to a lattice vector $z = z_1\boldsymbol{A}_1 + z_2\boldsymbol{A}_2$, such that z and $\tilde{\boldsymbol{c}}$ form a basis of the two-dimensional lattice. The area equality $|\boldsymbol{A}_1 \times \boldsymbol{A}_2| = |\tilde{\boldsymbol{c}} \times z|$ gives Diophantine equation $\tilde{n}_2 z_1 - \tilde{n}_1 z_2 = \pm 1$. Positive f is provided by -1 on the right, when the solutions (C.8) are

$$z_s = (z_{1s}, z_{2s}) = z_0 + s(\tilde{n}_1, \tilde{n}_2), \quad s = 0, \pm 1, \ldots \tag{9.3}$$

$$z_0 = \begin{cases} (0,1), & \text{if } \boldsymbol{c} = (n, 0), \\ (-1, 0), & \text{if } \boldsymbol{c} = (0, n), \\ \left(\tilde{n}_2^{\mathrm{Eu}(\tilde{n}_1)-1}, \dfrac{\tilde{n}_2^{\mathrm{Eu}(\tilde{n}_1)} - 1}{\tilde{n}_1}\right), & \text{otherwise.} \end{cases} \tag{9.4}$$

The series z_s is on the line parallel to the chiral vector, for $f = |\boldsymbol{A}_1 \times \boldsymbol{A}_2|/\tilde{c}$ away from it. All z_s correspond to the generators $(C_{Q_s}|f)$ of the variety of screw-axes, which combine with $\boldsymbol{C}_n$ into the same roto-helical group (Sect. 2.2.2); within convention C0 (2.11), among all z_s we chose z as a vector making minimal (but strictly positive) angle with the vector perpendicular to $\boldsymbol{c}$. Obviously, the rotation for $2\pi/Q$ corresponds to the projection of z onto $\boldsymbol{c}$, giving $Q = c^2/(\boldsymbol{c} \cdot z)$. Altogether, we finally get

$$\boldsymbol{L}^{(1)} = \boldsymbol{T}_Q(f)\boldsymbol{C}_n, \tag{9.5a}$$

$$n = \mathrm{GCD}(n_1, n_2), \tag{9.5b}$$

$$f = A_1 \frac{\sin\alpha}{\sqrt{\tilde{n}_1^2 X + \tilde{n}_2^2 + 2\tilde{n}_1\tilde{n}_2 Y}}, \tag{9.5c}$$

$$Q = n \frac{\tilde{n}_1^2 X + \tilde{n}_2^2 + 2\tilde{n}_1\tilde{n}_2 Y}{\tilde{n}_1 z_1 X + \tilde{n}_2 z_2 + (\tilde{n}_1 z_2 + \tilde{n}_2 z_1) Y}. \tag{9.5d}$$

---

[1] Recall that $\tilde{x} = x/n$ (division by the line group parameter n), while $\overline{x}$, $\underline{x}$ denote numerator and denominator of the rational $x = \overline{x}/\underline{x}$.

In conclusion, the roto-helical part $\boldsymbol{L}^{(1)}$ of the nanotube symmetry generates the whole tube from a single two-dimensional unit cell. The symmetry parameters f and $\tilde{Q} = Q/n$ depend only on $\tilde{\boldsymbol{c}}$ and thus they are the same for the ray of the nanotubes $n(\tilde{n}_1, \tilde{n}_2)$ differing by the order n of the principle axis.

9.1.2 Commensurability

As pointed out, nanotube is commensurate if Q is rational. However, instead of analyzing (9.5d), we directly check whether nanotube has a translational period. Obviously, if the translational vector $\boldsymbol{a} = a_1\boldsymbol{A}_1 + a_2\boldsymbol{A}_2$ exists, it is the minimal lattice vector (i.e., a_1 and a_2 are co-primes) orthogonal onto the chiral vector. This shows that the commensurability condition is the equation

$$\tilde{\boldsymbol{c}} \cdot \boldsymbol{a} = a_2\tilde{n}_2 + a_1\tilde{n}_1 X + (a_2\tilde{n}_1 + a_1\tilde{n}_2)Y = 0, \tag{9.6}$$

solvable in co-prime integers a_i. Period of such tube is the length of $\boldsymbol{a}$:

$$a = A_2\sqrt{a_1^2 X + a_2^2 + 2a_1a_2Y}. \tag{9.7}$$

Let q be the number of the layer lattice unit cells within a translational period a of a nanotube. Then the equality of the surface areas, $qA_1A_2\sin\alpha = ca$, gives

$$q = n\frac{\tilde{c}a}{A_2^2\sqrt{X - Y^2}}. \tag{9.8}$$

Recall that the real numbers may be viewed as an infinite dimensional vector space over the rational numbers. Therefore, as only X and Y may be irrational, commensurability condition (9.6) requires that 1, X, and Y are rationally dependent. In other words, either both X and Y are rational or there are rational w, x, and y (with $x \neq y$ as $X > Y$) and irrational J, such that $X = w + xJ$ and $Y = w + yJ$.

When both X and Y are rational, then (9.6) is a (rational) proportion between a_1 and a_2. Consequently, any nanotube (n_1, n_2) is commensurate with

$$q = n\frac{2\tilde{n}_1\tilde{n}_2\underline{X}\overline{Y} + \tilde{n}_1^2\overline{X}\underline{Y} + \tilde{n}_2^2\underline{XY}}{\mathrm{GCD}(\tilde{n}_1\underline{X}\overline{Y} + \tilde{n}_2\underline{XY}, \tilde{n}_2\underline{X}\overline{Y} + \tilde{n}_1\overline{X}\underline{Y})}, \tag{9.9}$$

$$\boldsymbol{a} = \frac{(\tilde{n}_1\underline{X}\overline{Y} + \tilde{n}_2\underline{XY}, -\tilde{n}_2\underline{X}\overline{Y} - \tilde{n}_1\overline{X}\underline{Y})}{\mathrm{GCD}(\tilde{n}_1\underline{X}\overline{Y} + \tilde{n}_2\underline{XY}, \tilde{n}_2\underline{X}\overline{Y} + \tilde{n}_1\overline{X}\underline{Y})}. \tag{9.10}$$

In the other case, rational and irrational parts of (9.6) give a system of two homogeneous equations in a_i:

$$a_1 w(\tilde{n}_1 + \tilde{n}_2) + a_2(\tilde{n}_1 w + \tilde{n}_2) = 0, \tag{9.11a}$$
$$a_1(\tilde{n}_1 x + \tilde{n}_2 y) + a_2 \tilde{n}_1 y = 0. \tag{9.11b}$$

It is solvable when the determinant of the system vanishes:

$$\tilde{n}_1^2 w(y - x) - \tilde{n}_1 \tilde{n}_2 x - \tilde{n}_2^2 y = 0. \tag{9.12}$$

This constraint on $\tilde{n}_1$ and $\tilde{n}_2$ singles out a subset of the chiral vectors yielding commensurate nanotubes. As $\boldsymbol{a}$ is fully determined by the reduced chiral vector $\tilde{\boldsymbol{c}}$, all the chiral vectors $n\tilde{\boldsymbol{c}}$ ($n = 1, 2, \ldots$) give commensurate nanotubes with period $\boldsymbol{a}$. Besides, when the roles of $\tilde{\boldsymbol{c}}$ and $\boldsymbol{a}$ are interchanged, a nanotube with the period $\tilde{\boldsymbol{c}}$ (orthogonal onto $\boldsymbol{a}$) is obtained. Hence, if such a lattice allows commensurate nanotubes, their chiral vectors lie on two perpendicular lines.

For $\tilde{n}_2 \neq 0$, the constraint (9.12) becomes

$$\frac{\tilde{n}_1}{\tilde{n}_2} = \frac{x \pm \sqrt{x^2 - 4wxy + 4wy^2}}{2w(y - x)} = \nu_\pm, \tag{9.13}$$

i.e., $\tilde{n}_1/\tilde{n}_2$ is rational only if $\sqrt{x^2 - 4wxy + 4wy^2}$ is. This singles out mutually orthogonal directions $\tilde{\boldsymbol{c}}^\pm = \boldsymbol{a}^\mp = (\overline{\nu}_\pm, \underline{\nu}_\pm)$ of commensurate nanotubes $\boldsymbol{c}^\pm = n\tilde{\boldsymbol{c}}^\pm$, with $q = n(\underline{\nu}_+ \overline{\nu}_- - \underline{\nu}_- \overline{\nu}_+)$ and $a^\pm = A_2\sqrt{\underline{\nu}_\mp^2 + \overline{\nu}_\mp^2 X + 2\underline{\nu}_\mp \overline{\nu}_\mp Y}$.

The case $\tilde{n}_2 = 0$ appears if and only if $w = 0$, meaning $Y = yX$ (i.e., $J = X$ and $x = 1$). Then the ray orthogonal onto $\tilde{\boldsymbol{c}}^+ = (1, 0)$ is obtained from (9.11b): $\tilde{\boldsymbol{c}}^- = (\overline{y}, \underline{y})$. The corresponding periods are $a^+ = A_1 \overline{y} |\tan\alpha|$ and $a^- = A_1$, while $q = n\underline{y}$.

Finally, as for all the commensurate nanotubes we have found q, n, and a, to complete the first family subgroup determination we use (9.5d) to calculate $r = q/Q$.

9.1.3 Additional Symmetries

Apart from the translational invariance, two-dimensional lattice has rotational $\boldsymbol{C}_2$ symmetry (rotation for π around the axis perpendicular to the layer). In addition, rhombic and rectangular lattices have vertical mirror and glide planes and also in rhombic rectangular and hexagonal lattices the order of the rotational axis is four and six, respectively. Particular atomic arrangements within the lattice unit cell may reduce the lattice symmetry group, and this way one of the 80 diperiodic groups [12] are obtained. Only some of these non-translational layer symmetries are preserved after layer folding into a nanotube: twofold rotational axis, mirror, and glide planes. When combined with the roto-helical group $\boldsymbol{L}^{(1)}$ (emerging from the lattice translations) given by (9.5), these additional symmetries yield line groups of the remaining 12 families.

Rotation C_2 of a layer becomes horizontal twofold axis, the U-axis, of the tube. Thus, whenever order of the principle axis of the layer is two, four, or six, symmetry of the nanotube is the fifth family line group $\boldsymbol{T}_Q(f)\boldsymbol{D}_n$ at least. Note that the higher order rotational symmetries of the layer do not give rise to the symmetry of nanotubes.

Vertical mirror (glide) plane is preserved in the nanotube only if the chiral vector is perpendicular onto it. When $\boldsymbol{c}$ is parallel to the plane, nanotube gets horizontal mirror (roto-reflectional) plane. All these transformations can be combined (Table 9.1) only with the roto-helical groups $\boldsymbol{T}(a)\boldsymbol{C}_n$ or $\boldsymbol{T}^1_{2n}(a)\boldsymbol{C}_n$ (i.e., $\tilde{q} = 1, 2$) of the achiral nanotubes.

First, we consider rectangular lattices, $\alpha = \pi/2$ (i.e., $Y = 0$). For irrational $X = J$, we have $w = y = 0$ (then $\underline{y} = 1$) and $x = 1$, yielding $\tilde{c}^+ = (1, 0)$ and $\tilde{c}^- = (0, 1)$, with $\tilde{q} = 1$, i.e., the helical factor reduces to the pure translational group. For X rational, from (9.9) we get $\tilde{q} = (\tilde{n}_1^2\overline{X} + \tilde{n}_2^2\underline{X})/\mathrm{GCD}(\tilde{n}_1, \underline{X})\mathrm{GCD}(\tilde{n}_2, \overline{X})$. Thus, for $X \neq 1$, the same result as for X irrational is achieved, while in the case $X = 1$ (square lattice) additionally $\tilde{q} = 2$ is obtained for $\tilde{c}^{\pm} = (\pm 1, 1)$.

Second, in the case of rhombic lattices $A_1 = A_2$ ($X = 1$) for Y irrational, taking $J = Y - 1$, we have $w = y = 1$ and $x = 0$ (then $\underline{x} = 1$), yielding $\tilde{c}^{\pm} = (\pm 1, 1)$ with $\tilde{q} = 2$, i.e., $\boldsymbol{L}^{(1)} = \boldsymbol{T}^1_{2n}(a/2)\boldsymbol{C}_n$. For rational Y, from (9.9) we get $\tilde{q} = (2\tilde{n}_1\tilde{n}_2\overline{Y} + (\tilde{n}_1^2 + \tilde{n}_2^2)\underline{Y})/\mathrm{GCD}(\tilde{n}_1\overline{Y} + \tilde{n}_2\underline{Y}, \tilde{n}_1\underline{Y} + \tilde{n}_2\overline{Y})$, allowing the same $\tilde{c}^{\pm}$ as for Y irrational. Only for $Y = 0$ and $Y = 1/2$ the additional pair, $\tilde{\boldsymbol{c}}^+ = (1, 0)$

Table 9.1 "Rolling-up" correspondence of the line and diperiodic groups. For each family F of the line groups after roto-helical subgroup $\boldsymbol{L}^{(1)}$ and the isogonal point group $\boldsymbol{P}^I_q$ (for irrational Q in the families 1 and 5, q is infinite), follow corresponding diperiodic groups enumerated according to [12]: for arbitrary chiral vector rolling gives either the first or the fifth family line group; only for special chiral vector(s) $\boldsymbol{a} = (n, 0)$, $\boldsymbol{b} = (0, n)$, $\boldsymbol{c} \in \{(n, 0), (0, n)\}$, $\boldsymbol{d} = (n, n)$, $\boldsymbol{e} = (-n, n)$, $\boldsymbol{f} \in \{(n, n), (-n, n)\}$, $\boldsymbol{g} \in \{(n, 0), (0, n), (-n, n)\}$, $\boldsymbol{h} \in \{(n, n), (-n, 2n), (-2n, n)\}$, $\boldsymbol{i} \in \{(n, 0), (0, n), (-n, n), (n, n), (-n, 2n), (-2n, n)\}$ the underlined groups (repeated after the corresponding vectors) give other line group families below

F	$\boldsymbol{L}^{(1)}$	$\boldsymbol{P}^I_q$	Diperiodic group
1	$\boldsymbol{T}_Q\boldsymbol{C}_n$	$\boldsymbol{C}_q$	1,2,4,5,8,9,10,$\underline{11}$,$\underline{12}$,$\underline{13}$,$\underline{14}$, $\underline{15}$,$\underline{16}$,$\underline{17}$,$\underline{18}$,$\underline{27}$,$\underline{28}$,$\underline{29}$,$\underline{30}$, $\underline{31}$,$\underline{32}$,$\underline{33}$,$\underline{34}$,$\underline{35}$,$\underline{36}$,65,66,67,68, $\underline{69}$,$\underline{70}$,$\underline{71}$,$\underline{72}$,74,$\underline{78}$,$\underline{79}$
5	$\boldsymbol{T}_Q\boldsymbol{C}_n$	$\boldsymbol{D}_q$	3,6,7,19,20,21,22,$\underline{23}$,$\underline{24}$,$\underline{25}$,$\underline{26}$, $\underline{37}$,$\underline{38}$,$\underline{39}$,$\underline{40}$,$\underline{41}$,$\underline{42}$,$\underline{43}$, $\underline{44}$,$\underline{45}$,$\underline{46}$,$\underline{47}$,$\underline{48}$, 49,50,51,52,53,54,$\underline{55}$,$\underline{56}$,$\underline{57}$,$\underline{58}$,$\underline{59}$,$\underline{60}$, $\underline{61}$,$\underline{62}$,$\underline{63}$,$\underline{64}$,73,75,76,$\underline{77}$,$\underline{80}$
2	$\boldsymbol{T}\boldsymbol{C}_n$	$\boldsymbol{S}_{2n}$	$\boldsymbol{a}$:17,33,34; $\boldsymbol{b}$:12,16,29,30
3	$\boldsymbol{T}\boldsymbol{C}_n$	$\boldsymbol{C}_{nh}$	$\boldsymbol{a}$:11,14,15,27,31,32; $\boldsymbol{b}$:28
4	$\boldsymbol{T}^1_{2n}\boldsymbol{C}_n$	$\boldsymbol{C}_{2nh}$	$\boldsymbol{e}$:13,18,35; $\boldsymbol{d}$:36; $\boldsymbol{h}$:69,72,78; $\boldsymbol{g}$:70,71,79
6	$\boldsymbol{T}\boldsymbol{C}_n$	$\boldsymbol{C}_{nv}$	$\boldsymbol{a}$:28; $\boldsymbol{b}$:11,14,15,27,31,32
7	$\boldsymbol{T}\boldsymbol{C}_n$	$\boldsymbol{C}_{nv}$	$\boldsymbol{a}$:12,16,29,30; $\boldsymbol{b}$:17,33,34
8	$\boldsymbol{T}^1_{2n}\boldsymbol{C}_n$	$\boldsymbol{C}_{2nv}$	$\boldsymbol{e}$:36; $\boldsymbol{d}$:13,18,35; $\boldsymbol{h}$:70,71,79; $\boldsymbol{g}$:69,72,78
9	$\boldsymbol{T}\boldsymbol{C}_n$	$\boldsymbol{D}_{nd}$	$\boldsymbol{a}$:42,45; $\boldsymbol{b}$:24,38,40
10	$\boldsymbol{T}\boldsymbol{C}_n$	$\boldsymbol{D}_{nd}$	$\boldsymbol{c}$:25,39,43,44,56,60,62,63
11	$\boldsymbol{T}\boldsymbol{C}_n$	$\boldsymbol{D}_{nh}$	$\boldsymbol{c}$:23,37,41,46,55,59,61,64
12	$\boldsymbol{T}\boldsymbol{C}_n$	$\boldsymbol{D}_{nh}$	$\boldsymbol{a}$:24,38,40; $\boldsymbol{b}$:42,45
13	$\boldsymbol{T}^1_{2n}\boldsymbol{C}_n$	$\boldsymbol{D}_{2nh}$	$\boldsymbol{f}$:26,47,48,55,56,57,58,61,62,63,64; $\boldsymbol{i}$:77,80

and $\tilde{c}^- = (0, 1)$ appear, giving $\tilde{q} = 1$ for the square lattice, and again $\tilde{q} = 2$ for the hexagonal lattice.

Additional symmetries of the layer reduce the number of the different nanotubes. For the layers with the principle axis order $n = 1, 2, 3, 4, 6$, the effective interval of the chiral angle is $[0, 2\pi/n')$, where $n' = \mathrm{LCM}(2, n) = 2, 2, 6, 4, 6$, respectively. Further, vertical mirror plane of the layer intertwines the chiral vectors of the optically isomeric tubes, enabling to halve this range to $[0, \pi/n']$. However, if there is not such a plane, the optical isomer of the (n_1, n_2) tube is obtained from the layer reflected in the mirror plane (perpendicular to $\boldsymbol{A}_1$) again as the (n_1, n_2) tube.

9.1.4 Symmetry-Based Common Characteristics of Nanotubes

It has been shown that symmetry of a nanotube rolled up from an arbitrary diperiodic layer and along any chiral vector is described by a line group. This has many important physical consequences, which depend not only on the geometry of the layer but also on the direction of the folding given by the chiral vector. While large diameter tubes (long chiral vectors) are in many aspects similar to the layers, many properties of small diameter tubes are determined by the chiral angle.

Depending on the two-dimensional lattice (but not on the particular diperiodic group corresponding to the layer) and on the chiral vector, nanotube may be incommensurate (then its line group is from the first or fifth family) or commensurate. Without optical isomers, i.e., achiral, are nanotubes with pure translational or zigzag helical factors; these commensurate structures are obtained for special chiral vectors from the rectangular and rhombic two-dimensional lattice, when also mirror/glide planes may be symmetries of the layer.

The conserved quantum numbers related to the roto-helical symmetries of nanotubes are *quasi-momenta* [8, 13, 9]: *helical*, $\tilde{k}$, from the helical *Brillouin zone* $(-\pi/f, \pi/f]$ and *remaining angular* $\tilde{m}$ (i.e., the part not included into the helical one), taking integer values from the interval $(-n/2, n/2]$. When U-axis, vertical, or horizontal mirror/glide planes are symmetries, the corresponding *parities* (Π_U, Π_v, and Π_h, taking values $+1$ and -1 for even and odd states, and 0 otherwise) are conserved. In commensurate nanotubes, one may alternatively use more conventional quantum numbers of linear and total angular quasi-momenta, k (from the Brillouin zone $(-\pi/a, \pi/a]$) and m (integers from $(-q/2, q/2]$), where q is the order of the principle axis of the isogonal point group. However, m is not conserved in the Umklapp processes.

These quantum numbers assigning energy bands of (quasi)particle spectra ($E^{\Pi}_{\tilde{m}}(\tilde{k})$ or $E^{\Pi}_{m}(k)$) correspond to the irreducible representations of the nanotube's line group. The dimension of irreducible representation is equal to the degeneracy of the band. Thus, for incommensurate nanotubes, degeneracy is either one or two, while for the commensurate also fourfold degeneracy is possible (families 9–13).

As an illustration we consider briefly inorganic metal chalcogenide nanotubes, MS_2 (M=Mo,W). The corresponding monolayer structure has diperiodic symmetry group DG78. As $X = 1$ and $Y = 1/2$, the corresponding nanotubes are always commensurate, with the roto-helical subgroup defined by (9.5). According to Table 9.1, all chiral nanotubes ($n_1 > n_2 > 0$) have symmetry groups of the first family. The symmetry of zigzag nanotubes $(n, 0)$, with additional vertical mirror plane, and armchair (n, n) ones, with additional horizontal mirror plane, is described by the line groups of the eight and fourth families, respectively. As all the groups from the first and eighth families are positive, the Brillouin zones of the chiral and zigzag inorganic nanotubes have no special points and consequently, there is no reduction at the Brillouin zone boundaries. Moreover, the bands of the chiral tubes are non-degenerate as their symmetry groups are abelian. On the other side, the zigzag tubes have a series of the double degenerate bands and a pair of a non-degenerate bands with well-defined vertical mirror parity (one band is odd and the other one is even). In contrast to this, all the bands of the armchair inorganic nanotubes are double degenerate (with no parity) reducing at the center of the Brillouin zone into two non-degenerate states with well-defined horizontal mirror parities.[2]

9.2 Carbon Nanotubes

Soon after the discovery of carbon nanotubes by Iijima [1] in 1991, they have became one of the most interesting objects of material science. Their unique properties investigated by researches from almost all fields of natural sciences, with applications important for diverse parts of technology, became a trademark of the new multidisciplinary field of nanoscience and nanotechnology (now well known by the acronym N&N). From the very beginning it was clear that carbon nanotubes are highly symmetric structures, and this was more or less explicitly used even in the early prediction of their conducting properties [8, 14], which was the result remarkable enough to start the period of N&N. Still, only in 1998, the full symmetry of these structures was found [9] and described by the line groups, which is latter on intensively used in the literature [15, 16]. In this chapter, as an illustration of the various applications of the line group symmetry in physics of quasi-one-dimensional crystals, we give a brief review of those properties of carbon nanotubes which are essentially related to their symmetry. For a more extended review see the references [11, 16].

Most of the experimentally grown carbon nanotubes are multi-walled, i.e., they have several coaxial single-wall tubes, each of them being a rolled up graphene layer, with the difference of radii of the adjacent walls being close to the distance between the graphite layers (3.44 Å).

---

[2] However, if the hamiltonian is real, then the time reversal must be taken into account to predict band degeneracy and topology, Sect. 8.1.2.

9.2.1 *Single-Wall Nanotubes*

Single-wall carbon nanotubes are the most interesting object of the contemporary solid state physics. From the point of view of symmetry, it is a single orbit system with highly nontrivial line group symmetry. This causes that symmetry gives insight to all properties of nanotubes.

9.2.1.1 Symmetry and Configuration

Graphene, quite recently synthesized [17] single layer of graphite, has symmetry of the symmorphic diperiodic group DG80, with the isogonal point group $\boldsymbol{D}_{6\mathrm{h}}$, and hexagonal lattice with both the periods $a_0 = 2.46\,\text{Å}$ (i.e., rhombic with the angle between periods $\alpha = 60°$). Due to the vertical mirror symmetry, according to the Sect. 9.1.3, the chiral angle of different tubes is in the range $[0, 30°]$, and the tubes with $\theta \in [30°, 60°]$ are their optical isomers. The limiting values $\theta = 0°, 30°$ correspond to the tubes $(n, 0)$ and (n, n), called *zigzag* ($\mathcal{Z}$) and *armchair* ($\mathcal{A}$). As it will be immediately justified, the latter two classes are *achiral*, while all others are *chiral* ($\mathcal{C}$) (Fig. 9.2).

The symmetry group of the nanotube (n_1, n_2) is extracted from Table 9.1: for chiral and achiral tubes it is from the fifth and thirteenth families, respectively:

$$\boldsymbol{L}_{\mathcal{C}} = \boldsymbol{T}_Q(f)\boldsymbol{D}_n, \quad \boldsymbol{L}_{\mathcal{ZA}} = \boldsymbol{T}_Q(f)\boldsymbol{D}_n, \tag{9.14}$$

with the first family subgroup parameters Q, n, and f to be found according to the prescription of the Sect. 9.1. As the necessary lattice parameters are $X = 1$ and $Y = 1/2$, the length of chiral vector is $c = a_0 n\sqrt{\tilde{n}_1^2 + \tilde{n}_2^2 + \tilde{n}_1\tilde{n}_2}$, giving also the tube diameter and chiral angle: $D = c/\pi$ and $\sin\theta = n_2 a_0/c$. All the tubes

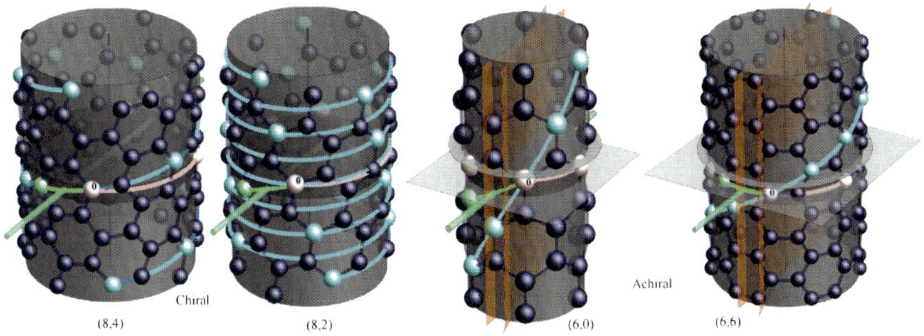

Fig. 9.2 Carbon nanotubes with symmetries. The symmetry groups of the depicted nanotubes (8,4), (8,2), (6,0), and (6,6) are $\boldsymbol{T}^9_{56}(0.8\,\text{Å})\boldsymbol{D}_4$, $\boldsymbol{T}^{11}_{28}(0.46\,\text{Å})\boldsymbol{D}_2$, $\boldsymbol{T}^1_{12}(2.14\,\text{Å})\boldsymbol{D}_{6\mathrm{h}}$, and $\boldsymbol{T}^1_{12}(1.24\,\text{Å})\boldsymbol{D}_{6\mathrm{h}}$, respectively. The symmetry elements are depicted by the action on the symcell atom C_{000} (indicated by 0). *Helix* and *circle* represent action of $\boldsymbol{T}^r_q$ and $\boldsymbol{C}_n$. Along x-axis is U-axis. *Parallelograms* are *vertical* and *horizontal* mirror planes, while zigzag plane and *circle* are glide and roto-reflectional plane

are commensurate because both X and Y are rational. From (9.9) we find isogonal group principle axis order q. Then, using (9.3) we get z, which according to (9.5d) gives $Q = q/r$; as q is already known, r directly follows, while f is given by (9.5c) (alternatively, one applies $f = a/\tilde{q}$, with a obtained from (9.7)). This completes search for the first family subgroup:

$$n = \mathrm{GCD}(n_1, n_2); \tag{9.15a}$$

$$f = \frac{a_0}{\sqrt{\tilde{n}_1^2 + \tilde{n}_2^2 + \tilde{n}_1\tilde{n}_2}}, \quad a = \frac{a_0\sqrt{3}\sqrt{\tilde{n}_1^2 + \tilde{n}_2^2 + \tilde{n}_1\tilde{n}_2}}{\mathcal{R}}; \tag{9.15b}$$

$$q = 2n\frac{\tilde{n}_1\tilde{n}_2 + \tilde{n}_1^2 + \tilde{n}_2^2}{\mathcal{R}}, \tag{9.15c}$$

$$r_1 = \frac{n_1 + 2n_2 - (\frac{n_2}{n})^{\mathrm{Eu}(\frac{n_2}{n})-1} q\mathcal{R}}{n_1\mathcal{R}}, \quad r \stackrel{\tilde{q}}{=} r_1. \tag{9.15d}$$

Here, $\mathcal{R} = \mathrm{GCD}(\tilde{n}_1 + 2\tilde{n}_2, \tilde{n}_2 + 2\tilde{n}_1)$, which is equal to 3 if $\tilde{n}_1 - \tilde{n}_2$ is divisible by 3, and 1 otherwise; r_1 is helicity parameter subdued to the convention C1 of (2.14). Thus, the transformations of the symmetry group of chiral and achiral tubes are, respectively:

$$\ell_{tsu} = (C_Q|f)^t C_n^s U^u, \quad \ell_{tsuv} = (C_Q|f)^t C_n^s U^u \sigma_{\mathrm{v}}^v. \tag{9.16}$$

All tubes are with nontrivial helical axis, because from (9.15c) it follows that $\tilde{q} = 2\frac{\tilde{n}_1\tilde{n}_2 + \tilde{n}_1^2 + \tilde{n}_2^2}{\mathcal{R}}$ is even. In fact, it can be represented [18] as $\tilde{q} = 12K + 2$, with $K = 0, 1, \ldots$. Its minimal value $\tilde{q} = 2$ singles out achiral helical groups, corresponding to achiral tubes. As the other values are much larger, elementary cell of chiral tubes contains many monomers. Different tubes have different roto-helical symmetries, i.e., chiral indices and triples (Q, n, f) are biuniquely related. Even more, all chiral tubes have different pairs (Q, n), and only in the achiral tubes $(n, 0)$ and (n, n), we get the groups $\boldsymbol{T}^1_{2n}(\frac{\sqrt{3}}{2}a_0)\boldsymbol{C}_n$ and $\boldsymbol{T}^1_{2n}(\frac{a_0}{2})\boldsymbol{C}_n$ differing only in the fractional translations (also $\mathcal{R}$ is 1 for zigzag and 3 for armchair tubes).

The isogonal point group is $\boldsymbol{D}_q$ for chiral and $\boldsymbol{D}_{2n\mathrm{h}}$ for achiral tubes. Its principle axis order q is equal to the half of the number of the atoms in the elementary cell.

The graphene elementary cell contains two carbon atoms, which are mutually connected by rotation C_2. Thus, pure translations and this rotation generate whole graphene from a single atom. After folding, graphene translations become roto-helical transformations $\boldsymbol{L}^{(1)}$, which together with U-axis coming from C_2 symmetry of the layer, generate nanotube from a single atom. This means that all single-wall carbon nanotubes are mono-orbit systems: symcell contains an arbitrary chosen atom, while the transversal is the fifth family subgroup of the symmetry group, i.e., $\boldsymbol{L}_C$ for both chiral and achiral tubes.

Conveniently, we define *nanotube reference frame* with z-axis being the nanotube axis, and x-axis coinciding with the U-axis through center of carbon hexagons.

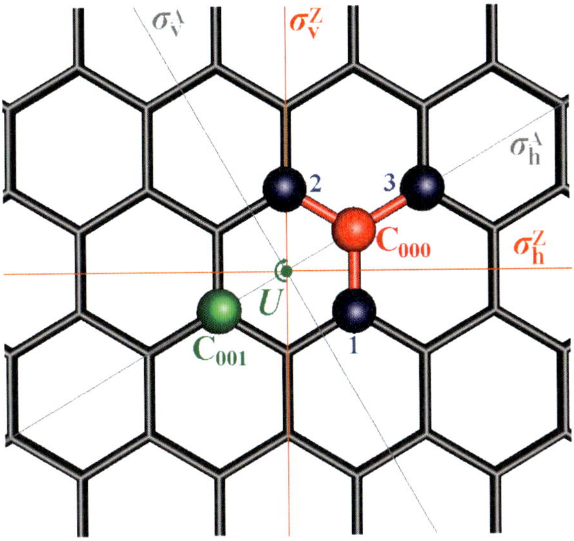

Fig. 9.3 Symcell atom C_{000} and its nearest neighbors 1,2, and 3 at honeycomb. *Perpendicular* to the plane C_2-axis (*dot* with *arrow*), maps C_{000} to C_{001} and becomes U-axis of the nanotube (along x-axis). Lines depict the graphene mirror planes becoming σ_h and σ_v in the cases of zigzag (superscript Z) and armchair (A) tubes, when the chiral vector is in σ_h and *perpendicular* to σ_v

Further, as the symcell atom we chose the atom C_{000} at graphene honeycomb (Fig. 9.3). After rolling, its cylindrical coordinates are

$$\boldsymbol{r}_{000} = \left(\frac{D}{2}, \phi_0, z_0\right), \quad \phi_0 = 2\pi \frac{n_1 + n_2}{nq\mathcal{R}}, \quad z_0 = \frac{n_1 - n_2}{\sqrt{6nq\mathcal{R}}} a_0. \tag{9.17a}$$

Acting by the transversal element ℓ_{tsu} of (9.16) on C_{000} we get coordinates of any other atom C_{tsu}:

$$\boldsymbol{r}_{tsu} = \ell_{tsu}\boldsymbol{r}_{000} = \left(\frac{D}{2}, (-1)^u\varphi_0 + 2\pi\left(\frac{rt}{q} + \frac{s}{n}\right), (-1)^u z_0 + t\frac{n}{q}a\right). \tag{9.17b}$$

In particular, the nearest neighbors 1, 2, and 3 (Fig. 9.3) are atoms $C_{t_i s_i u_i}$ (i = 1, 2, 3) with

$$t_{1/2} = \mp\frac{n_{2/1}}{n}, \; t_3 = t_1 + t_2, \; s_{1/2} = \frac{2n_{1/2} + (1 \pm r_1\mathcal{R})n_{2/1}}{q\mathcal{R}}, \; s_3 = s_1 + s_2, \; u_i = 1. \tag{9.18}$$

In chiral cases the stabilizer of any atom is trivial $\boldsymbol{C}_1$ (the identity element only), and the whole nanotube is generic orbit a_1 of the fifth family line groups (Fig. 3.1, Table 3.2). For the achiral tubes the transversal $\boldsymbol{L}_C$ is a halving subgroup of the symmetry group $\boldsymbol{L}_{\mathcal{ZA}}$. Therefore, the stabilizer is with two elements: orbit type and stabilizer of C_{000} are b_1 and $\{e, C_n\sigma_x\}$ for zigzag, and d_1 and $\{e, \sigma_h\}$ for armchair tubes. Also, it follows that carbon nanotubes belong to fifth conformation class.

9.2.1.2 Relaxation

As a single-wall nanotube is a mono-orbit system, its configuration is determined by the coordinates of the orbit representative atom $\boldsymbol{r}_{000} = (D/2, \varphi_{000}, z_{000})$ and symmetry group. For the fifth family line groups (transversals of all nanotubes) $\boldsymbol{T}_Q(f)\boldsymbol{D}_n$, the continual parameters are Q and f. Hence, there are altogether five parameters to be varied. However, the increased symmetry $\boldsymbol{T}^1_{2n}(a/2)\boldsymbol{D}_{n\mathrm{h}}$ prevent variation of $Q = 2n$. In addition, for the zigzag tubes, with atoms being in the vertical mirror planes, φ_{000} is fixed, likewise z_{000} for the armchair tubes, with atoms in the horizontal mirror planes; thus, reduces number of the relaxation parameters to three. In Table 9.2 configuration parameters for simply rolled and relaxed nanotubes are compared.

9.2.1.3 Band Topology

The conclusions of the Sect. 8.1.2 enable to derive some quite general characteristics of the band structures of carbon nanotubes. Since the dihedral axis reverses the both momenta, i.e., $(k, m) \to (-k, -m)$ or $(\tilde{k}, \tilde{m}) \to (-\tilde{k}, -\tilde{m})$, it suffices to consider only the irreducible domain $k \in [0, \pi/a]$ or $\tilde{k} \in [0, \pi/f]$.

For chiral nanotubes, the energy bands are double degenerate within interior of the irreducible domain. However, if simultaneously $\tilde{k} = 0, \pi/f$ and $\tilde{m} = 0, n/2$ the corresponding states are physically the same and at the edge of the Brillouin zone singlet states, even and odd with respect to the dihedral axis appear. All other states are doublets and those with opposite $\tilde{m}$ meet at the center or at the Brillouin zone edge. Thus, only the bands $\tilde{m} = 0, n/2$ (i.e., when $\tilde{m} = -\tilde{m}$) are symmetric around the center and the edge of the Brillouin zone, with extremes in the density of states at $\tilde{k} = 0$ and $\tilde{k} = \pi/f$; these edge point states are singlets, even or odd with respect to the U transformation.

Achiral carbon nanotubes have in addition vertical σ_v and horizontal σ_h mirror symmetries. Leaving k invariant while reversing m, vertical mirror symmetry causes additional (altogether fourfold) degeneracy, for the bands characterized by $m = 1, \dots, n-1$. However, the bands $m = 0$ and $m = n$ remain double degenerate, but with the well-defined σ_v parity. At the Brillouin zone center there are eight singlets characterized by different combinations of the angular momentum quantum numbers $m = 0, n$, and two types of the mirror parities (σ_h and σ_v). The remaining states ($m = 1, \dots, n-1$) are double degenerate, even or odd with respect to σ_h. At another Brillouin zone edge, the doublets $m = 0, n$ have σ_v parity, while if n is even the corresponding $m = n/2$ doublet has σ_h parity. Hence, z-reversal parity characterizes all the states at the edges of the bands causing van Hove singularities in the density of states.

Table 9.2 Relaxed nanotubes of the least radii. For each nanotube (column NT), after the family number F and group parameters q, r, and n, the non-relaxed (purely rolled) and relaxed configuration parameters and their difference (in percents) are presented. It turned out that the $Q = q/r$ remained fixed during relaxation. Therefore only fractional translation f and cylindrical coordinates of the initial atom are given. For the achiral tubes the parameter which is not varied is denoted as/in the corresponding columns

NT	Group				Nonrelaxed				Relaxed				Difference [%]			
	F	q	r	n	f	ρ	φ	z	f	ρ	φ	z	f	ρ	φ	z
(4,0)	13	8	1	4	2.13	1.57	0.79	0.71	2.10	1.68	/	0.70	1.51	−7.17	/	2.10
(3,2)	5	38	15	1	0.49	1.71	0.83	0.16	0.49	1.79	0.83	0.16	−0.83	−4.88	−0.15	2.74
(4,1)	5	14	11	1	0.47	1.79	0.74	0.47	0.47	1.89	0.75	0.46	−0.09	−5.16	−0.15	0.66
(5,0)	13	10	1	5	2.13	1.96	0.63	0.71	2.14	2.07	/	0.72	−0.47	−5.29	/	−1.56
(3,3)	13	6	1	3	1.23	2.03	0.70	0	1.25	2.11	0.70	/	−1.38	−3.50	−0.59	/
(4,2)	5	28	9	2	0.80	2.07	0.67	0.27	0.80	2.16	0.68	0.27	0.15	-4.36	−0.30	−0.99
(5,1)	5	62	51	1	0.38	2.18	0.61	0.51	0.38	2.25	0.61	0.51	−0.25	−3.33	−0.13	0.44
(6,0)	13	12	1	6	2.13	2.36	0.52	0.71	2.13	2.43	/	0.70	0.24	−3.43	/	0.83
(4,3)	5	74	21	1	0.35	2.38	0.59	0.12	0.35	2.44	0.60	0.12	−0.97	−2.24	−0.66	1.21
(5,2)	5	26	11	1	0.35	2.45	0.56	0.34	0.34	2.53	0.56	0.34	−0.43	−3.54	−0.06	1.53
(6,1)	5	86	73	1	0.33	2.57	0.51	0.54	0.32	2.67	0.51	0.54	0.35	−3.88	−0.11	−0.37
(4,4)	13	8	1	4	1.23	2.71	0.52	0	1.23	2.79	0.53	/	0.39	−2.96	−0.75	/

9.2.1.4 Electronic Bands and Conductivity

We consider the simplest tight-binding dynamical model of electrons in carbon nanotubes. This spin-independent model is based on graphene sp^2 bonding: carbon $1s$ orbital is occupied by the two localized core electrons, while the bonds with the nearest neighbors are realized by three bonding hybridized orbitals ($2s$ and two in-plane $2p$ orbitals) occupied by three electrons per atom. Therefore, the relevant state space is spanned by the remaining $p^{\perp}$-orbital [19] perpendicular to the tube surface, half-filled by a single electron per atom. Such an orbital at the site C_{tsu} is denoted as $\mid tsu\rangle$. It is generated from the $p^{\perp}$-orbital $\mid 000\rangle$ of the reference atom C_{000}, with wave function:

$$\langle \boldsymbol{r} | 000\rangle = \chi_{000}(\rho, \varphi, z) = \sqrt{\frac{2Z_{\text{eff}}^7}{15\pi a_B^7}} |\boldsymbol{r} - \boldsymbol{r}_{000}| \left(\rho \cos\varphi - \frac{D}{2} \cos\varphi_0 \right) \mathrm{e}^{-\frac{Z_{\text{eff}}}{a_{\text{B}}} |\boldsymbol{r}-\boldsymbol{r}_{000}|} \tag{9.19}$$

(a_{B} is Bohr radius and $Z_{\text{eff}} \approx 3.81$). The orbitals from different sites are considered to be orthogonal. Hamiltonian is built within nearest neighbor approximation, assuming that they are symmetrically distributed (this is also an approximation; it partly neglects curvature effects). Therefore, denoting three nearest neighbors of the atom C_{tsu} by $C_{(tsu);i}$ ($i = 1, 2, 3$), the nonvanishing hamiltonian matrix elements are diagonal ones, $\langle tsu|\ H\ |tsu\rangle = E_{\text{C}}$, equal to the carbon atom p-orbital energy, and $\langle tsu|\ H\ |(tsu); i\rangle = V \approx -3\,\text{eV}$.

As carbon nanotubes are mono-orbit systems, we use the algorithm explained in Sect. 8.5.1, with the transversal $\boldsymbol{Y} = \boldsymbol{L}_C$ and $\boldsymbol{P}$ being the stabilizer of the initial atom C_{000}. The pulled down hamiltonian is[3]

$$H^{\downarrow\lambda} = \sum_{i=0}^{3} \langle 000|\ H\ |t_i s_i u_i\rangle D^{(\lambda)^T}(\ell_{t_i s_i u_i}). \tag{9.20}$$

Here, for $i = 0$ the on-site matrix element is assumed, i.e., $\ell_{t_0 s_0 u_0} = \ell_{000}$ is the identity, while for $i > 0$ we use $\ell_{t_i s_i u_i}$ given by (9.18). This matrix should be found for each irreducible representation $D^{(\lambda)}$ of the symmetry group of the nanotube (Table 4.5 for chiral and Table 4.13 for achiral tubes).

For chiral tubes and representations ${}_{\tilde{k}}E_{\tilde{m}}$ we get

$$H^{\downarrow(\tilde{k}\tilde{m})} = E_{\text{C}}\left(\begin{smallmatrix}1 & 0\\ 0 & 1\end{smallmatrix}\right) + V \begin{pmatrix} 0 & \sum_{i=1}^{3} \mathrm{e}^{-\mathrm{i}\tilde{\psi}_i} \\ \sum_{i=1}^{3} \mathrm{e}^{\mathrm{i}\tilde{\psi}_i} & 0 \end{pmatrix}, \quad \tilde{\psi}_3 = \tilde{\psi}_2 - \tilde{\psi}_1,$$

$$\tilde{\psi}_1 = -\tilde{k}a\frac{n_2}{q} + 2\pi\tilde{m}\frac{2n_1 + (1 + r_1\mathcal{R})n_2}{qn\mathcal{R}}, \quad \tilde{\psi}_2 = \tilde{k}a\frac{n_1}{q} + 2\pi\tilde{m}\frac{(1 - r_1\mathcal{R})n_1 + 2n_2}{qn\mathcal{R}}.$$

---

[3] As in the considered case there is a single orbit and single orbital per atom, the term $\mid A, \psi_A\rangle\langle B, \psi_B \mid$ determining block $H^{\downarrow\lambda}_{AB}$ in the pulled down hamiltonian matrix (8.22) reduces to the superfluous factor, projector $|000\rangle\langle 000|$.

Solving the eigenproblem of this hamiltonian we easily get the energy bands,

$$\varepsilon_{\tilde{m}}^{\pm}(\tilde{k}) = E_{\mathrm{C}} \pm |V| \sqrt{\sum_{i=1}^{3}(1 + 2\cos\tilde{\psi}_i)}, \tag{9.21}$$

and the corresponding eigenvectors in the generalized Bloch form (Sect. 8.1):

$$|\tilde{k}\tilde{m}0;\pm\rangle = \frac{1}{\sqrt{|\boldsymbol{L}_C|}}\sum_{ts}\left(\mathrm{e}^{-\mathrm{i}\psi_{\tilde{m}}^{\tilde{k}}(t,s)}\,|ts0\rangle \pm \mathrm{e}^{\mathrm{i}(h_{\tilde{m}}^{\tilde{k}} - \psi_{\tilde{m}}^{\tilde{k}}(t,s))}\,|ts1\rangle\right), \tag{9.22a}$$

$$|-\tilde{k},-\tilde{m},0;\pm\rangle = \frac{1}{\sqrt{|\boldsymbol{L}_C|}}\sum_{ts}\left(\pm\,\mathrm{e}^{\mathrm{i}(h_{\tilde{m}}^{\tilde{k}} + \psi_{\tilde{m}}^{\tilde{k}}(t,s))}\,|ts0\rangle + \mathrm{e}^{\mathrm{i}\psi_{\tilde{m}}^{\tilde{k}}(t,s)}\,|ts1\rangle\right); \tag{9.22b}$$

here, $\tilde{\psi}^{(\tilde{k},\tilde{m})}(t,s) = \frac{\tilde{k}a}{q}t + \frac{2\pi\tilde{m}}{n}s$ and $h_m^k = \arg\{V(\mathrm{e}^{\mathrm{i}\psi_1} + \mathrm{e}^{\mathrm{i}\psi_2} + \mathrm{e}^{\mathrm{i}(\psi_1+\psi_2)})\}$.

Analogously, we can get the energies and states for all the representations of chiral and achiral nanotubes [20]. Also, for linear quantum numbers, the bands have the same form when $\tilde{\psi}_i$ is substituted by

$$\psi_1 = -ka\frac{n_2}{q} + 2\pi m\frac{2n_1+n_2}{qn\mathcal{R}}, \quad \psi_2 = ka\frac{n_1}{q} + 2\pi m\frac{n_1+2n_2}{qn\mathcal{R}}, \quad \psi_3 = \psi_2 - \psi_1. \tag{9.23}$$

Obviously, for each $\tilde{m}$ there are two bands (9.21), symmetric with respect to E_{C}. Taking into account that only half of the states are occupied (at zero temperature), we see that Fermi level E_{F} coincides with E_{C} (therefore, according to the convention $E_{\mathrm{F}} = 0$ it is usual to put $E_{\mathrm{C}} = 0$ in (9.21)). The bands $\varepsilon_{E_{\tilde{m}}}^{+}(\tilde{k})$ are conducting, while the valence bands are $\varepsilon_{E_{\tilde{m}}}^{-}(\tilde{k})$. Consequently, there is a *gap* between conducting and valence bands (and the tube is semiconducting) unless for some values $\tilde{k}$ and $\tilde{m}$ the square root vanishes. Simple calculation shows that this occurs only for the tubes with $n_1 - n_2$ divisible by three, i.e., only these tubes are conducting. Namely, for all tubes [21] Fermi level is at $\tilde{k}_{\mathrm{F}} = \frac{2q\pi}{3na}$ on the band $\varepsilon_{E_{\tilde{m}_{\mathrm{F}}}}^{-}(\tilde{k})$ with $\tilde{m}_{\mathrm{F}} \stackrel{\circ}{=} \frac{2}{3}nr_1\mathcal{R}$ (equality modulo interval $(-n/2, n/2]$). So, only when $n_1 - n_2$ is divisible by three, the square root vanishes and the gap disappears since $\varepsilon_{E_{\tilde{m}_{\mathrm{F}}}}^{-}(\tilde{k}_{\mathrm{F}}) = \varepsilon_{E_{\tilde{m}_{\mathrm{F}}}}^{+}(\tilde{k}_{\mathrm{F}}) = E_{\mathrm{F}}$ (Fig. 9.4).

However, Landau non-crossing rule [22] for the bands reveals more subtle symmetry-based detail. Indeed, having all the quantum numbers identical, the bands $\varepsilon_{E_{\tilde{m}_{\mathrm{F}}}}^{-}(\tilde{k})$ and $\varepsilon_{E_{\tilde{m}_{\mathrm{F}}}}^{+}(\tilde{k})$ cannot cross, i.e., in a more precise model a small *secondary gap* appears. Still, for armchair tubes $\mathcal{R} = 3$, giving $\tilde{m}_{\mathrm{F}} = 0$ for which there are two representations of the opposite σ_{v}-parity, and it is easily checked that just the bands with different parity are crossed. To summarize, there are three types of nanotubes: armchair ones are conducting, the remaining tubes with $n_1 - n_2$ divisible by 3 are quasi-conducting (due to small gap of 0.01 eV they are conducting at the room

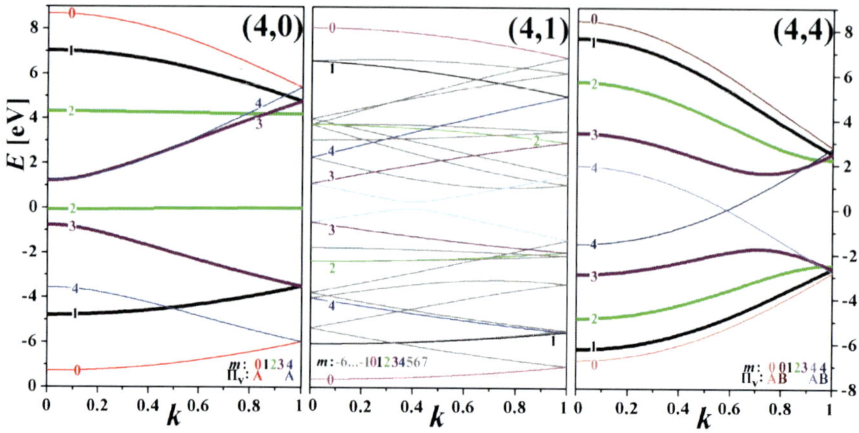

Fig. 9.4 Electronic bands of a chiral (quasi-metallic), zigzag (semiconducting) and armchair (metallic) nanotubes. *Thin* and *thick* bands are twofold and fourfold, respectively. Band quantum numbers are listed at the *bottom*

temperature), and all other tubes are semiconducting (with gaps of order of 1 eV). Calculations show that gap decreases with tube diameter and vanishes in the infinite limit, when graphene, being semi-metal, is obtained.

9.2.1.5 Bloch States and Electronic Density

In the interior of the helical Brillouin zone energy $\varepsilon^{\pm}_{\tilde{m}}(\tilde{k})$ is double degenerate with the multiplet of states (9.22) transforming according to the irreducible representation ${}_{\tilde{k}}E_{\tilde{m}}$ (U-parity for these representations is $\Pi_U = 0$ and then $\tilde{\Pi}^U = 1$). The corresponding Bloch wave functions $\Psi^{(\tilde{k},\tilde{m})1} = \Psi^{\tilde{k},\tilde{m}}(\boldsymbol{r})$ and $\Psi^{(\tilde{k},\tilde{m})2} = \Psi^{-\tilde{k},-\tilde{m}}(\boldsymbol{r})$ are of the type (8.2) and may be expanded over the symmetry-adapted basis of the Bloch functions (5.26), with the representative functions (Table 5.2):

$$\Psi^{\pm\tilde{k},\pm\tilde{m}}_{IKM}(\boldsymbol{r}) = \mathrm{e}^{\mp \mathrm{i}(\tilde{m}\varphi + (\tilde{k} - \frac{2\pi\tilde{m}r_1}{f})z)} R^M_{IK}(\rho) H^M_K(\varphi, z), \tag{9.24}$$

where $H^M_K(\varphi, z)$ are the fifth family harmonics (Table 5.1).

The tight-binding model (Sect. 8.5.1) with n atomic orbitals $\ell_{tsu}\chi_i(\boldsymbol{r})$ ($i = 1, \ldots, n$) per atom, with $\chi_i(\boldsymbol{r}) = \langle \boldsymbol{r} \mid 000; i\rangle$ being orbitals of C_{000}, gives the electronic eigenfunctions in the inductive form (8.21):

$$\Psi^{(\tilde{k}\tilde{m})l}(\boldsymbol{r}) = \frac{1}{\sqrt{|\boldsymbol{L}^{(5)}|}} \sum_{tsu} \sum_{l'i} c_i^{(\tilde{k}\tilde{m})l'} D^{(\tilde{k}\tilde{m})^*}_{ll'}(\ell_{tsu})\ell_{tsu}\chi_i(\boldsymbol{r}). \tag{9.25}$$

Here, $c_i^{(\tilde{k}\tilde{m})l'} = c_{000,i}^{(\tilde{k}\tilde{m})l'}$ are the coefficients associated to the atomic orbitals of C_{000} in the expansion of $\Psi^{(\tilde{k}\tilde{m})l}(\boldsymbol{r})$. To find the amplitudes $\alpha^M_{IK} = (\Psi^{(\tilde{k}\tilde{m})l}_{IKM}(\boldsymbol{r}), \Psi^{(\tilde{k}\tilde{m})l}(\boldsymbol{r}))$ in the expansion (5.28), we substitute (9.25), and instead of acting on $\chi_i(\boldsymbol{r})$ by the unitary operators $D(\ell_{tsu})$, we apply their inverses $D(\ell_{tsu})^\dagger = D(\ell_{tsu}^{-1})$ on the symmetry-adapted functions on the left; this by (8.1) gives

$$\alpha_{IK}^{M} = \sum_{l'l''i} \left(\frac{1}{\sqrt{|\boldsymbol{L}^{(5)}|}} \sum_{tsu} D_{ll'}^{(\tilde{k}\tilde{m})^*}(\ell_{tsu}) D_{ll''}^{(\tilde{k}\tilde{m})}(\ell_{tsu}) \right) c_i^{(\tilde{k}\tilde{m})l'} \left(\Psi_{IKM}^{(\tilde{k}\tilde{m})l''}(\boldsymbol{r}), \chi_i(\boldsymbol{r}) \right).$$

The orthogonality theorem [23] reduces the braced factor to $\delta_{l',l''}/|\lambda|$, and Bloch eigenstates expressed through the amplitudes of the atomic orbitals are obtained

$$\alpha_{IK}^{M} = \frac{1}{2} \sum_{l'i} c_i^{(\tilde{k}\tilde{m})l'} \left(\Psi_{IKM}^{(\tilde{k}\tilde{m})l'}(\boldsymbol{r}), \chi_i(\boldsymbol{r}) \right). \tag{9.26}$$

Figure 9.5 shows the expansion (performed numerically) of the state with quantum numbers $\tilde{k} = \frac{\pi}{10f}$, $\tilde{m} = 1$, $U = 0$ (corresponding to the energy $E = -0.285$ eV of the tube (4,2)). Also, the electronic density of this tube is shown. It is sum

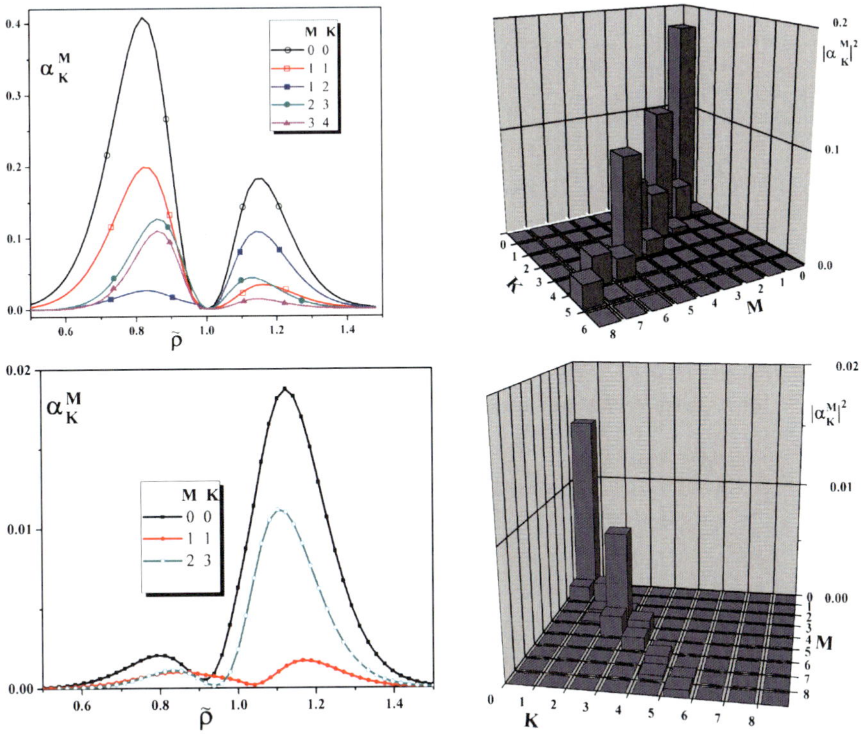

Fig. 9.5 Bloch eigenstate $\Psi^{(\tilde{k},\tilde{m})}(\boldsymbol{r})$ with $\tilde{k} = \pi/10f = 0.389993\,\text{Å}^{-1}$ and $\tilde{m} = 1$, corresponding to eigenenergy $E = -0.285$ eV (*top panels*) and total electronic density (*bottom panels*) of the carbon nanotube (4,2) with symmetry group $\boldsymbol{L} = \boldsymbol{T}_{28}^{9}(f = 0.8\,\text{Å})\boldsymbol{D}_2$ and diameter $D = 4.14$ Å. *Left*: non-negligible expansion coefficients α_K^M plotted in the significant range $0 < \tilde{\rho} < 1.5$ of the reduced radial coordinate $\tilde{\rho} = 2D/\rho$. *Right*: Harmonic expansions at radii $\tilde{\rho} = 1.15$

$\varrho = \sum |\Psi^{(\tilde{k}\tilde{m})l}(\boldsymbol{r})|^2$ of the densities $|\Psi^{(\tilde{k}\tilde{m})l}(\boldsymbol{r})|^2$ of all the filled states, i.e., the ones below Fermi level. At zero temperature it is an invariant function and can be expanded over the harmonics H_K^M.

9.2.1.6 Optical Properties

Having at disposal electronic bands and Bloch eigenstates assigned by the complete set of conserved quantum numbers, we can straightforwardly apply the prescription given in Sect. 8.3.1 to predict optical properties of carbon nanotubes. The allowed optical transitions are singled out by the selection rules (8.8). As for the parities, for chiral tubes only U-parity applies; since in the interior of the Brillouin zone the Bloch states have not defined parity, i.e., $\Pi_U = 0$. The nontrivial restrictions appear only at the edges of the zone. On the other side, for achiral tubes the bands with $m = 0, n$ have σ_v parity throughout the zone, while the z-reversal parities are effective only at $k = 0, \pi/a$. The allowed transitions are illustrated in Fig. 9.6.

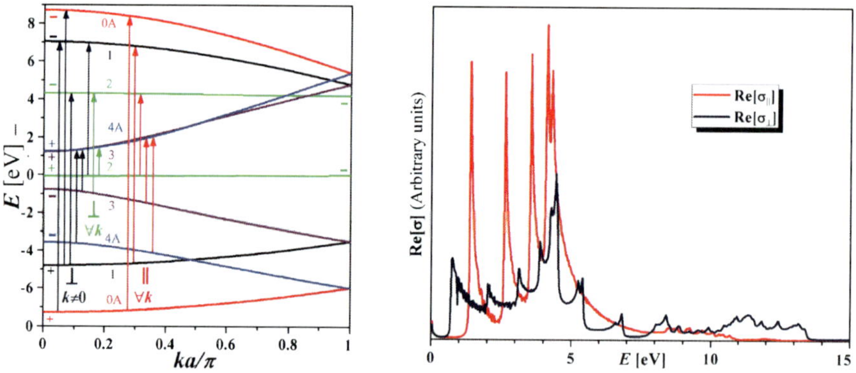

Fig. 9.6 *Left*: Optical transitions allowed by the selection rules for the tube (4,0). Bands are assigned by quantum numbers m and $\Pi_v = A$. The allowed transitions differ for electrical field parallel to nanotube (*red*) and perpendicular to it (*green*); in the last case the transitions prevented by Π_h parity in $k = 0$ (indicated by + or − sign on the *left*) are indicated by *blue* color. *Right: Parallel* and *perpendicular* components of the optical conductivity tensor for nanotube (10,10)

Taking into account all the selection rules, we apply (8.9) to calculate *optical conductivity*. The main features are shown in Fig. 9.6: strong anisotropy and strong chirality dependence of the absorption.

9.2.1.7 Ion Dynamics

As each carbon nanotube is a single orbit of a line group, the dynamical representation for the chiral tubes can be read from Table 7.5 (row a_1), and from Table 7.13 for the achiral nanotubes (rows b_1 and d_1 for zigzag and armchair cases, respectively). There are six bands for each m. For the purpose of the further discussion we write here only the $k = 0$ components:

$$D_{0C}^{\text{dyn}} = 3({}_0A_0^+ + {}_0A_0^- + {}_0A_{q/2}^+ + {}_0A_{q/2}^-) + 6\sum_{m=1}^{q/2-1} {}_0E_m; \tag{9.27a}$$

$$D_{0\mathcal{Z}}^{\text{dyn}} = D_{\mathcal{Z}}^+ + D_{\mathcal{Z}}^-, \quad D_{\mathcal{Z}}^{\pm} = 2{}_0A_0^{\pm} + 2{}_0A_n^{\pm} + {}_0B_0^{\pm} + {}_0B_n^{\pm} + 3\sum_{m=1}^{n-1} {}_0E_m^{\pm}; \tag{9.27b}$$

$$D_{0\mathcal{A}}^{\text{dyn}} = 2D_{\mathcal{A}}^+ + D_{\mathcal{A}}^-, \quad D_{\mathcal{A}}^{\pm} = {}_0A_0^{\pm} + {}_0B_0^{\pm} + {}_0A_n^{\pm} + {}_0B_n^{\pm} + 2\sum_{m=1}^{n-1} {}_0E_m^{\pm}. \tag{9.27c}$$

The phonon bands are usually obtained within force constant model. An example of the band structure is given in Fig. 9.7. There are four acoustic modes: two translational in the perpendicular plane are multiplet of the representation ${}_0E_1$ for chiral and ${}_0E_1^+$ for achiral tubes; the longitudinal one corresponds to ${}_0A_0^-$, while the irreducible representation of the twisting mode is again ${}_0A_0^-$ for chiral, but ${}_0B_0^+$ for achiral tubes. The slopes of the acoustic bands at $k = 0$ determine the transversal, longitudinal, and twisting sound velocities: $v_{\text{T}} = 9.41$ km/s, $v_{\text{L}} = 20.37$ km/s, $v_{\text{TW}} = 14.98$ km/s.

The infrared active modes transform according to the irreducible representations ${}_0A_0^-$ and ${}_0E_1$ for chiral tubes and ${}_0A_0^-$ and ${}_0E_1^+$ for achiral. Then, using (9.27), from (8.11), with $N_{\text{ac}}^{\text{ir}} = 4, 3, 3$ for chiral, zigzag, and armchair tubes, respectively, we conclude that there are 11, 5, and 6 infrared active modes, respectively. However, some of them are multiplets of ${}_0E_1$ or ${}_0E_1^+$ representations, thus having the same energies; counting this, we may expect at most 6 and 3 lines in the spectra of chiral and achiral tubes, respectively.

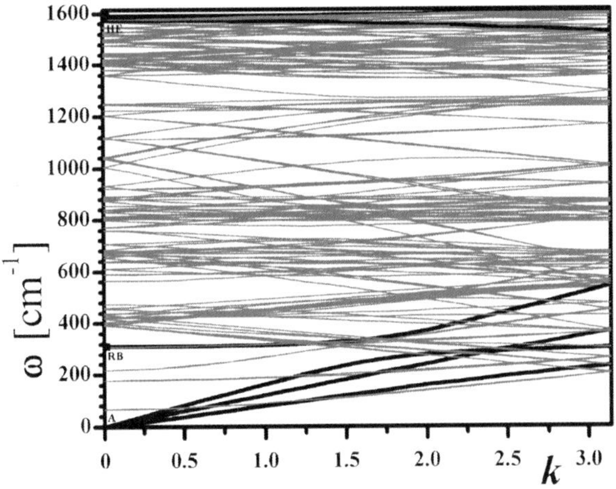

Fig. 9.7 Phonon dispersions of the tube (8,2). *Bolded branches* are beginning with acoustic (A), radial-breathing (RB), and high energy (HE) modes

Similarly, for the Raman active modes we find the irreducible representations ${}_0A_0^+, {}_0A_0^-, {}_0E_1, {}_0E_2$ for chiral and ${}_0A_0^+, {}_0B_0^+, {}_0E_1^-, {}_0E_2^+$ for achiral tubes. Consequently, the number of the Raman active modes is 26, 14, and 15 for chiral, zigzag, and armchair tubes, respectively ($N_{ac}^{R} = 4, 1, 1$), but as some of them are double degenerate, there may be at most 15, 8, and 9 different frequencies in the Raman spectra.

As the Raman and infrared measurements are among the basic tools in the characterization of materials, the frequencies of the active phonons are extensively calculated [15, 24] for nanotubes, in order to get their dependence on the tube diameter and chirality. However, it is experimentally found that the totally symmetric modes, corresponding to the representation ${}_0A_0^+$, are much more intensive in the Raman spectra than the others. The lowest among them is called *breathing mode*, because the vibrations of the atoms are almost radial, and in the limiting case of infinite diameter it becomes transversal acoustic graphene mode. The other two (for chiral) or one (for achiral tubes) symmetric modes are in the high-energy region.

9.2.1.8 Diffraction

Diffraction amplitude [25] of a single-wall carbon nanotube is proportional to the geometrical factor for its single orbit, i.e., for the transversal $\boldsymbol{T}_q^r(f)\boldsymbol{D}_n$ (Table 8.2):

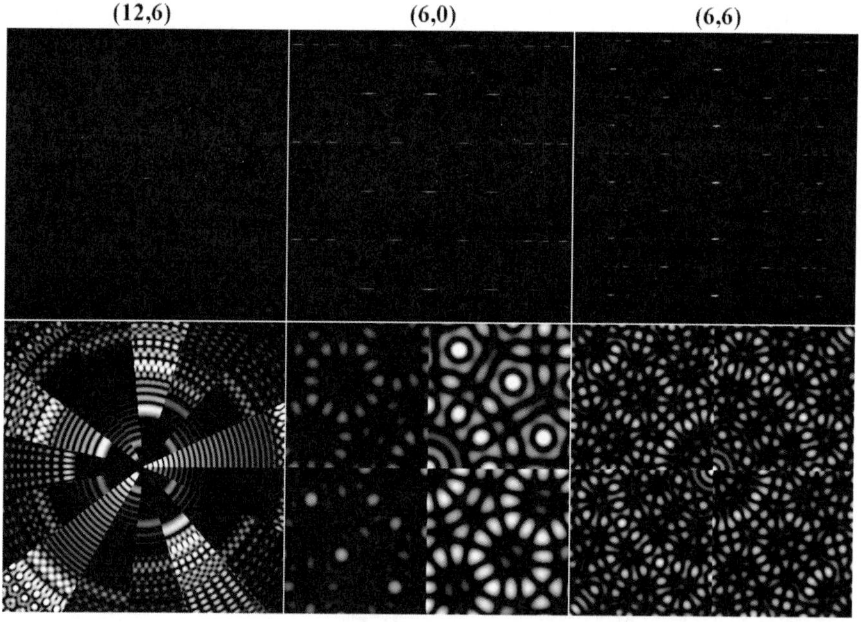

Fig. 9.8 Diffraction patterns of carbon nanotubes. Below the patterns for normal incident wave ($\Phi = 0$) are layer lines for chiral, zigzag, and armchair tubes

$$G_A(\boldsymbol{k}_K) = 2\sum_M \mathrm{i}^{Mq-Kp} J_{Mq-Kp}(D\pi k_\perp)\mathrm{e}^{\mathrm{i}(Kp+Mq)\Phi}$$
$$\cos\left[(Mq - Kp)\varphi_{000} + 2\pi K \frac{z_{000}}{a}\right].$$

Diffraction patterns are illustrated in Fig. 9.8. As for carbon nanotubes q is even, the symmetry of the diffraction space is $\boldsymbol{D}_{q\mathrm{h}}$. According to general discussion (Sect. 8.4), layer lines are spaced by $1/a$, and every $\tilde{q}$th one includes the same Bessel functions multiplied by different cosine factors. The layer lines $K = 0, \pm\tilde{q}, \pm 2\tilde{q}, \ldots$ are with central peak, while other ones have *intensity gap*, which is minimal for $K = r$ or $K = \tilde{q} - r$. The leading Bessel functions in the layer lines $K = \pm 1$ are J_p or J_{q-p}.

Note that for zigzag tubes the atom C_{101} is in the xz-plane; thus $\varphi_{101} = 0$, and if C_{101} is taken for the orbit representative the cosine factor becomes M-independent. Thus there are only two different patterns (for odd and even K) of the layer lines, though the intensities are K-dependent. This manifests that this orbit belongs also to conformation class $\boldsymbol{Y}^{(4)}$ (as chosen for b_1 in Table 3.2). Analogously, for armchair tubes $z_{000} = 0$ gives the geometrical factor of the conformation class $\boldsymbol{Y}^{(8)}$. Chiral tubes are characterized by the properties of the conformation class $\boldsymbol{Y}^{(5)}$ only.

9.2.1.9 Potentials

The total potential produced by a tube at the point $\boldsymbol{r}$ is

$$V(\boldsymbol{r}) = \sum_{tsu} v(\boldsymbol{r}, \boldsymbol{r}_{tsu}), \tag{9.28}$$

where $v(\boldsymbol{r}, \boldsymbol{r}_{tsu})$ is the potential at $\boldsymbol{r}$ produced by the atom C_{tsu}. It is invariant (Sect. 5.2.1), and its general form is given by (5.28). Recall that if the terms with $|K| = M = 1$ vanish, the potential has larger symmetry than the system. We analyze here several examples of the expansions of the potentials over harmonics.

Coulomb potential of ions, $v(\boldsymbol{r}, \boldsymbol{r}_{tsu}) = \frac{1}{|\boldsymbol{r}-\boldsymbol{r}_{tsu}|}$ may be treated analytically. In fact, due to the *Poisson equation* $\Delta V(\boldsymbol{r}) = -4\pi \sum_{tsu} \delta(\boldsymbol{r} - \boldsymbol{r}_{tsu})$, the expansion over the basis (5.19) is straightforwardly found

$$V(\boldsymbol{r}) = \sum_{KM} \int_0^\infty \alpha_K^{Mb} U_{Kb}^M(b\rho, \varphi, z)db, \quad \alpha_K^{Mb} = \frac{8\pi}{D} \frac{U_{Kb}^{M*}(\frac{D}{2}, \varphi_{000}, z_{000})}{b^2 + (2\pi \frac{2K-M}{a})^2}. \tag{9.29}$$

The amplitudes $\alpha_K^M(\rho)$ are obtained in terms of modified Bessel functions $K_m(x)$ and $I_m(x)$. E.g., for the achiral tubes we get for $K = M = 0$, $K = M/2$ ($M \neq 0$ even), and otherwise, respectively:

$$\alpha_0^0(\rho) = H_0^{0^*}(\varphi_{000}, z_{000}) \begin{cases} 0, & \rho \le \frac{D}{2}; \\ -4\pi \ln\left(\frac{2\rho}{D}\right), & \rho > \frac{D}{2}, \end{cases} \tag{9.30a}$$

$$\alpha_{\frac{M}{2}}^M(\rho) = \frac{2\pi}{nM} H_{\frac{M}{2}}^{M^*}(\varphi_{000}, z_{000}) \begin{cases} \left(\frac{2\rho}{D}\right)^{nM}, & \rho \le \frac{D}{2}; \\ \left(\frac{2\rho}{D}\right)^{-nM}, & \rho > \frac{D}{2}, \end{cases} \tag{9.30b}$$

$$\alpha_K^M(\rho) = 4\pi H_K^{M^*}(\varphi_{000}, z_{000}) \begin{cases} K_{nM}\left(\pi \frac{|2K-M|D}{a}\right) I_{nM}\left(2\pi \frac{|2K-M|}{a}\rho\right), & \rho \le \frac{D}{2}; \\ I_{nM}\left(\pi \frac{|2K-M|D}{a}\right) K_{nM}\left(2\pi \frac{|2K-M|}{a}\rho\right), & \rho > \frac{D}{2}. \end{cases} \tag{9.30c}$$

Several harmonics of low order in K and M are plotted in Fig.9.9.

Another example is Van der Waals interaction between the layers in graphite and walls of multi-wall nanotubes. The pairwise potential is well fitted [26] to the interaction of the layers in graphite by Lenard–Jones form:

$$v(\boldsymbol{r}) = -\frac{18.5426}{|\boldsymbol{r}|^6} + \frac{29000.4}{|\boldsymbol{r}|^{12}}. \tag{9.31}$$

We numerically calculate total potential $V(\boldsymbol{r})$ and expand it over harmonics. The amplitudes $\alpha_K^M(\rho)$ for the radius $\rho_\pm = \frac{D}{2} \pm 3.44$ Å, corresponding to the adjacent layer in a multi-wall nanotube, are depicted in Fig. 9.10. These results will be used in Sect. 9.2.2 to discuss layer–layer interaction in double-wall tubes.

An important common characteristics of all the discussed expansions is that amplitudes $\alpha_K^M(\rho)$ rapidly decrease both with M and K.

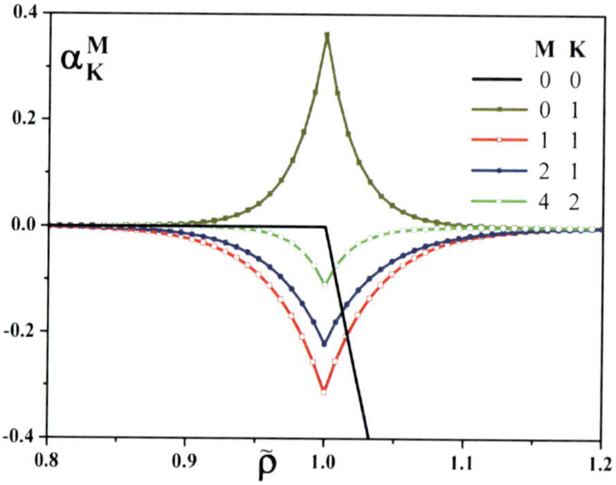

Fig. 9.9 Total potential produced by the carbon nanotube (13,13) (symmetry group $\boldsymbol{L} = \boldsymbol{T}_{26}^1(f = 1.23\,\text{Å})\boldsymbol{D}_{13\text{h}}$, diameter $D = 17.64$ Å), resulting from the Coulomb potential of the atoms. Non-negligible expansion coefficients $\alpha_K^M(\tilde{\rho})$ are plotted in the significant range $0.75 < \tilde{\rho} < 1.2$ of the reduced radial coordinate $\tilde{\rho} = 2\rho/D$

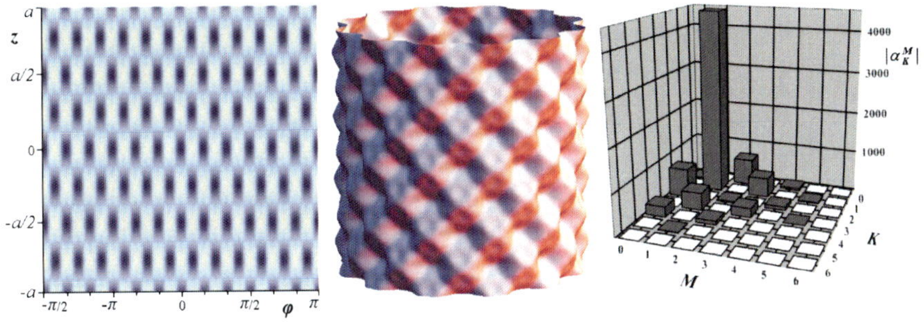

Fig. 9.10 Van der Waals potential $V(\rho, \varphi, z)$ of the tube (12,12) (symmetry group $\boldsymbol{L} = \boldsymbol{T}^1_{24}(f = 1.23\,\text{Å})\boldsymbol{D}_{12\text{h}}$ and diameter $D = 16.28$ Å) at $\rho = 7.2$ Å. *Right*: coefficients α^M_K (arbitrary units) in the expansion (5.28) over line group harmonics. *Left* and *middle*: density and cylindrical (φ and z, radius of the surface at φ and z equals to $V(\rho, \varphi, z)$ added to the nanotube radius) plots of V

9.2.2 Double- and Multi-Wall Nanotubes

Although in this section we primarily consider double-wall carbon nanotubes, the methods and results presented can be straightforwardly generalized to multi-wall nanotubes. The double-wall nanotube $W@W' = (n_1, n_2)@(n'_1, n'_2)$ consists of two coaxially arranged single-wall nanotubes, denoted as $W = (n_1, n_2)$ and $W' = (n'_1, n'_2)$. Since the axes z and z' of the walls coincide, their relative position is determined by angle Φ and vertical displacement Z: x-axis of the inner wall, defined by (9.17a), should be rotated for Φ and upraised for Z to match the x'-axis of the outer wall. Therefore, in the frame of the interior wall, its atoms have coordinates (9.17), while the positions $\boldsymbol{r}'_{tsu}$ of the outer wall atoms are given by the same expression, only with Φ and Z added to ϕ'_0 and z'_0, respectively.

9.2.2.1 Symmetry

Symmetry group of a multi-wall carbon nanotube is intersection of the symmetry groups (9.14) of the walls. Therefore we apply Theorem 3, which for double-wall carbon nanotubes gives the following general result [27]. As both Q and Q' are rational, the commensurability conditions reduce to that of fractional translations or equivalently to the commensurability of periods. When it is satisfied, the double-wall tube is commensurate with the period $a_\cap = \hat{a}'a = \hat{a}a'$ (the second equality defines co-primes $\hat{a}$ and $\hat{a}'$). However, using (9.15b), we find $a/a' = \sqrt{\tilde{q}\mathcal{R}'/\tilde{q}'\mathcal{R}}$, implying that whenever $\mathcal{R} \neq \mathcal{R}'$ tubes are incommensurate. Thus, for commensurate tubes $a/a' = \sqrt{\tilde{q}/\tilde{q}'}$, meaning that commensurability requires

$$\hat{a} = \sqrt{\tilde{q}/\text{GCD}(\tilde{q}, \tilde{q}')}, \quad \hat{a}' = \sqrt{\tilde{q}'/\text{GCD}(\tilde{q}, \tilde{q}')}, \tag{9.32}$$

i.e., that the square roots are co-prime integers. When it is satisfied then the first family subgroup of the symmetry group of the double-wall tube is $\boldsymbol{L}^{(1)}_\cap = \boldsymbol{T}^{r_\cap}_{q_\cap}(a_\cap)\boldsymbol{C}_{n_\cap}$, with

$$n_\cap = \mathrm{GCD}(n, n'), \quad q_\cap = n_\cap \mathrm{GCD}(r\hat{a}'\frac{n'}{n_\cap} - r'\hat{a}\frac{n}{n_\cap}, \sqrt{\tilde{q}\tilde{q}'}), \quad a_\cap = \hat{a}'a = \hat{a}a'. \tag{9.33}$$

Helicity parameter $r_\cap$ is determined by the integer $r_\cap = (r\hat{a}\tau + \frac{q}{n}s_0)q_\cap/q$ with $\tau = \sqrt{\tilde{q}\tilde{q}'}/\mathrm{GCD}(r\hat{a}'\frac{n'}{n_\cap} - r'\hat{a}\frac{n}{n_a}, \sqrt{\tilde{q}\tilde{q}'})$ and $s_0 = \tau(r\hat{a}q' - r'\hat{a}'q)(\hat{n}^{\mathrm{Eu}(\hat{n}')} - 1)/n'\tilde{q}\tilde{q}'$; then $r_{1\cap} = r_\cap + jq_\cap/n_\cap$, where j is the minimal nonnegative integer for which $r_{1\cap}$ and $q_\cap$ are the co-primes.

The remaining symmetry elements, parities, are present only for special relative positions of the walls, when some of their U-axes and/or mirror planes coincide [9]. Recalling again the theorem of Abud and Sartori [28], we conclude that just such configurations, having maximal symmetry, are stable. Therefore, U-axis is always present, while when both walls are achiral, mirror planes appear, too.

Leaving aside more detailed study of particular double-wall tubes, we only stress out that the walls may be commensurate or incommensurate. In the later case symmetry is described by a finite axial point group. Even in the commensurate cases $\boldsymbol{L}_\cap^{(1)}$ is small in comparison to the single-wall tubes. Both facts manifest the incompatibility of the symmetries of the walls and have strong impact on the interaction of the walls, as we shall see immediately.

9.2.2.2 Interwall Interaction

The interaction of the walls W and W′ is known to be of the van der Waals type. With the model of the Lenard–Jones interatomic potential (9.31), the total interaction between the walls is

$$V(\Phi, Z) = \sum_{t's'u'} \sum_{tsu} v(\boldsymbol{r}'_{t's'u'}, \boldsymbol{r}_{tsu}) = \sum_{t',s',u'} V_{\mathrm{in}}(\boldsymbol{r}'_{t's'u'}) = \sum_{t,s,u} V_{\mathrm{out}}(\boldsymbol{r}_{tsu}). \tag{9.34}$$

Here, $V_{\mathrm{in}}(\boldsymbol{r}) = \sum_{t's'u'} v(\boldsymbol{r}'_{t's'u'}, \boldsymbol{r})$ and $V_{\mathrm{out}}(\boldsymbol{r}) = \sum_{tsu} v(\boldsymbol{r}_{tsu}, \boldsymbol{r})$ are the potentials (9.28) of the interior and outer wall, respectively. It is obvious that $V(\Phi, Z)$ is invariant under transformations of the $\boldsymbol{L}$ of the interior and $\boldsymbol{L}'$ of the outer wall atoms. Therefore, the total symmetry of the potential is the product $\boldsymbol{L}_\times = \boldsymbol{L}\boldsymbol{L}' = \boldsymbol{L} \otimes \boldsymbol{L}'/\boldsymbol{L} \cap \boldsymbol{L}'$ of these groups. Notice that if there was no interaction between walls, the symmetry of the double-wall tube would be the direct product $\boldsymbol{L} \otimes \boldsymbol{L}'$. In other words, the interaction breaks the symmetry to $\boldsymbol{L}_\cap = \boldsymbol{L} \cap \boldsymbol{L}'$ by the factor $\boldsymbol{L}_\times$, i.e., the symmetry of the interaction is exactly the *symmetry breaking* group.

The breaking group $\boldsymbol{L}_\times$ is an ordinary line group only if the commensurability condition (9.32) is fulfilled, while otherwise it is a bihelical one (Sect. E.2). In the commensurate case, the parameters of the first family subgroup $\boldsymbol{L}_\times^{(1)} = \boldsymbol{T}_{q_\times}^{r_{1\times}}(a_\times)\boldsymbol{C}_{n_\times}$ of the breaking group are [27]

$$n_\times = \frac{\mathrm{LCM}(n,n')\sqrt{\tilde{q}\tilde{q}'}}{\mathrm{GCD}(r_1\hat{a}'n'/n - r_1'\hat{a}n/n, \sqrt{\tilde{q}\tilde{q}'})}, \qquad q_\times = \mathrm{LCM}(q,q'), \tag{9.35}$$

$$a_\times = \frac{aa'\mathrm{GCD}(r_1\hat{a}'n'/n_\cap - r_1'\hat{a}n/n_\cap, \sqrt{\tilde{q}\tilde{q}'})}{A\,\mathrm{GCD}(q,q')}, \tag{9.36}$$

$$\frac{qq'r_{1\times}}{\overline{q,q'}} = \frac{(r_1\hat{a}'q' - r_1'\hat{a}q)\hat{a}^{\mathrm{Eu}(\hat{a}')} + r_1'q\hat{a}}{\hat{a}\hat{a}'} \quad (\mathrm{mod}\ \underline{n_\cap\tilde{q}\tilde{q}', r_1\hat{a}'q' - r_1'\hat{a}q}). \tag{9.37}$$

By convention, $r_{1\times}$ is the unique solution of the last equation which is less than $q_\times$ and co-prime with $q_\times$.

The mentioned incompatibility of $\boldsymbol{L}$ and $\boldsymbol{L}'$ yields simultaneously low symmetry of double-wall nanotubes and high symmetry of their interaction. This implies fine periodicity (Fig. 9.11) of the interaction potential along and around the z-axis, i.e., small periods $a_\times$ and $2\pi/n_\times$. Precisely, if V_{in} in (9.34) is expanded over harmonics of $\boldsymbol{L}$, the remaining summation over $\boldsymbol{L}'$ cancels all the terms which are not also

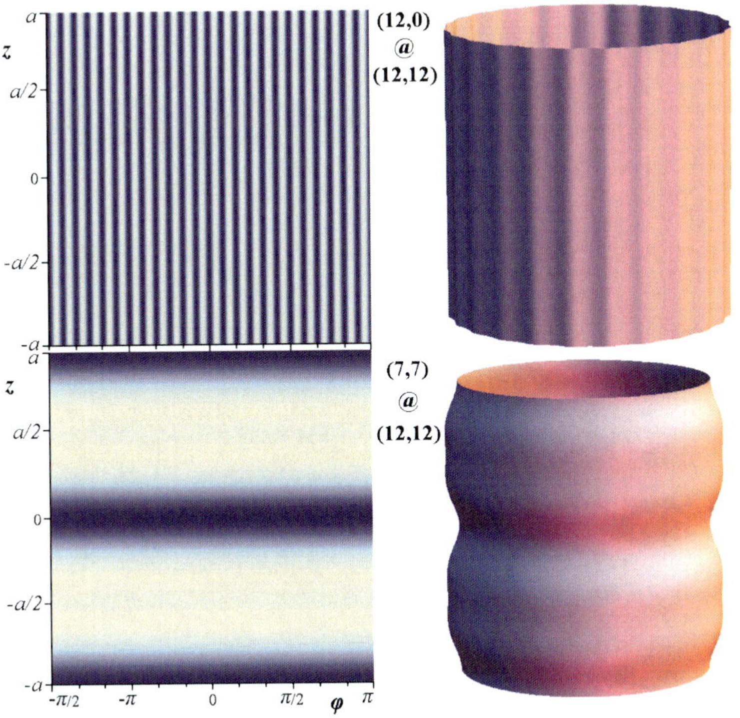

Fig. 9.11 Interwall potential $V(\Phi, Z)$ (density and cylindrical plot). Nanotube (12,12) is taken as the emitter, and potential and units are as in Fig. 9.10. *Up*: (12,0)@(12,12), incommensurate, thus super slippery along z-axis, with rotational symmetry $\boldsymbol{C}_{24}$. *Down*: (7,7)@(12,12), commensurate, roto-translational symmetry $\boldsymbol{T}(a)\boldsymbol{C}_{168}$) (almost rotationally super slippery due to large n)

harmonics of $\boldsymbol{L}'$. Hence, only a sparse distribution of the common harmonics which contribute to the interaction is left: the first common harmonics will be with large K and M. In the view of the rapid decrease of the amplitudes α_K^M (Fig. 9.10), this means that the interaction of the walls is generally very small.

In particular, for incommensurate walls, when $\boldsymbol{L}_\cap$ is finite, the breaking group is bihelical, with broken translational commensurability condition, meaning that the orbits of $\boldsymbol{L}_\times$ are quasi-continual along Z, i.e., for any Z, there is an arbitrary close $Z + \varepsilon$ that is obtained from Z by a transformation from $\boldsymbol{L}_\times$. However, due to the invariance of V, values $V(\Phi, Z)$ and $V(\Phi, Z+\varepsilon)$ are equal, and as V is continuous function, this implies that it is constant along Z. Hence, no energy is needed for the coaxial relative translation of the incommensurate walls, i.e., the walls are super slippery [29].

Let us only mention here that super slippery effect cannot be realized for the coaxial rotations of the walls, because from the beginning it was assumed that Q and Q' are rational. Still, if for some other types of nanotubes Q and Q' were irrational it would be possible to build up a super slippery double-wall nanotube, an ideal component of nano-machines.

References

1. S. Iijima, Nature **354**, 56 (1991)
2. R. Tenne, L. Margulis, M. Genut, G. Hodes, Nature London **360**, 444 (1992)
3. P. Hoyer, Langmuir **12**, 1411 (1996)
4. N.G. Chopra, R.G. Luyken, K. Cherrey, V.H. Crespi, M.L. Cohen, S.G. Louie, A. Zettl, Science **269**, 966 (1995)
5. J. Sha, J. Niu, X. Ma, J. Xu, X. Zhang, O. Yang, D. Yang, Adv. Matter. **14**, 1219 (2002)
6. J.C. Hutleen, K.B. Jirage, C.R. Martin, J. Am. Chem. Soc. **1420**, 6603 (1998)
7. Z.W. Pan, Z.R. Dai, Z.L. Wang, Science **291**, 1947 (2001)
8. J.W. Mintmire, B.I. Dunlap, C.T. White, Phys. Rev. Lett. **68**, 631 (1992)
9. M. Damnjanović, I. Milošević, T. Vuković, R. Sredanović, Phys. Rev. B **60**, 2728 (1999)
10. Y. Li, S.V. Rotkin, U. Ravaioli, Nano Lett. **3**, 183 (2003)
11. M. Damnjanović, I. Milošević, E. Dobardžić, T. Vuković, B. Nikolić, in *Applied Physics of Nanotubes: Fundamentals of Theory, Optics and Transport Devices*, ed. by S.V. Rotkin, S. Subramoney (Springer-Verlag, Berlin-Heidelberg-New York, 2005), pp. 41–88
12. V. Kopsky, D. Litvin, *Subperiodic Groups, International Tables for Crystallography*, vol. E (Kluwer, Dordrecht, 2003)
13. M. Dresselhaus, G. Dresselhaus, R. Saito, Phys. Rev. B **45**, 6234 (1992)
14. N. Hamada, S. Sawada, A. Oshiyama, Phys. Rev. Lett. **68**, 1579 (1992)
15. S. Reich, C. Thomsen, J. Maultzsch, *Carbon Nanotubes — Basic Concepts and Physical Properties* (Wiley-VCH, Weinheim, 2004)
16. E.B. Barros, A. Jorio, G.G. Samsonidze, R.B. Capaz, A.G.S. Filho, J.M. Filho, G. Dresselhaus, Physics Reports **431**, 261–302 (2006)
17. K.S. Novoselov, A.K. Geim, S.V. Morozov, D. Jiang, Y. Zhang, S.V. Dubonos, I.V. Grigorieva, A.A. Firsov, Science **306**, 666 (2004)
18. M. Damnjanović, I. Milošević, T. Vuković, R. Sredanović, J. Phys. A **32**, 4097 (1999)
19. S. Tasaki, K. Maekawa, T. Yamabe, Phys. Rev. B **57**, 9301 (1998)
20. M. Damnjanović, T. Vuković, I. Milošević, J. Phys. A **33**, 6561 (2000)
21. M. Damnjanović, T. Vuković, I. Milošević, Solid State Comm. **116**, 265 (2000)

22. L.D. Landau, E.M. Lifshitz, *Quantum Mechanics* (Elsevier, Amsterdam, 1980)
23. L. Jansen, M. Boon, *Theory of Finite Groups. Applications in Physics* (North-Holland, Amsterdam, 1967)
24. E. Dobardžić, I. Milošević, B. Nikolić, T. Vuković, M. Damnjanović, Phys. Rev. B **68** (2003)
25. T. Vuković, M. Damnjanović, Nanotechnology **18**, 375708 (2007)
26. R. Saito, R. Matsuo, G.D. T. Kimura, M.S. Dresselhaus, Chem. Phys. Lett. **348**, 187 (2001)
27. M. Damnjanović, I. Milošević, T.V. E. Dobardžić, B. Nikolić, J. Phys. A **36**, 10349 (2003)
28. H. Abud, G. Sartori, Ann. Phys. **150**, 307 (1983)
29. M. Damnjanović, T. Vuković, I. Milošević, Eur. Phys. J. B **25**, 131 (2002)

Appendix A
Koster–Seitz Notation

Abstract Necessary details on notation of Euclidean transformations which are widely used in three-dimensional crystallography and solid state physics are summarized.

Arbitrary transformation in the Euclidean space can be seen as a successive application of an orthogonal transformation X (rotation, inversion, or their combination) and a translation for some vector $\boldsymbol{f}$. Thus, it can be expressed by the symbol $(X|\boldsymbol{f})$ defined as follows:

$$(X|\boldsymbol{f})\boldsymbol{r} = X\boldsymbol{r} + \boldsymbol{f}, \qquad \text{(A.1)}$$

where $\boldsymbol{r}$ is a radius vector. In an orthonormal frame $\{e_x, e_y, e_z\}$, vectors $\boldsymbol{r}$ and $\boldsymbol{f}$ are columns of their coordinates ($\boldsymbol{r} = (x, y, z)^T$ and $\boldsymbol{f} = (f_x, f_y, f_z)^T$) while X is a 3×3 matrix ($X = (X)_{ij}$). These matrices give the coordinate representation of (A.1). The action (A.1) can also be given in a four-dimensional notation: vector $\underline{\boldsymbol{r}} = (x, y, z, 1)^T$ is used instead of $\boldsymbol{r}$, and the transformation (A.1) is represented by a 4×4 matrix:

$$[X|\boldsymbol{f}] = \begin{pmatrix} X_{11} & X_{12} & X_{13} & f_x \\ X_{21} & X_{22} & X_{23} & f_y \\ X_{31} & X_{32} & X_{33} & f_z \\ 0 & 0 & 0 & 1 \end{pmatrix}. \qquad \text{(A.2)}$$

Then, the transformed vector is obtained after omitting the last coordinate 1 in $[X|\boldsymbol{f}]\underline{\boldsymbol{r}}$.

Obviously, when the translation vector $\boldsymbol{f}$ is equal to 0, the transformation reduces to the orthogonal part: $(X|0) = X$. Analogously, pure translations are obtained for orthogonal part X equal to the identical matrix I. Thus, $(I|0)$ is the identity transformation. Multiplication rule and inverse transformation follow from (A.1) or from its matrix form (A.2):

$$(X|\boldsymbol{f})(X'|\boldsymbol{f}') = (XX'|\boldsymbol{f} + X\boldsymbol{f}'), \quad (X|\boldsymbol{f})^{-1} = (X^{-1}| - X^{-1}\boldsymbol{f}). \qquad \text{(A.3)}$$

Damnjanović, M., Milošević: . Lect. Notes Phys. **801**, 171–190 (2010)
DOI 10.1007/978-3-642-11172-3

Conventionally, rotation for angle $\varphi = \frac{2\pi}{Q}$ around z-axis is denoted by R_φ or C_Q, while σ and U denote mirror plane and rotation for π around an axis perpendicular to z; if necessary, subscript specifies position of these symmetry elements: $\sigma_{\rm h}$ is horizontal mirror plane, U_x coincides with x-axis, $\sigma_{\rm v}$ is vertical mirror plane, and σ_x such a plane containing z-axis. The product $C_{2n}\sigma_{\rm h}$ is denoted as S_{2n}. When translation is along z-axis, $\boldsymbol{f} = f\boldsymbol{e}_z$, we write $(R|f)$ and $[R|f]$. Finally, the Lie algebra generators (Sect. E.1), as being derivatives of transformations $(R(x)|\boldsymbol{f}(x))$ over some parameter x, we write $(R(x)|\boldsymbol{f}(x))'_x = \frac{\partial}{\partial x}(R(x)|\boldsymbol{f}(x))|_{x=0}$, and $[R(x)|\boldsymbol{f}(x)]'_x$ analogously. In particular, z-components of the linear and angular momenta are $p_z = -\mathrm{i}\hbar(I|f)'_f$ and $l_z = -\mathrm{i}\hbar(R_\varphi|f)'_\varphi$.

Appendix B
Rod Groups

Abstract Among (infinitely many) line groups there are 75 *rod groups*, with the order q of the principle axis of rotation of the isogonal group taking crystallographic values $q = 1, 2, 3, 4, 6$ or $q = 1, 2, 3$ for the line groups with roto-reflections (i.e., for the families 2, 9, and 10).

Rod groups are listed in Table B.1 using international notation [1, 2] which is obtained substituting $\boldsymbol{L}$ by $\boldsymbol{p}$ in the symbol of the corresponding line groups.

Table B.1 Rod groups (international and factorized notation) listed by families F of the line groups according to the order q of the principle axis of isogonal group

F	$q=1$	$q=2$	$q=3$	$q=4$	$q=6$
1	$L1 = \boldsymbol{T}\boldsymbol{C}_1$	$L2 = \boldsymbol{T}\boldsymbol{C}_2$	$L3 = \boldsymbol{T}\boldsymbol{C}_3$	$L4 = \boldsymbol{T}\boldsymbol{C}_4$	$L6 = \boldsymbol{T}\boldsymbol{C}_6$
		$L2_1 = \boldsymbol{T}_2\boldsymbol{C}_1$	$L3_1 = \boldsymbol{T}_3\boldsymbol{C}_1$	$L4_1 = \boldsymbol{T}_4\boldsymbol{C}_1$	$L6_1 = \boldsymbol{T}_6\boldsymbol{C}_1$
			$L3_2 = \boldsymbol{T}_3^2\boldsymbol{C}_1$	$L4_2 = \boldsymbol{T}_4\boldsymbol{C}_2$	$L6_2 = \boldsymbol{T}_6\boldsymbol{C}_2$
				$L4_3 = \boldsymbol{T}_4^3\boldsymbol{C}_1$	$L6_3 = \boldsymbol{T}_6\boldsymbol{C}_3$
					$L6_4 = \boldsymbol{T}_6^2\boldsymbol{C}_2$
					$L6_5 = \boldsymbol{T}_6^5\boldsymbol{C}_1$
2	$L\bar{1} = \boldsymbol{T}\boldsymbol{S}_2$	$L\bar{4} = \boldsymbol{T}\boldsymbol{S}_4$	$L\bar{3} = \boldsymbol{T}\boldsymbol{S}_6$		
3	$L\bar{2} = \boldsymbol{T}\boldsymbol{C}_{1h}$	$L2/m = \boldsymbol{T}\boldsymbol{C}_{2h}$	$L\bar{6} = \boldsymbol{T}\boldsymbol{C}_{3h}$	$L4/m = \boldsymbol{T}\boldsymbol{C}_{4h}$	$L6/m = \boldsymbol{T}\boldsymbol{C}_{6h}$
4		$L2_1/m = \boldsymbol{T}_2\boldsymbol{C}_{1h}$		$L4_2/m = \boldsymbol{T}_4^1\boldsymbol{C}_{2h}$	$L6_3/m = \boldsymbol{T}_6^1\boldsymbol{C}_{3h}$
5	$L12 = \boldsymbol{T}\boldsymbol{D}_1$	$L222 = \boldsymbol{T}\boldsymbol{D}_2$	$L32 = \boldsymbol{T}\boldsymbol{D}_3$	$L422 = \boldsymbol{T}\boldsymbol{D}_4$	$L622 = \boldsymbol{T}\boldsymbol{C}_6$
		$L2_122 = \boldsymbol{T}_2\boldsymbol{C}_1$	$L3_12 = \boldsymbol{T}_3\boldsymbol{C}_1$	$L4_122\boldsymbol{T}_4\boldsymbol{D}_1$	$L6_122 = \boldsymbol{T}_6\boldsymbol{D}_1$
			$L3_22 = \boldsymbol{T}_3^2\boldsymbol{D}_1$	$L4_222 = \boldsymbol{T}_4\boldsymbol{D}_2$	$L6_222 = \boldsymbol{T}_6\boldsymbol{D}_3$
				$L4_322$	$L6_322 = \boldsymbol{T}_6\boldsymbol{D}_3$
					$L6_422 = \boldsymbol{T}_6^2\boldsymbol{D}_2$
					$L6_522 = \boldsymbol{T}_6^5\boldsymbol{D}_1$
6	$L1m = \boldsymbol{T}\boldsymbol{C}_{1v}$	$L2mm = \boldsymbol{T}\boldsymbol{C}_{2v}$	$L3m = \boldsymbol{T}\boldsymbol{C}_{3v}$	$L4mm = \boldsymbol{T}\boldsymbol{C}_{4v}$	$L6mm = \boldsymbol{T}\boldsymbol{C}_{6v}$
7	$L1c = \boldsymbol{T}'\boldsymbol{C}_1$	$L2cc = \boldsymbol{T}'\boldsymbol{C}_2$	$L3c = \boldsymbol{T}'\boldsymbol{C}_3$	$L4cc = \boldsymbol{T}'\boldsymbol{C}_4$	$L6cc = \boldsymbol{T}'\boldsymbol{C}_6$
8		$L2_1mc = \boldsymbol{T}_2\boldsymbol{C}_{1v}$		$L4_2mc = \boldsymbol{T}_4\boldsymbol{C}_{2v}$	$L6_3mc = \boldsymbol{T}_6\boldsymbol{C}_{3v}$
9	$L\bar{1}m = \boldsymbol{T}\boldsymbol{D}_{1d}$	$L\bar{4}2m = \boldsymbol{T}\boldsymbol{D}_{2d}$	$L\bar{3}m = \boldsymbol{T}\boldsymbol{D}_{3d}$		
10	$L\bar{1}c = \boldsymbol{T}'\boldsymbol{S}_2$	$L\bar{4}2c = \boldsymbol{T}'\boldsymbol{S}_4$	$L\bar{3}c = \boldsymbol{T}'\boldsymbol{S}_6$		
11	$L\bar{2}2m = \boldsymbol{T}\boldsymbol{D}_{1h}$	$L2/mmm = \boldsymbol{T}\boldsymbol{D}_{2h}$	$L\bar{6}2m = \boldsymbol{T}\boldsymbol{D}_{3h}$	$L4/mmm = \boldsymbol{T}\boldsymbol{D}_{4h}$	$L6/mmm = \boldsymbol{T}\boldsymbol{D}_{6h}$
12	$L\bar{2}2c = \boldsymbol{T}'\boldsymbol{C}_{1h}$	$L2mcc = \boldsymbol{T}'\boldsymbol{C}_{2h}$	$L\bar{6}2c = \boldsymbol{T}'\boldsymbol{C}_{3h}$	$L4mcc = \boldsymbol{T}'\boldsymbol{C}_{4h}$	$L6mcc = \boldsymbol{T}'\boldsymbol{C}_{6h}$
13		$L2_1/mcm = \boldsymbol{T}_2\boldsymbol{D}_{1d}$		$L4_2/mcm = \boldsymbol{T}_4\boldsymbol{D}_{2d}$	$L6_3/mcm = \boldsymbol{T}_6\boldsymbol{D}_{3d}$

Appendix C
Elements of the Number Theory

Abstract Elementary number theory notation and terminology necessary in analysis of the line groups is introduced.

For each real number x, the greatest integer being less than or equal to x is called *integer part* of x and denoted by $[x]$. The difference $\{x\} = x - [x]$ is called *fractional part*.

The *greatest common divisor* of the integers x and x' is denoted as $\underline{x, x'}$ or $\mathrm{GCD}(x, x')$. Analogously, for the *least common multiple* we use either $\overline{x, x'}$ or $\mathrm{LCM}(x, x')$. We say that x and x' are *coprimes* if $\underline{x, x'} = 1$. For any x and x', we define coprimes $\hat{x}$ and $\hat{x}'$ by

$$x = \hat{x}\mathrm{GCD}(x, x'), \quad x' = \hat{x}'\mathrm{GCD}(x, x'). \tag{C.1}$$

It is useful to note that

$$\overline{x, x'} = \hat{x}\hat{x}'\underline{x, x'} = \hat{x}x' = \hat{x}'x, \quad xx' = \hat{x}\hat{x}'\underline{x, x'}^2 = \underline{x, x'}\,\overline{x, x'}. \tag{C.2}$$

Especially, when $x = ab$ and $x' = a'b'$, it follows that

$$\underline{x, x'} = \underline{a, a'}\,\underline{b, b'}\,\underline{\hat{a}, \hat{b}'}\,\underline{\hat{a}', \hat{b}}, \quad \overline{x, x'} = \underline{a, a'}\,\underline{b, b'}\frac{\hat{a}\hat{a}'\hat{b}\hat{b}'}{\underline{\hat{a}, \hat{b}'}\,\underline{\hat{a}', \hat{b}}} = \frac{\overline{a, a'}\,\overline{b, b'}}{\underline{\hat{a}, \hat{b}'}\,\underline{\hat{a}', \hat{b}}}, \tag{C.3}$$

$$\hat{x} = \frac{\hat{a}\hat{b}}{\underline{\hat{a}, \hat{b}'}\,\underline{\hat{a}', \hat{b}}}, \quad \hat{x}' = \frac{\hat{a}'\hat{b}'}{\underline{\hat{a}, \hat{b}'}\,\underline{\hat{a}', \hat{b}}}, \quad \frac{\hat{x}}{\hat{x}'} = \frac{\hat{a}\hat{b}}{\hat{a}'\hat{b}'}. \tag{C.4}$$

The numbers x and y are *equal modulo* positive integer x' if $\{\frac{x}{x'}\} = \{\frac{y}{x'}\}$, i.e., if there is an integer i such that $x = y + ix'$; this is denoted as $x \stackrel{x'}{=} y$ or $x = y \pmod{x'}$. In particular, the minimal integer y equal x modulo x' is $y = x'\{\frac{x}{x'}\}$.

The *inverse* of x modulo x', denoted as $x^{-1}_{(x')}$, is the minimal integer y such that $xy \stackrel{x'}{=} 1$, i.e., such that $xy = 1 + Ax$ (A integer). It exists only if x and x' are co-primes (since $x'\{\frac{xy}{x'}\} = \underline{x, x'}\hat{x}'\{\frac{\hat{x}y}{\hat{x}'}\}$ is multiple of $\underline{x, x'}$), and then the Euler theorem shows that

$$y = x_{(x')}^{-1} \overset{x'}{=} x^{\mathrm{Eu}(x')-1}, \tag{C.5}$$

where $\mathrm{Eu}(x')$ is Euler function giving the number of the co-primes with x' being less than or equal to x'. Note that $A = (x^{\mathrm{Eu}(x')} - 1)/x'$ is integer and co-prime with both x and $x_{(x')}^{-1}$ and that when $x < x'$, x is the inverse of its inverse. In some occasions we take a benefit of a fact that the parities of x', x, $x_{(x')}^{-1}$, and A can be combined in only six ways which are given in Table C.1.

Table C.1 Possible combinations of the parities ("o" and "e" stand for odd and even) of x', x, y, and A related by $xy = 1 + Ax'$

x'	x	y	A	xy	x'	x	y	A	xy	x'	x	y	A	xy
1	o	o	o	e o	3	o	e	o	o e	5	e	o	o	o o
2	o	o	e	o e	4	o	e	e	o e	6	e	o	o	e o

Diophantine equation in unknown integer x is

$$ax \overset{b}{=} c, \tag{C.6}$$

where a, b, and c are integers. The above analysis shows that it is solvable if and only if c is a multiple of $d = \mathrm{GCD}(a, b)$. Especially, when a and b are co-primes (i.e., $d = 1$), particular (minimal positive) solution of (C.6) is $x_0 \overset{b}{=} ca^{\mathrm{Eu}(b)-1}$, while general solution is $x_z = x_0 + ib$, $z \in \mathbb{Z}$. Otherwise, when $d \geq 1$ (and divides c), there are d particular solutions of (C.6): $x_i = x_0 + i\frac{b}{d}$ ($i = 0, \ldots, d-1$), with $x_0 \overset{b/d}{=} \frac{c}{d}(\frac{a}{d})^{\mathrm{Eu}(\frac{b}{d})-1}$ (solving the original equation divided by d); then the general solution is $x_{iz} = x_i + zb$.

Diophantine equation in two unknowns x and y is

$$ax \mp by = c, \tag{C.7}$$

where a, b, and c are integers. When a and b are co-primes, its particular solution is a pair (x_0, y_0):

$$x_0 = ca^{\mathrm{Eu}(b)-1}, \quad y_0 = \pm\frac{c}{b}(a^{\mathrm{Eu}(b)} - 1),$$

while the general one consists of the pairs (x_z, y_z):

$$x_z = ca^{\mathrm{Eu}(b)-1} + zb, \quad y_z = \pm\frac{c}{b}(a^{\mathrm{Eu}(b)} - 1) \pm za, \quad z \in \mathbb{Z}. \tag{C.8}$$

Generally, the Diophantine equation (C.7) is solvable if and only if c is a multiple of GCD(a, b). Then there is a unique particular solution (x_0, y_0) such that $0 \leq x_0 < \frac{a}{\underline{a,b}}$, and the general solution is

$$x_z = x_0 + \frac{b}{\underline{a, b}} z, \quad y_z = y_0 \pm \frac{a}{\underline{a, b}} z. \tag{C.9}$$

Appendix D
Construction of the Representations

Abstract Due to the structure of the line groups, construction of their irreducible representations and co-representations can be straightforwardly performed with a help of the basic group theoretical considerations, which are briefly reviewed here.

D.1 Irreducible Representations

The first family line groups are direct products of cyclic groups, which allows to find their representations. On the other side, induction of the representations from halving subgroup gives a method to construct irreducible representations of other families.

D.1.1 Cyclic Groups

As cyclic groups are abelian their irreducible representations are, according to Schur lemma, one-dimensional. For unitary representations this means that the generator is represented by $D^{(\lambda)}(g) = \mathrm{e}^{\mathrm{i}\lambda}$, for γ restricted to $(-\pi, \pi]$. If group is of the finite order n, then g^n is the identity, and $1 = D^{(\lambda)}(g^n) = \mathrm{e}^{in\lambda}$. Thus, different representations are counted by $\lambda = 2m\pi/n$ for m integer taking values from $(-n/2, n/2]$. In particular, this gives form of the irreducible representations of the helical groups $\boldsymbol{T}_Q(f)$ and pure rotational point factor $\boldsymbol{C}_n$ used in Sect. 4.1.1.

D.1.2 Direct Product

Complete set of the irreducible representations of the direct product group $\boldsymbol{L} = \boldsymbol{A} \otimes \boldsymbol{B}$ is

$$D^{(\alpha\beta)}(\boldsymbol{L}) = D^{(\alpha)}(\boldsymbol{A}) \otimes D^{(\beta)}(\boldsymbol{B}), \text{ i.e. } D^{(\alpha\beta)}(ab) = D^{(\alpha)}(a) \otimes D^{(\beta)}(b), \quad \text{(D.1)}$$

where $D^{(\alpha)}(\boldsymbol{A})$ and $D^{(\beta)}(\boldsymbol{B})$ run over the complete sets of the irreducible representations of $\boldsymbol{A}$ and $\boldsymbol{B}$. This immediately gives the irreducible representations of the first family groups (Sect. 4.1.1), being the direct product of the cyclic groups $\boldsymbol{T}_Q(f)$ and $\boldsymbol{C}_n$.

D.1.3 Induction from a Halving Subgroup

Knowing the irreducible representations of a halving subgroup, we construct the irreducible representations by the inductive method [3]. Let $\boldsymbol{L}$ be a group, and $\boldsymbol{L}'$ its halving subgroup with Δ being the complete set of the irreducible representations $\Delta^{(\mu)}(\boldsymbol{L}')$. In $\boldsymbol{L}$ there is a unique coset of $\boldsymbol{L}'$; choosing an arbitrary element ℓ of $\boldsymbol{L}$ outside $\boldsymbol{L}'$ as *coset representative*, we decompose the group:

$$\boldsymbol{L} = \boldsymbol{L}' + \ell\boldsymbol{L}'.$$

Halving subgroup $\boldsymbol{L}'$ is invariant subgroup and contains squares of all the elements of the group; in particular, ℓ^2 and $\ell^{-1}\ell'\ell$ for any ℓ' from $\boldsymbol{L}'$ are from $\boldsymbol{L}'$. Therefore, if $\Delta^{(\lambda)}(\boldsymbol{L}')$ is an irreducible matrix representation of $\boldsymbol{L}'$, the matrix $\Delta^{(\lambda)}_{\ell}(\ell') = \Delta^{(\lambda)}(\ell^{-1}\ell'\ell)$ is one of the matrices of the representation $\Delta^{(\lambda)}(\boldsymbol{L}')$, but not necessarily the same as $\Delta^{(\lambda)}(\ell')$. The set of matrices $\Delta^{(\lambda)}_{\ell}(\boldsymbol{L}')$ is irreducible (as simply reordered set $\Delta^{(\lambda)}(\boldsymbol{L}')$) and a representation of $\boldsymbol{L}'$ since $\Delta^{(\lambda)}_{\ell}(\ell'\ell'') = \Delta^{(\lambda)}_{\ell}(\ell')\Delta^{(\lambda)}_{\ell}(\ell'')$. So, $\Delta^{(\lambda)}(\boldsymbol{L}')$ and $\Delta^{(\lambda)}_{\ell}(\boldsymbol{L}')$ are two irreducible representations of the same dimension. We check whether they are equivalent, i.e., whether there is a nonsingular matrix A such that for each ℓ' holds the relation $A^{-1}\Delta^{(\lambda)}(\ell')A = \Delta^{(\lambda)}_{\ell}(\ell')$. Depending on the outcome the irreducible representation(s) of $\boldsymbol{L}$ are constructed from $\Delta^{(\lambda)}(\boldsymbol{L}')$ in one of the two following ways:

Theorem 1 *Let Δ be the set of all nonequivalent irreducible matrix representations $\Delta^{(\lambda)}(\boldsymbol{L}')$ ($\lambda = 1, 2, \ldots$). For each λ we find $\Delta^{(\lambda)}_{\ell}(\boldsymbol{L}')$ and the equivalent to it representation from Δ we denote by $\Delta^{(\lambda_\ell)}$. Two cases are possible:*

1. $\Delta^{(\lambda)}(\boldsymbol{L}') = \Delta^{(\lambda_\ell)}(\boldsymbol{L}')$.
   Then two non-equivalent irreducible representations $D^{(\lambda\pm)}$ of $\boldsymbol{L}$ are obtained:

$$D^{(\lambda\pm)}(\ell') \stackrel{\text{def}}{=} \Delta^{(\lambda)}(\ell'), \quad D^{(\lambda\pm)}(\ell) \stackrel{\text{def}}{=} \pm A, \tag{D.2a}$$

   where A is a nonsingular matrix satisfying:

   a. $A^{-1}\Delta^{(\lambda)}(\ell')A = \Delta^{(\lambda)}(\ell^{-1}\ell'\ell)$ *for each ℓ', and*
   b. $A^2 = \Delta^{(\lambda)}(\ell^2)$.

2. $D^{(\lambda)}(\boldsymbol{L}') \neq D^{(\lambda_\ell)}(\boldsymbol{L}')$. *Then the pair $D^{(\lambda)}(\boldsymbol{L}')$ and $D^{(\lambda_\ell)}(\boldsymbol{L}')$ give single irreducible representation $D^{(\lambda,\lambda_\ell)}$ of $\boldsymbol{L}$:*

$$D^{(\lambda,\lambda_\ell)}(\ell') \stackrel{\text{def}}{=} \begin{pmatrix} \Delta^{(\lambda)}(\ell') & 0 \\ 0 & \Delta_\ell^{(\lambda)}(\ell') \end{pmatrix}, \quad D^{(\lambda,\lambda_\ell)}(\ell) \stackrel{\text{def}}{=} \begin{pmatrix} 0 & \Delta^{(\lambda)}(\ell^2) \\ 1 & 0 \end{pmatrix}. \tag{D.2b}$$

The set of the obtained irreducible representations of **L** *(two for each* λ *such that* $\lambda_\ell = \lambda$*, and one for each pair of nonequal* λ *and* λ_ℓ*) is the complete set of nonequivalent unitary irreducible representations of* **L**.

The representations listed in Tables 4.1–4.13 are obtained by the three-step procedure, according to the following scheme:

1					$\boldsymbol{L}^{(1)}$				
ℓ	$C_{2n}\sigma_\mathrm{h}$	σ_h	σ_h	U	σ_v		$(\sigma_\mathrm{v}\vert\frac{a}{2})$		σ_v
2	$\boldsymbol{L}^{(2)}$	$\boldsymbol{L}^{(3)}$	$\boldsymbol{L}^{(4)}$	$\boldsymbol{L}^{(5)}$	$\boldsymbol{L}^{(6)}$		$\boldsymbol{L}^{(7)}$		$\boldsymbol{L}^{(8)}$
ℓ					U_d	U_d	U	U	U
3					$\boldsymbol{L}^{(9)}$	$\boldsymbol{L}^{(10)}$	$\boldsymbol{L}^{(11)}$	$\boldsymbol{L}^{(12)}$	$\boldsymbol{L}^{(13)}$

(D.3)

The first step is to find representations of the first family line groups. In the second step we construct the representations of the families 2, …, 8, applying the above theorem to the irreducible representations of the first family subgroup, with the coset representative ℓ indicated in the second row. Finally, in the third step Theorem 1 is used to find representations of the remaining families 9, …, 13 (the applied coset representatives and the halving subgroups, being from the positive families 6, 7, or 8 (Sect. 2.3.3), are listed in (D.3)).

D.2 Co-representations of the Magnetic Groups

Irreducible co-representations of the magnetic group $\boldsymbol{L}^*(\boldsymbol{L}') = \boldsymbol{L}' + \Theta\ell^*\boldsymbol{L}'$ are constructed by the induction procedure similar to the one described in Sect. D.1.3 for the ordinary representations. We again assume that the irreducible representations of the halving subgroup $\boldsymbol{L}'$ (purely geometrical transformations) are known. Then, for each such representation $\Delta^{(\lambda)}(\boldsymbol{L}')$ we define the set of matrices $\Delta_{\ell'}^{(\lambda)^*}(\ell^*) = \Delta^{(\lambda)^*}(\ell^{*-1}\ell\ell^*)$, which is again an irreducible representation. According to the relationship between $\Delta^{(\lambda)}(\boldsymbol{L})$ and $\Delta_{\ell^*}^{(\lambda)^*}(\boldsymbol{L})$, Wigner defined three types of irreducible representations of $\boldsymbol{L}$, each giving an irreducible co-representation of $\boldsymbol{L}^*(\boldsymbol{L})$ in a particular way.

Theorem 2 *Let* Δ *be the set of all the nonequivalent irreducible matrix representations* $\Delta^{(\lambda)}(\boldsymbol{L}')$ $(\lambda = 1, 2, \dots)$ *of* $\boldsymbol{L}'$*. For each* λ *we find* $\Delta_{\ell^*}^{(\lambda)^*}$ *and equivalent to it representation* $\Delta^{(\lambda^*_{\ell^*})}$ *from* Δ*. Then there are three kinds of irreducible representations of* $\boldsymbol{L}'$*, and for each of them the corresponding irreducible co-representation is constructed as follows.*

1. *If* $\Delta^{(\lambda)}(\boldsymbol{L}') = \Delta^{(\lambda^*_{\ell*})}(\boldsymbol{L}')$ *and* $AA^* = \Delta^{(\lambda)}(\ell^{*2})$ *for a nonsingular operator* A *such that* $\Delta^{(\lambda)^*}(\ell^{*-1}\ell\ell^*) = A^{-1}\Delta^{(\lambda)}(\ell)A$ *for each* ℓ *in* $\boldsymbol{L}'$, *then we obtain irreducible co-representation* $\bar{D}^{(\lambda)}(\boldsymbol{L}*(L'))$ *of* $\boldsymbol{L}^*(L')$ *of the same dimension as* $\Delta^{(\lambda)}(\boldsymbol{L}')$*:*

$$\bar{D}^{(\lambda)}(\ell) \stackrel{\text{def}}{=} \Delta^{(\lambda)}(\ell), \quad \bar{D}^{(\lambda)}(\ell^*\ell) \stackrel{\text{def}}{=} A\Delta^{(\lambda)^*}(\ell). \tag{D.4a}$$

2. *If* $\Delta^{(\lambda)}(\boldsymbol{L}') = \Delta^{(\lambda^*_{\ell*})}(\boldsymbol{L}')$ *and* $AA^* = -\Delta^{(\lambda)}(\ell^{*2})$ *for a nonsingular operator* A *such that* $\Delta^{(\lambda)^*}(\ell^{*-1}\ell\ell^*) = A^{-1}\Delta^{(\lambda)}(\ell)A$ *for each* ℓ *in* $\boldsymbol{L}'$, *then we obtain the irreducible co-representation* $\bar{D}^{(\lambda)}(\boldsymbol{L}^*(\boldsymbol{L}'))$ *of* $\boldsymbol{L}^*(\boldsymbol{L}')$ *of the doubled dimension:*

$$\bar{D}^{(\lambda)}(\ell) \stackrel{\text{def}}{=} \begin{pmatrix} \Delta^{(\lambda)}(\ell) & 0 \\ 0 & \Delta^{(\lambda)^*}_{\ell*}(\ell) \end{pmatrix}, \quad \bar{D}^{(\lambda)}(\ell^*\ell) \stackrel{\text{def}}{=} \begin{pmatrix} 0 & \Delta^{(\lambda)}(\ell^{*^2})\Delta^{(\lambda)}_{\ell*}(\ell) \\ \Delta^{(\lambda)^*}(\ell) & 0 \end{pmatrix}. \tag{D.4b}$$

3. *If* $\Delta^{(\lambda)}(\boldsymbol{L}') \neq \Delta^{(\lambda^*_{\ell*})}(\boldsymbol{L}')$, *then* $\Delta^{(\lambda)}(\boldsymbol{L}')$ *and* $\Delta^{(\lambda^*_{\ell*})}(\boldsymbol{L}')$ *give by* (D.4b) *single irreducible co-representation* $\bar{D}^{(\lambda)}(\boldsymbol{L}^*(L'))$ *of* $\boldsymbol{L}^*(L')$ *of the doubled dimension.*

The set of the obtained irreducible co-representations of $\boldsymbol{L}^*(L')$ *(one for each* λ *such that* $\lambda^*_{\ell*} = \lambda$, *and one for each pair of nonequivalent* λ *and* $\lambda^*_{\ell*}$*) is complete.*

Appendix E
Generalizations of the Line Groups

Abstract The structure of the line groups allows some generalizations, which have physical applications. By emphasizing the factorization of the line groups as the crucial for their construction and consequently for their very definition, the possibility to widen the scope of the factors $\boldsymbol{Z}$ and $\boldsymbol{P}$ is opened, namely, we had tacitly assumed that the both factors were discrete and that the point factor was finite, although this was not required by the factorization itself. Therefore we here briefly discuss the groups having the same form of factorization $\boldsymbol{L} = \boldsymbol{Z}\boldsymbol{P}$ but without the above-mentioned restrictions.

The most straightforward way to the forthcoming generalizations is to consider the symmetry group $\boldsymbol{L}_{\text{cyl}} = \boldsymbol{T}^1\boldsymbol{C}_\infty$ of the homogeneous cylinder. Leaving the parities aside, it contains all translations along z-axis, all rotations around it, and their combinations; in particular, it contains all the first family line groups. This group is a subgroup of the Euclidean group $\boldsymbol{T}^3\boldsymbol{SO}(3)$, comprising all the translations and rotations of the three-dimensional Euclidean space.[1] Further, $\boldsymbol{L}_{\text{cyl}} = \boldsymbol{T}^1\boldsymbol{C}_\infty$ is a two-dimensional Lie group, with Lie generators (Appendix A)

$$P = (I|z)'_z, \quad L = (R_\varphi|0)'_\varphi. \tag{E.1}$$

These are essentially z-components of *linear* and *angular momenta*. However, the same Lie group is formed by arbitrary two independent linear combinations $\alpha_i P + \beta_i L$ $(i = 1, 2)$ of these generators. Note that such a linear combination with both nontrivial coefficients is proportional to some of the continual helical generators

$$\tilde{P}_h = (R_\varphi|\frac{h}{2\pi}\varphi)'_\varphi = \frac{h}{2\pi}P + L. \tag{E.2}$$

The helical momentum introduced in (2.8) is proportional to this generator: $\tilde{p}_h = \frac{\hbar}{h}P_h$. Thus, for any two different h_1 and h_2 the group $\boldsymbol{L}_{\text{cyl}}$ is generated by P_{h_1} and P_{h_2}.

[1] Note that $\boldsymbol{SO}(2) = \boldsymbol{C}_\infty$.

E.1 Continual Line Groups

At first, we consider cases when one of the generators remains discrete, while the other one becomes continual. When the discrete rotational group $\boldsymbol{P} = \boldsymbol{C}_n$ is substituted by $\boldsymbol{C}_\infty$, five continual line groups, being one-dimensional Lie groups (Table E.1) are obtained. Further, allowing continual helical generator and retaining discrete $\boldsymbol{P}$, we get two additional one-dimensional Lie groups.

Table E.1 Continual line groups. For each family of the continual line groups the international symbol, different factorizations, generators, and the isogonal point group $\boldsymbol{P}_\text{I}$ are given. Continual subgroup is point factor (containing $\boldsymbol{C}_\infty$ generated by L) for groups 1–5, and generalized translational group ($\boldsymbol{T}^1$, with Lie generator P) for groups 6 and 7. $Z_0 = (I|a)$ is written instead of the conventional Z_n^1

	International symbol	Factorizations	Generators	$\boldsymbol{L}^{(1)}$	$\boldsymbol{P}_\text{I}$
1	$\boldsymbol{L}\infty(a)$	$\boldsymbol{T}(a)\otimes\boldsymbol{C}_\infty$	Z_0, L	$\boldsymbol{T}(a)\otimes\boldsymbol{C}_\infty$	$\boldsymbol{C}_\infty$
2	$\boldsymbol{L}\infty/m(a)$	$\boldsymbol{T}(a)\wedge\boldsymbol{C}_{\infty\text{h}}$	Z_0, L, σ_h	$\boldsymbol{T}(a)\otimes\boldsymbol{C}_\infty$	$\boldsymbol{C}_{\infty\text{h}}$
3	$\boldsymbol{L}\infty2(a)$	$\boldsymbol{T}(a)\wedge\boldsymbol{D}_\infty$	Z_0, L, U	$\boldsymbol{T}(a)\otimes\boldsymbol{C}_\infty$	$\boldsymbol{D}_\infty$
4	$\boldsymbol{L}\infty m(a)$	$\boldsymbol{T}(a)\otimes\boldsymbol{C}_{\infty\text{v}}=\boldsymbol{C}_{\infty\text{v}}\wedge\boldsymbol{T}'(a)$ $=\boldsymbol{C}_\infty\wedge\boldsymbol{T}'(a/2)$	Z_0, L, σ_v	$\boldsymbol{T}(a)\otimes\boldsymbol{C}_\infty$	$\boldsymbol{C}_{\infty\text{v}}$
5	$\boldsymbol{L}\infty/mmm(a)$	$\boldsymbol{T}(a)\wedge\boldsymbol{D}_{n\text{h}}=\boldsymbol{T}'(a)\boldsymbol{D}_{n\text{h}}$ $=\boldsymbol{T}'(a/2)\boldsymbol{C}_{n\text{h}}=\boldsymbol{T}'(a/2)\boldsymbol{D}_n$	$Z_0, L, U, \sigma_\text{v}$	$\boldsymbol{T}(a)\otimes\boldsymbol{C}_\infty$	$\boldsymbol{D}_{\infty\text{h}}$
6	$\boldsymbol{L}n(0)$	$\boldsymbol{T}^1\otimes\boldsymbol{C}_n$	P, C_n	$\boldsymbol{T}^1\otimes\boldsymbol{C}_n$	$\boldsymbol{C}_n$
7	$\boldsymbol{L}n2(0)$	$\boldsymbol{T}^1\otimes\boldsymbol{D}_n$	P, C_n, U	$\boldsymbol{T}^1\otimes\boldsymbol{C}_n$	$\boldsymbol{D}_n$

E.2 Bihelical Line Groups

Finally, we briefly consider discrete groups with two helical generators, $(C_{Q^+}|f^+)$ and $(C_{Q^-}|f^-)$, and possibly other finite order generators. Obviously, any pair of such generators commute, implying that independently of other generators, such group contains subgroup $\boldsymbol{T}_{Q^+}(f^+)\boldsymbol{T}_{Q^-}(f^-)$. Let us start with analysis of this product.

As obviously all the elements of these groups commute, the resulting product is the factor group

$$\boldsymbol{T}_{Q^+,Q^-}(f^+,f^-) = \frac{\boldsymbol{T}_{Q^+}(f^+)\otimes\boldsymbol{T}_{Q^-}(f^-)}{\boldsymbol{T}_{Q^+}(f^+)\cap\boldsymbol{T}_{Q^-}(f^-)}. \tag{E.3}$$

Being intersection of two cyclic groups, $\boldsymbol{T}_{Q^+}(f^+)\cap\boldsymbol{T}_{Q^-}(f^-)$ is also cyclic, generated by a roto-helical transformation $(C_{Q^\cap}|f^\cap)$. Therefore it is either infinite helical group $\boldsymbol{T}_{Q_\cap}(f_\cap)$ or a trivial group (comprising the identity transformation only, i.e., $Q_\cap = 1$, $f_\cap = 0$). Let us examine the nontrivial case.

Theorem 3 *The product group* $\boldsymbol{T}_{Q^+,Q^-}(f^+,f^-)$ *is line group* $\boldsymbol{T}_Q(f)\boldsymbol{C}_n$ *if and only if the intersection group* $\boldsymbol{T}_{Q_\cap}(f_\cap) = \boldsymbol{T}_{Q^+}(f^+)\cap\boldsymbol{T}_{Q^-}(f^-)$ *is nontrivial helical group, which is equivalent to the* commensurability condition*: there exists irrational number* J *such that* $\frac{1}{Q^\pm} = x^\pm + J^\pm y^\pm$*, with rational* $x^\pm$ *and* $y^\pm$ *such that ratios*

f^+/f^- and y^+/y^- are equal and rational, i.e., when $\frac{f^+}{f^-} = \frac{y^+}{y^-} = \frac{F^+}{F^-}$ for some co-primes $F^\pm$. Then $\frac{1}{Q_\cap} = x_\cap + Jy_\cap$ and $\frac{1}{Q} = x + Jy$, with:

$$f = \frac{f^\pm}{F^\pm}, \qquad y = \frac{y^\pm}{F^\pm}, \tag{E.4}$$

$$f^\cap = nF^+F^-f, \; y_\cap = nF^+F^-y, \tag{E.5}$$

where n is determined (together with integer Z) as the minimal positive integer satisfying $\frac{x^+}{F^+} - \frac{x^-}{F^-} = \frac{Z}{\tau F^+F^-}$; finally, Q and $Q_\cap$ (i.e., x and $x_\cap$) are found (together with integers $Z^\pm$ and $Z^\pm_\cap$) from the systems of equations

$$x = \frac{x^\pm}{F^\pm} - \frac{Z^\pm}{nF^\pm}, \qquad Z^+ - Z^- = Z, \tag{E.6}$$

$$x_\cap = nF^\mp x^\pm - Z^\pm_\cap, \; Z^+_\cap - Z^-_\cap = Z. \tag{E.7}$$

In order to prove the theorem we first note that helical group $\boldsymbol{T}_Q(f)$ is a subgroup of $\boldsymbol{T}_{Q'}(f')$ if and only if there are positive integers F' and Z' such that $f = F'f'$ and $\frac{1}{Q'} - \frac{F'}{Q} = Z'$. Indeed, as the both groups are cyclic, the subgroup condition actually means that there is positive integer F' satisfying $(C_{Q'}|f')^{F'} = (C_Q|f)$.

Now we consider the intersection. As $\boldsymbol{T}_{Q_\cap}(f_\cap)$ is the maximal (cyclic) subgroup of both $\boldsymbol{T}_{Q^+}(f^+)$ and $\boldsymbol{T}_{Q^-}(f^-)$, its generator satisfies $(C_{Q^+}|f^+)^{F^-} = (C_{Q_\cap}|f_\cap) = (C_{Q^-}|f^-)^{F^+}$, and according to the lemma, it is nontrivial if and only if there are reals $Q^\cap \geq 1$ and $f^\cap > 0$ such that the two pairs of equations (conveniently, $\check{x}$ denotes $1/x$)

$$f^\cap = F^\mp_\cap f^\pm, \quad F^\mp_\cap \check{Q}^\pm - \check{Q}_\cap = Z^\mp_\cap, \tag{E.8}$$

are simultaneously solvable in the integers $Z^\pm_\cap$ and $F^\pm_\cap > 0$. The first condition means that the fractional translations $f^\pm$ are commensurate (their ratio is rational). Then there are co-prime (thus minimal) integers $F^\pm$ such that $f^+/f^- = F^+/F^-$; all other pairs are their multiples, in particular $F^\pm_\cap = \tau F^\pm$. Consequently, $f^\pm$ are both multiples of the (maximal) length $f = f^\pm/F^\pm$, while $f_\cap = \tau F^+F^-f$. Substituting this in the last pair of (E.8) we get $F^\mp_\cap \check{Q}^\pm - \check{Q}_\cap = Z^\pm_\cap$. When these conditions are subtracted, one gets equation $F^-_\cap \check{Q}^+ - F^+_\cap \check{Q}^- = Z^+_\cap - Z^-_\cap$, showing that if 1, $\check{Q}^+$ and $\check{Q}^-$ are not rationally independent, the intersection is trivial (as the only solution is $F^-_\cap = F^+_\cap = 0$). Otherwise there is irrational J and rational $x^\pm$ and $y^\pm$ such that $\check{Q}^\pm = x^\pm + Jy^\pm$ (if both $\check{Q}^+$ and $\check{Q}^-$ are rational, then this is automatically satisfied for arbitrary J, $x^\pm = \check{Q}^\pm$, and $y^\pm = 0$). Irrational part of the subtracted equations becomes commensurability condition on $y^\pm$, with the same ratio as that of the fractional translations: $y^+/y^- = F^+/F^-$, i.e., there is (maximal) rational y such that $y = y^\pm/F^\pm$. However, the rational part

$$\frac{x^+}{F^+} - \frac{x^-}{F^-} = \frac{Z^+_\cap - Z^-_\cap}{\tau F^+F^-} \tag{E.9}$$

is automatically satisfied: its left side simple to calculate; being rational, it is quotient of two co-primes and can be (by simultaneous multiplication of numerator and denominator by the minimal appropriate factor) adapted to the form $Z/\tau F^+F^-$, directly giving τ. Then (E.8) means that $\breve{Q}$ is of the form $\breve{Q} = x_\cap + Jy_\cap$ (with rational $x_\cap$ and $y_\cap$). Taking into account previous relations, its irrational part yields $y_\cap = \tau F^+F^-y$, while the rational part, being (E.7), is automatically solvable in $x_\cap$ and $Z^\pm_\cap$ (due to (E.9)) and with $Z = Z^+_\cap - Z^-_\cap$ gives the values of all these quantities.

Finally, we examine whether $\boldsymbol{T}_{Q^+,Q^-}(f^+,f^-)$ is one of the ordinary line groups. If it is, it must be from the first family, as only roto-helical operations are involved. Thus, we search for Q, n, and f such that for each t^+ and t^- there are s and t satisfying

$$(C_{Q^+}|f^+)^{t^+}(C_{Q^-}|f^-)^{t^-} = (C_Q|f)C_n^s. \qquad \text{(E.10a)}$$

This gives equations

$$t^+f^+ + t^-f^- = tf, \quad t^+\breve{Q}^+ + t^-\breve{Q}^- = t\breve{Q} + s\breve{n}. \qquad \text{(E.10b)}$$

The first condition is equivalent to the first equation of (E.4), being necessary for nontrivial intersection, as it has already been proved. Including this in the second condition, the requirement on the existence of a real Q and integers Z^+, Z^-, and n such that

$$\breve{Q}^\pm - F^\pm\breve{Q} = Z^\pm\breve{n} \qquad \text{(E.11)}$$

is obtained. Again, subtracting these conditions multiplied by $F^\mp$, we find the requirement that 1, $\breve{Q}^+$, and $\breve{Q}^-$ are rationally dependent, which is, coincident with the condition necessary for the nontrivial intersection (exactly the second equation of (E.4)). Further, x^+ and x^- are related by (E.9) with $\tau = n$. Introducing this into the first (or second) condition we find that $\breve{Q}$ is also rationally dependent on the same set, with the irrational part (if not zero) being y: $\breve{Q} = x + Jy$. Finally, we determine x (together with $Z^\pm$) from the rational part (E.6).

Therefore, if the commensurability conditions of Theorem 3 are satisfied, the product of two helical groups is the first family line group. Otherwise, it is an abelian group, but not one of the ordinary line groups, and we call it *bihelical line group*. The orbits of bihelical groups are quasi-continual: if translational commensurability condition (for $f^\pm$) is broken, but the rotational one (for $y^\pm$) is preserved, the orbits are quasi-continual along the z-axis, while in the opposite case along the coordinate φ; if the both conditions are broken, the orbits are quasi-continual in both directions.

It remains to investigate whether additional transformations may be incorporated into the bihelical groups. It turns out that $\boldsymbol{C}_n$ and U-axis are always compatible, while vertical and horizontal mirror and glide (roto-helical) planes may appear only when both helical groups are achiral, but with incommensurable (fractional) translations.

Appendix F
Modified Group Projector Technique

Abstract One of the most frequent tasks in physics is to solve eigenproblem of the hermitian operator H (hamiltonian) in the state space $\mathcal{S}$ of some system. Symmetry group $\boldsymbol{L}$, acting in $\mathcal{S}$ through representation $D(\boldsymbol{L})$ with the operators $D(\ell)$ commuting with H, may be applied to simplify this problem. Well-known Bloch theorem is a special case of the group projector technique, applicable for translational group.

Namely, Wigner's group projector technique [4] is aimed to find symmetry-adapted basis $|\ \mu m, t\rangle$ ($t = 1, \dots, f^{\mu}$; here f^{μ} is the frequency number of the irreducible component $D^{(\mu)}(\boldsymbol{L})$ in $D(\boldsymbol{L})$) satisfying (8.1), which is also eigenbasis of H. However, the original technique involves summation of the representative matrices over the group, preventing its direct application to the crystalline systems, when infinite groups and infinite dimension of the state space occur. Modified group projector technique [5–7] avoids both obstacles using the product structure of the relevant groups and inductive nature of the state space, thus enabling numerical implementations of full symmetry.

First, we construct auxiliary space $\mathcal{S} \otimes \mathcal{S}^{(\mu)^*}$, where $\mathcal{S}^{(\mu)*}$ is the space of the conjugated irreducible representation $D^{(\mu)}$, with the standard vectors (5.37) $|\mu^* m\rangle$. The representation $D^{\mu}(\boldsymbol{L}) = D(\boldsymbol{L}) \otimes D^{(\mu)^*}(\boldsymbol{L})$ acts in this space. The basis $|\mu, t\rangle$ of the fixed points determines basis $|\mu m, t\rangle$ in $\mathcal{S}$ by the partial scalar product:

$$|\mu m, t\rangle = \langle \mu m | \mu, t\rangle, \quad (m = 1, ..., |\mu|). \tag{F.1}$$

Thus, one needs to find this set of fixed points (Fig. F.1), which is the range of the Wigner projector

$$L^{\mu} = \frac{|\mu|}{|\boldsymbol{L}|} \sum_{\ell \in \boldsymbol{L}} D(\ell) \otimes D^{(\mu)^*}(\ell) \tag{F.2}$$

on the multiple irreducible space of the identity representation; this is *modified projector* of $\boldsymbol{L}$ for the representations $D(\boldsymbol{L})$ and $D^{(\mu)}(\boldsymbol{L})$. However, this subspace coincides with the space spanned by the common fixed points, i.e., eigenvectors for eigenvalue 1, of the operators $D(\ell) \otimes D^{(\mu)^*}(\ell)$ representing group generators only.

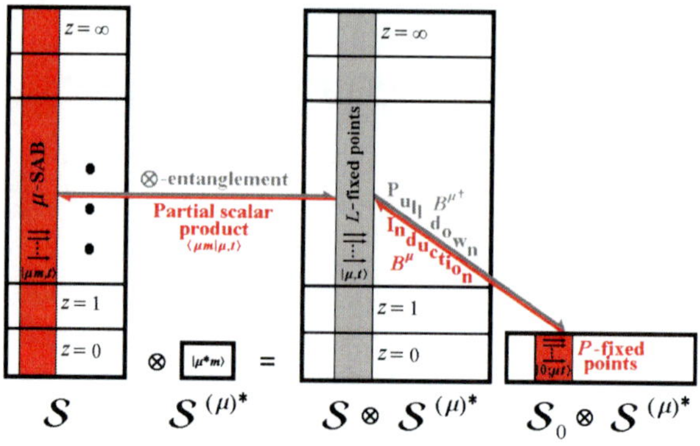

Fig. F.1 Scheme of the modified group projector technique in the induced spaces. The symmetry-adapted basis $|\mu m, t\rangle$ looked for in the state space $\mathcal{S}$ is effectively found in the monomer space $\mathcal{S}_0$: the fixed points $|0;\mu t\rangle$ of the stabilizer group $\boldsymbol{P}$ in this space are induced to auxiliary space $\mathcal{S}\otimes\mathcal{S}^{(\mu)*}$ (where they are fixed points $|\mu, t\rangle$ of the whole group $\boldsymbol{L}$), and then by the partial trace mapped to $|\mu m, t\rangle$

So, instead of the summation over the whole group, one solves finite system of linear equations.

The second difficulty, infinite dimension of the involved operators, is in the studies of crystals resolved by the convenient algebraic properties of the modified projector. Let $\boldsymbol{L} = \boldsymbol{ZP}$, where $\boldsymbol{P}$ is finite subgroup of $\boldsymbol{L}$, and $\boldsymbol{Z}$ the group of the generalized translations.[1] The Lagrange decomposition onto cosets of $\boldsymbol{P}$ is: $\boldsymbol{L} = \bigcup_z z\boldsymbol{P}$. Simultaneously, action of the transversal $\boldsymbol{Z}$ makes partition of the system into orbits S_P of $\boldsymbol{Z}$. Taking a representative atom s_P from each orbit one gets the "initial monomer" M. Assume that the representatives are chosen such that M is invariant under $\boldsymbol{P}$, i.e. $\boldsymbol{P}$ is its stabilizer. Further we assume that the total space $\mathcal{S}$ is spanned over orbitals associated to atoms, i.e. each atom contributes by its orbitals. Due to symmetry, atoms from the same orbit have the same orbitals. Orbitals from the atoms of M span finite dimensional *monomer space* $\mathcal{S}_0$, and $\mathcal{S}$ is decomposed as $\mathcal{S} = \sum_z D(z)\mathcal{S}_0$.

When ℓ is from $\boldsymbol{P}$, monomer space $\mathcal{S}_0$ is invariant under $D(\ell)$, i.e. $D(\boldsymbol{P})$ reduces in $\mathcal{S}_0$ to the representation $D^{\downarrow}(\boldsymbol{P})$ of the stabilizer (precisely, $D(\boldsymbol{L})$ is *induced* representation [8] of $D^{\downarrow}(\boldsymbol{P})$). This enables to construct the modified projector in the *finite dimensional space* $\mathcal{S}_0 \otimes \mathcal{S}^{(\mu)^*}$ related to the stabilizer only:

$$L^{\mu\downarrow} = \frac{|\mu|}{|\boldsymbol{P}|}\sum_{\ell\in\boldsymbol{P}} D^{\downarrow}(\ell)\otimes D^{(\mu)^*}(\ell). \tag{F.3}$$

[1] Quite generally, the technique uses [7] arbitrary transversal $\boldsymbol{Z}$ and the corresponding stabilizer group $\boldsymbol{P}$, and instead to the monomer calculations are reduced the to the *symcell*. However, when transversal is not a subgroup, some expressions are cumbersome, and for simplicity we use generalized translations instead of transversal. Consequently, whenever transversal is a group, it may be used as $\boldsymbol{Z}$ without any change in the forthcoming results.

The most important fact is that the operator $B^{\mu} = \frac{1}{\sqrt{|\mathbf{Z}|}} \sum_z I_0 \otimes D^{(\mu)^*}(z)$ (I_0 is $|\mathcal{S}_0|$ dimensional identity matrix) is partial isometry interrelating modified projectors as

$$L^{\mu\downarrow} = B^{\mu^\dagger} L^{\mu} B^{\mu}. \tag{F.4}$$

Thus, B^{μ} biuniquely relates the fixed points of L^{μ} in $\mathcal{S}$ to the fixed points of $L^{\mu\downarrow}$ in $\mathcal{S}_0$. Precisely, to any basis $|0;\mu t\rangle$ in the range of $L^{\mu\downarrow}$ correspond vectors

$$|\mu, t\rangle = B^{\mu}\ |0;\mu t\rangle, \tag{F.5}$$

yielding by (F.1) a *symmetry-adapted basis* in $\mathcal{S}$.

Technically, arbitrary basis $|\,0;\mu t\rangle = \sum_{mA\psi_A} c_{\psi}^{(\mu m;t)}\ |\,A\psi_A\rangle\langle\mu m\,|$ in the range of $L^{\mu\downarrow}$ may be used in (F.5); as A and ψ_A count atoms in the monomer and their orbitals, states $|\,A\psi_A\rangle$ are basis in $\mathcal{S}_0$. The vectors $|\,\mu, t\rangle$ are then given by the coefficients $c_{A\psi_A}^{(\mu m,t)}$. The component of the vector $|\,\mu, t\rangle$ corresponding to the monomer zM is according to (F.5):

$$|z;\mu m, t\rangle = \frac{1}{\sqrt{|\mathbf{Z}|}} \sum_{m' A\psi_A} c_{A\psi_A}^{(\mu m',t)} D_{mm'}^{(\mu)^*}(z)\ |zA, \psi_A\rangle. \tag{F.6}$$

In particular $|0;\mu m, t\rangle = \frac{1}{\sqrt{|\mathbf{Z}|}} \sum_{A\psi_A} c_{A\psi_A}^{(\mu m,t)}\ |A, \psi_A\rangle$, since for the initial monomer z is identity.

To find a symmetry-adapted eigenbasis of the hamiltonian, we construct in the same finite dimensional space $\mathcal{H}_0 \otimes \mathcal{H}^{(\mu)^*}$ the *pulled down hamiltonian*

$$H^{\mu\downarrow} = B^{\mu^\dagger} H \otimes I_{\mu} B^{\mu} \tag{F.7}$$

(I_{μ} is identity matrix in $\mathcal{S}^{(\mu)}$). As it commutes with $L^{\mu\downarrow}$, there are common eigenvectors. Those among them which are fixed points of $L^{\mu\downarrow}$ are to be chosen as $|0;\mu t\rangle$ to generate, through (F.5) and (F.1), the part of symmetry-adapted eigenbasis corresponding to μth irreducible representation. The eigenvalues of $H^{\mu\downarrow}$ corresponding to $|\,0;\mu t\rangle$ are exactly the eigenvalues (with degeneracy $|\mu|$) of H corresponding to $|\mu m, t\rangle$ for all $m = 1, \ldots, |\mu|$.

To construct $H^{\mu\downarrow}$ in the introduced atomic basis $|A, \psi_A\rangle$ we assume that monomer atom A interacts with N_B^A atoms from the orbit of another monomer atom B. When the hamiltonian matrix elements $\langle zA, \psi_A \mid H \mid z'B, \psi_B\rangle$ are introduced in (F.7), we obtain $H^{\mu\downarrow}$ with submatrices representing interaction of the orbits of A and B:

$$H_{AB}^{\mu\downarrow} = \sum_{z=1}^{N_B^A} \sum_{\psi_A\psi_B} \langle A, \psi_A|\ H\ |zB, \psi_B\rangle\ |B\psi_B\rangle\langle A\psi_A|\ \otimes D^{(\mu)^T}(z). \tag{F.8}$$

Note that for the finite range interactions the sum over the transversal elements z reduces to the finite number of neighbors.

References

1. B.K. Vainshtein, *Modern Crystallography: Fundamentals of Crystals, Symmetry and Methods of Structural Crystallography*, vol. 1 (Springer, Berlin, 1994)
2. V. Kopsky, D. Litvin, *Subperiodic Groups, International Tables for Crystallography*, vol. E (Kluwer, Dordrecht, 2003)
3. L. Jansen, M. Boon, *Theory of Finite Groups. Applications in Physics* (North-Holland, Amsterdam, 1967)
4. E.P. Wigner, *Group Theory and its Applications to the Quantum Mechanics of Atomic Spectra* (Academic Press, New York, 1959)
5. M. Damnjanović, I. Milošević, J. Phys. A **27**, 4859 (1994)
6. M. Damnjanović, I. Milošević, J. Phys. A **28**, 1669 (1995)
7. M. Damnjanović, T. Vuković, I. Milošević, J. Phys. A **33**, 6561 (2000)
8. S.L. Altmann, *Induced Representations in Crystals and Molecules* (Academic Press, London, 1977)

Index